ERWIN ROMMEL FIRST WAR:
A NEW LOOK AT INFANTRY ATTACKS

ERWIN ROMMEL FIRST WAR:
A NEW LOOK AT INFANTRY ATTACKS

by Erwin Rommel & Zita Steele

Edited & Translated
from the German by Zita Steele

FLETCHER & CO. PUBLISHERS
www.fletcherpublishers.com

NOTE: The photographs in this collection belonged to Erwin Rommel. Some were seized by U.S. military forces during the final stages of World War II. Others were part of his private collection of First World War photos that are part of his estate in Germany.

DISCLAIMER: This book does not in any way promote Nazi ideology. In addition, Rommel was never a member of the Nazi party, nor was he responsible for any war crimes or genocide.

Erwin Rommel First War: A New Look at Infantry Attacks

By Erwin Rommel & Zita Steele
Fletcher & Co. Publishers
© August 2023, Fletcher & Co. Publishers LLC.

All rights reserved, including the right to reproduce this book, or portions thereof, in any form without written permission except for the use of brief quotations embodied in critical articles and reviews.

Cataloging-in-Publication data for this book is available from the Library of Congress.
Library of Congress Catalog Number: 2023940389
Photography: Erwin Rommel
Edited, Annotated & Translated from the German: Zita Steele
Original illustrations & writing: Zita Steele
Cover design: Zita Steele with assistance from Brian Walker
Interior design & photo of Zita Steele: Noël-Marie Fletcher

First Edition
Published in the United States of America

Cataloging information
ISBN-13 978-1-941184-33-2

Acknowledgements

I'd like to extend my thanks to the following professional soldiers for their support of my military history research and writing:

- Col. Jerry D. Morelock, Ph.D., U.S. Army, Retired
- Lt. Col. John Rice, Royal Regiment of Fusiliers, British Army, Retired
- Lt. Col. Ed Sherwood, 101st Airborne, U.S. Army, Retired

Another shout-out to Jerry for his most kind efforts to help me make this book the best it could be. Thank you, Jerry, for your help in editing this book and for sharing your great wisdom—and most of all, for being such an inspiring person and wonderful friend.

A special thanks to my very talented colleague Brian Walker, Group Design Director at HistoryNet, for his guidance in restoring the maps, for restoring some particularly tough images for me and for helping me with cover design. Despite a very busy schedule he was willing to take time to share his expertise with me, and helped me bring this years-long project to a swifter and more visually appealing conclusion. Thank you, Brian!

I'd also like to thank archivist John E. Taylor of the U.S. National Archives and Records Administration, who is no longer with us. A military intelligence expert, Mr. Taylor introduced me to Rommel's photo collection when I was a teenager. His enthusiasm for war history was contagious, and his encouragement helped set me on the path of becoming a military historian.

And last but definitely not least I would like to thank my mother Noël-Marie Fletcher for always encouraging me to follow my dreams, for always being in my corner, and for being so positive and motivating about this project when things were difficult.

Contents
— By Zita Steele

- About this Project—10
- My Approach to Rommel's First War—12
  - *New English Translation*
  - *My Commentary*
  - *Rommel's Post-War Battle Lessons*
  - *His Lost Notes*
  - *About Rommel's Images*
  - *My Approach to His Maps*
  - *Editing & Style*
  - *Rommel's Phrases Added to His Chapter Titles*
- Rommel as a Soldier & Author—20
  - *Rommel the Mountain Trooper: About German Army Mountain Rangers*
  - *A Note About Storming: Demystifying the Vague German Tactic*
  - *Rommel as a Writer*
- Young Rommel on the Eve of Battle—28
  - *Lack of Maturity*
  - *Girl Trouble*
  - *A Death in the Family*
  - *An Unexpected Child*
  - *Going Off to War*
- Conclusion: Rommel at War's End—378
  - *Rommel the Veteran*
  - *Rommel's Insomnia and Post-Traumatic Stress Disorder*
  - *The Ultimate Betrayal—Not Once, But Twice*
  - *My Views on Rommel*
    - Rommel & Nazism
    - Rommel's Strengths & Weaknesses
    - Rommel & the Resistance
    - A Reflection about Rommel
- Selected Sources and Reference Materials—392
- About the Author—393

By Erwin Rommel

ROMMEL'S FOREWORD—42

CHAPTER 1
MOBILE WARFARE IN BELGIUM AND NORTHERN FRANCE — 44

- *"Chocolate & Bread Rolls"*—Marching Out, 1914
- *"Soldier Songs"*—At the Border
- *"I Had to be Tough"*—Reconnaissance against Longwy and Preparation for the First Fight
- *"A Horrific Sight"*—Combat at Bleid
- *"Cries of the Wounded"*—At the Meuse, Engagements at Mont and in the Forest of Doulcon
- *"Under My Command"*—Combat at Gesnes
- *"A Hailstorm"*—Pursuit through the Argonne, Fighting Near Pretz
- *"Death Reaped Harvest"*—Attack on the Defuy Woods
- *"An Absolute Hell"*—At the Defuy Woods
- *"Torment"*—Night Attack of Sept. 9–10, 1914
- *"Blacking Out"*—Countermarch through the Argonne
- *"Monstrous Bitterness"*—Assault at Montblainville, Storming of the Bouzon Forest
- *"Hit by a Bullet"*—Forest Battle along the Roman Road

CHAPTER 2
BATTLES IN THE ARGONNE 1915 — 110

- *"Whiz Bangs"*—The Company Sector in the Charlotte Valley
- *"Bitter Fighting"*—Storming on Jan. 29, 1915
- *"Blood Drenched Soil"*—In Front of Central and Bagatelle
- *"One Last Handclasp"*—Storm Attack on Central
- *"Fire Magic"*—Storm Attack on Sept. 8, 1915

Chapter 3
Static Warfare in the High Vosges 1916 & Mobile Warfare in Romania 1916–17

144

- *"In Good Spirits"*—The New Formation
- *"Extraordinary Difficulties"*—Assault Troop Operation Pinetree Peak
- *"Horrifying Night"*—At Skurduk Pass
- *"Like Rushing Water"*—The Storming of Lesului
- *"Terrifying Minutes"*—Battle at Kurpenul-Valarii
- *"Storm into the Unknown"*—Hill 1001, Magura Odobesti
- *"Bitter Cold"*—Gagesti
- *"The Air of Winzenheim"*—At Vidra

Chapter 4
Battles in the Southeast Carpathians, August 1917

194

- *"Surprise Attack"*—March to the Carpathian Front
- *"An Especially Valorous Deed"*—Attack on Trassenknie on Aug. 9, 1917
- *"Exhausted from Blood Loss"*—The Attack on Aug. 10, 1917
- *"The Situation Looked Desperate"*—The Storming of Mount Cosna on Aug. 11, 1917
- *"Better to be a Hammer than an Anvil"*—The Combat Activity on Aug. 12, 1917
- *"Maelstrom"*—On the Defensive, Aug. 13–18, 1917
- *"Into our Own Artillery Fire"*—The 2nd Storming of Mount Cosna on Aug. 19, 1917
- *"For the Rest of My Life"*—Again on Active Defense

CHAPTER 5
THE TOLMEIN OFFENSIVE 1917   260

- *"Unfavorable Circumstances"*—Deployment and Preparations for the 12th Battle of the Isonzo
- *"To Try My Luck"*—The 1st Day of the Attack: Hevnik, Hill 114
- *"Stouthearted Warriors"*—The 2nd Day of the Attack, Oct. 25, 1917. Surprising Breach into the Kolovrat Position.
- *"That Was Fun!"*—The Attack on Kuk, the Valley of Luico-Savogna Is Cut Off and the Luico Pass opened
- *"Nothing Ventured, Nothing Gained"*—The Storming of Mount Cragonza
- *"We Risked the Attack"*—The Capture of Hill 1192 and the Mrzli Peak (Hill 1356), and the Storming of Mount Matajur

CHAPTER 6
PURSUIT ACROSS TAGLIAMENTO AND PIAVE 1917–1918   330

- *"Seething with Fury"*—Masseris, Compeglio, Torrente Torre, Tagliamento, Klautana Pass
- *"Utterly Fatigued"*—Pursuit to Cimolais
- *"Could I Keep My Promise?"*—Attack on the Italian Positions West of Cimolais
- *"Blown Up into the Air"*—Pursuit across Erto and through the Vajont Gorge
- *"Not to My Taste"*—Battle at Longarone
- *"To My Great Pain"*—Battles in the Mount Grappa Area

ABOUT THIS PROJECT

I have been researching Erwin Rommel for decades since I was in high school. My research led me to become the first historian to publish an analysis of Rommel's wartime photography in my *"Erwin Rommel: Photographer"* book series. This book is very different from others in the series. This project took me over five years to complete and is unlike anything I have done before.

It is a new evaluation of Rommel as a young soldier and a general based on his experiences during the First World War. The book includes a new translation I have made of Rommel's famous book, *"Infantry Attacks!"* as well as photos from Rommel's private collection. The photos consist of images I obtained from his estate in Germany and from the U.S. National Archives.

This book is more than just a new translation—I have combined my own writings, including detailed commentary and notes in each chapter, to provide a new perspective on Rommel in World War I as well as introductory material and final assessments to analyze the war's impact on his character and future generalship.

I've learned a lot about Rommel over all these years, but what I found out when I took a deeper look into Rommel's World War I experiences—and especially how he expressed these experiences in his writings, drawings and photos—deepened my knowledge about him and caused me to view many aspects of his life in a different light.

In creating this book, I hope to share this knowledge with others and make a lasting contribution to history. I hope this book will enhance readers' knowledge of Rommel and his campaigns, and especially help people "get to know" Rommel better—whether they are trying to understand his military tactics, leadership or simply Rommel as a person.

My Approach to Rommel's First War

New English Translation

Being able to "hear" Rommel recount his war experiences in German was key to unlocking the secrets of his experiences from this time in his life. I located an original 1938 German copy of Rommel's book, *"Infantry Attacks!"* from the U.S. National Archives.

The text of the book appears in *Fraktur* typeface which was common in Germany until the 20th century. Sometimes people unfamiliar with *Fraktur* can find it very difficult to read. In my enthusiasm for German studies, I acquired the skill of reading *Fraktur* while I was still in college and put my skills to use here.

When I compared the original German version to the most widely used English translation from 1944—which was produced by the U.S. Army—I noticed some striking differences.

I should point out that the U.S. Army translation, which was made while World War II was still ongoing, is the most well-known translation that exists in the English-speaking world. Censorship—which I will discuss in more detail a few paragraphs down—was a major problem. I found that many important details shedding light on Rommel's character and actions were omitted. Even aside from this, the U.S. Army translation is extremely dry and hard to chew through. Anyone who picks up that translation for the first time may come away thinking that Rommel was a boring writer and didn't have much to say apart from his insights on tactics.

Sample text from the 1938 German copy of Rommel's book using the Fraktur typeface which is antiquated today.

The original German version is a colorful, exciting book. Rommel writes with a distinctly South German swing, peppering his narrative with soldierly slang. It includes some very emotional passages about fallen comrades and mixes in some humor as well. It reminded me of veterans' memoirs that I read today.

I decided to produce a new translation and include it in this book because I thought the world needed one. I have studied German for many years and am fluent in it; one of my talents is German-to-English translation. I have often used this skill in my professional life as a journalist and historian, and also in having ordinary conversations with German people. I have often been "pressganged" into service as an interpreter by Germans who want to get something across in English, and native German speakers have often complimented me about communicating what they meant exactly right. So I wanted to share my experience of hearing Rommel's "voice" in his memoir with readers. I am confident in my translation of Rommel and hope readers enjoy it. Here are the steps I took:

- **Vividly Accurate German:** Contrary to common stereotypes about German being harsh or clinical, German is a beautiful and extremely poetic language. It is also a very precise language. Making a good English translation of German is tricky—many translators take too broad of an approach and smudge away the fine details of it, or they get so narrowly focused on precision that they wring all the poetry and moodiness out. Rommel had a unique and colorful writing style. I've used my love of German and all my powers of translating to "get across" what Rommel was saying as accurately as possible—his tone, his slang, the military maneuvers he describes and more.

- **Restoring Lost Information—Censorship:** The U.S. Army cut a lot of detail out of Rommel's book—sometimes entire passages and anecdotes. It is interesting what they chose to get rid of: colorful adjectives, grief at the loss of comrades, words of empathy or acts of compassion for captured enemies, descriptions of bloody scenes and events, sometimes Rommel's words of cool calculation or bravado in the face of enemies. I found the 1944 translation to have been scrubbed fairly clean not only of interesting details in the writing but also of Rommel's humanity. There was also one notable instance in which text was changed for political reasons—when Rommel expressed not being afraid of Russians, the reference was changed to say "Romanians" instead of "Russians." I have restored the lost information and included notes to point out to readers where significant U.S. Army omissions occurred.

- **Correcting Errors:** Sometimes the U.S. Army translation introduced errors into the text, for example failures to understand or communicate certain German idioms or references to equipment. I have corrected these.

- **Notes on Context:** Rommel colored his narrative with many references tied to German culture, such as inside jokes and references to songs. I felt that this background information was important. I have included explanatory notes about these interesting highlights in my commentary and created footnotes to point them out where appropriate.

My Commentary

I have written commentary throughout the book which includes my observations, analysis and interesting details that I point out for readers. Commentary is included in each chapter and has been designed to accompany Rommel's writing. In some chapters the commentary is short; in others it is longer. Commentary is absent from some subsections where I felt it was not needed. As I mentioned before, this project is more than just a translation of Rommel's writing. It is also an interpretation of Rommel's evolution as a leader. My commentary appears in gray-colored sections. Setting it apart from the main text gives readers the opportunity to read or skip it as they please.

I have also written introductory material to establish a clear picture of Rommel's maturity and state of affairs at the time the war broke out. This sets the stage for his development as a soldier and future general. I offer my assessments of Rommel's personal life, the unique identity he attained as a *Gebirgsjäger* mountain ranger and German "storming" tactics.

I conclude the book by discussing Rommel's postwar years. I also address some controversies about Rommel and offer a final analysis with some definitive statements about Rommel as I see him.

Rommel's Post-War Battle Lessons

The book includes Rommel's commentary as well. This is in the form of his post-battle lessons, which he called his "Considerations"—points that formed food for thought. He typically included these at the end of subsections where he felt it was needed. Sometimes he offered none. I have also commented on Rommel's assessments of himself and his various battlefield performances.

The result is a unique synthesis between past and present. Engaging with the writings of commanders from long ago proves that history is a living thing and we can continue to have dialogue with the past to learn new things.

His Lost Notes

One of the most exciting things I found out about while working on this project is what I refer to as "Rommel's Lost Notes." This is a secret of the original Fraktur typeface which apparently nobody noticed before. Rommel chose to emphasize certain parts of the text—he highlighted words to underscore certain things and make them stand out. In Fraktur, this emphasis is made by adding extra space between letters. This is easy to miss for those who are unfamiliar with *Fraktur*.

The most important "highlights" were those he made in his tactical lessons. These are truly significant. They show us exactly what words and phrases Rommel wanted students of warfare to focus on. These shed new light on his military thinking and hone in on the kernels of military knowledge he really wanted to impart to people.

These "lost notes" are one of the most exciting discoveries I have made in my life, and I am very happy to share them with people; I hope that others find them just as illuminating. The exact words and phrases that Rommel wanted to highlight now appear in bold font.

About Rommel's Images

I saw Rommel's first war as a mirror to his future. His early experiences foreshadowed events in his later career.

This is the story not only of a young soldier but of the making of a general. That is why I have chosen to show images of Rommel from both before and during World War I and from World War II.

I have included the World War II images to remind readers of the commander that Rommel was growing into.

I have included Rommel's personal photos from his private albums. As with the other photos in my Rommel Photography book series, I have restored some photos, as needed, that have been damaged or discolored with age to bring out more details in the images.

My Approach to His Maps

Some English-language versions of Rommel's book do not include all of his maps. This book includes every single map that Rommel created to illustrate his World War I actions—a total of 84. Many of his original sketches came out quite dark because he produced them for display on a projector screen to teach his students. There is no reason for us to confine these maps to the limits of 1930s technology. To allow readers to see the details of the maps more clearly, I have removed the black borders around them and brightened them up for clarity. I have also left the original German type on the maps.

Rommel's maps were an integral part of his work. During the course of his narrative, he describes himself drawing his own maps for battle plans and showing them to his men and to senior officers.

Cleaning up the maps helped to deepen my understanding of Rommel as a visual thinker. The maps show Rommel's exacting precision, his creativity and the details he took pains to get exactly right. I believe including the complete collection of World War I maps was a valuable component of this project.

Editing & Style

All military ranks are given in German. This ensures accuracy and will enable readers with an interest in the German Army during this time period to research details on their own.

Times are given in German format (that is, according to the 24 hour clock). This is of course the way that Rommel had written them.

The names of some units (such as the *Alpenkorps*) are given in German. Others have been rendered in English for convenience.

I have preserved Rommel's German references to location names and included notes with references to modern-day place names for clarity when appropriate.

German is at times more repetitive than English, with the result that verbatim translations can sometimes become eye-wateringly complex or difficult to read; German also regularly runs away what we know as "run-ons," among other unusual grammatical things. Since my goal as translator was, of course, to enable people to actually comprehend what Rommel says, I have edited the text for clarity where appropriate.

Rommel's Phrases Added to His Chapter Titles

    The main chapters are titled and arranged exactly as Rommel had arranged them—in six parts. Rommel was not interested in trying to produce a record of specific months or years; he seems to have been focused on chunking specific "types" of war experiences into six neat capsules and then dividing these into subsections. I have preserved his style of organization.

    For all his military prowess, Rommel was not a good headline writer. In order to bring out Rommel's war experiences more in each chapter, I have added phrases to Rommel's original subheadings. I did not invent them. They are drawn from Rommel's own words from the subsection in question. For example, a subsection originally called *Hill 1001, Magura Odobesti* has become *"Storm Into the Unknown": Hill 1001, Magura Odobesti*. The "storm into the unknown" phrase is a direct quote from Rommel's narrative described by the headline. I have done this for each section in the book. These phrases capture the spirit of what occurs in each passage.

Rommel as a Soldier & Author

Rommel, shown here in Goslar in the 1930s, was proud of his service as a German army mountain ranger in World War I. He remained an avid skier for the rest of his life.

Rommel the Mountain Trooper: About German Army Mountain Rangers

During the war, Rommel served as an Army ski trooper and mountain ranger—called a *Gebirgsjäger*—in the Württemberg Mountain Battalion. In German, *Gebirgsjäger* literally means "mountain hunter" but is arguably better translated as "mountain ranger." These elite Alpine troops are equivalent to Special Forces units. To be a mountain ranger in the German Army was and is prestigious. A successful ranger must be an expert in hiking and combat exercises. During World War I, feats like skiing, mountain-climbing, and maneuvering pack animals across rugged terrain were combined with shooting and stealth maneuvers. Rangers also had to transport howitzers and supplies on skis, since they were obliged to haul their gear up and down steep slopes. *Gebirgsjäger* rangers are usually sent to difficult combat areas.

Additional social prestige is associated with the *Gebirgsjäger* because Germans have great cultural esteem for mountains and hiking sports. Mountains represent beauty and strength in German culture, and are commonly associated with the concept of the German homeland. Hiking activities are prized as healthy and uplifting. Many residents of Germanic countries take vacations in mountainous areas to climb or engage in sports. A mountaineering soldier, therefore, is culturally entitled to a special type of admiration in Germany.

The Edelweiss flower has always been the symbol of German army mountain rangers. The name Edelweiss means "noble white" in German. The flower is embroidered on *Gebirgsjäger* caps and uniforms and worn with great pride. It is an emblem of courage. The tiny white bloom only grows on very high mountain peaks. It flourishes in rocky crags where other flowers die due to lack of oxygen and nourishment. The pale, simple Edelweiss can often be seen sprouting from the most barren summits or from the sides of sheer cliffs. Thus, for German-speaking people, the Edelweiss is a symbol of strength and resilience. It is considered a privilege for soldiers to wear it.

Many *Gebirgsjäger* soldiers have been drawn from mountainous regions. Rommel was a native of the Swabian Alps, distinguished by very steep hills, crisp air, forests and snowy winters. Rommel's training as a mountain ranger left him with great physical toughness.

A Note About Storming: Demystifying the Vague German Tactic

Before delving into Rommel's narrative, I would like to offer a brief explanation of a tactic that Rommel and other German soldiers often used: an assault technique referred to in German as "storming." English speakers sometimes have difficulty understanding what this term means, sometimes associating it with a natural storm or with Nazi propaganda. It is used in different contexts in German, which adds to the confusion. My attempt here to clarify the German concept of "storming," will hopefully give readers a better understanding not only of Rommel's tactics, but of the term used in a general German-speaking context.

In German, the word *Sturm* literally means "storm," and has some of the same connotations we have in English—it can be used to describe a natural event such as a thunderstorm, and can also be used as a verb to describe an invasive action. English speakers usually encounter trouble when Germans use the word "storm" in this latter context. They often interpret it as an "assault," or just simply as a "storm" without explaining what the action really means. In German, a "storm" action is basically a sudden infantry charge. As Rommel's writing shows, a typical German "storm" usually takes place after a period of stealth or quiet observation. It occurs quite suddenly—a commander gives a signal, and all members of his team rush forward in unison. The German "storm" can accurately be described as a "thundering" swarm of violence or, more figuratively, as a "cloudburst" of forceful action. This is different from what might simply be called an "assault." Although an assault denotes aggressive action, it is a vague term, and might encompass varying tactics—for example, an assault can begin with fire upon the enemy preceding a charge, and can also include coordinated maneuvers to take ground. A German "storm," however, is always a direct physical charge. It is a violent rush, like a stampede, typically accompanied by loud shouting and noise. The sudden, combined effect of noise and concentrated violence strikes like a "storm" and basically shocks and overwhelms the enemy.

Elements of the German military known for this tactic might accurately be described as "shock troops." However, the German "storm" tactic can be executed by anybody familiar with the concept. For example, in Germany, "storming" is performed not only by military forces, but also by police and fire brigades. It can be used as a surprise attack or an invasive technique to break through areas in a surge. Military troops can "storm" an enemy position, while police or firemen can "storm" a house or building.

In a German military context, the act of "storming" an enemy is perceived as a bold and daring action, and therefore has a certain aura of heroic glamor. During the 1930s, right-wing radicals tried to hijack the "storm" concept for political use. Seeking to fashion themselves as heroes or fighters, they tried to associate the notion of "storming" with their ideology. One can see this in the rise of Hitler's so-called S.A. "storm troopers," who were not elite soldiers, but merely thugs who attacked and harassed unarmed people and committed vandalism. Another notable example is Julius Streicher's vile publication, *"Der Stürmer,"* which purveyed hateful Nazi propaganda and venomous racist tirades. The name of this publication is often translated literally as, "The Stormer." A more apt translation could be "The Berserker"—Streicher was trying to invoke the military concept of "storming" as an aggressive, head-on attack at opposition. As Rommel's narrative demonstrates, the "storming" tactic was not a Nazi invention, but a bold technique developed and used by Germans long before the Nazi era.

Rommel as a Writer

Rommel had a unique writing style. Reading his work in the original German reveals a lot about his character and outlook as a veteran. Rommel the author had a very South German way of writing. He used colloquialisms. Like many soldiers, he also used slang when talking about fighting – like "gunning down" opponents, "ripping through" positions, etc. Yet Rommel's writings had other characteristics that truly stand out.

Self-Denial

Rommel's writing style tends to be extremely self-effacing. The German language allows people to take an impersonal tone when describing actions or events, for example: "One found it cold," or "Dinner was eaten." This passive and self-deflecting style is not abnormal in German. However, the extent to which Rommel makes use of this style is excessive and highly unusual. His attempts to deflect attention from himself is at times overpowering—especially in cases when he describes harrowing or uncomfortable situations. Although Rommel does mention himself specifically, such as when he gets wounded or particularly in matters of decision-making, he tends to avoid references to himself in many cases when describing the grim realities of war. He opts for a "group" approach, referring to "we" and "us" rather than himself when describing hard times. He regularly makes use of German grammar to circumvent having to mention himself, even in long passages when it is grammatically awkward to do so. (For clarity, I have at times had to put Rommel back into his own sentences to make the narrative more understandable in English.)

This is very different from other German authors. Germany is a nation of poets and intellectuals, and Germans are opinionated people; it was perfectly acceptable for a German to express an individual viewpoint in a war memoir at this time period. One notable example is the German edition of World War I veteran Ernst Jünger's famed memoir, *In Stahlgewittern* (known as "Storm of Steel"). Jünger wasn't afraid to place himself at the front and center of his own narrative. He described his feelings looking into a dead man's eyes and the sight of corpses; his writing is gritty precisely because readers see the world of war through Jünger's lofty and pitiless gaze. His war memoir is considerably darker and gorier than Rommel's, and has a much more nationalist tone.

Rommel's self-effacing style is something I notice when I read articles by veterans, particularly of the Vietnam War, describing their war experiences. Although they want to record a battle they witnessed for posterity, these men focus on everybody else around them; they often put themselves last, choosing not to dwell too much on their own feelings or experiences. Sometimes these men nearly erase themselves from their own war stories because they seem to feel they "don't deserve" attention compared to other comrades who they believe suffered more than they did.

Optimism and Human Life

Rommel's writing on war is sometimes bloody but never jaded or gruesome. In many ways, his memoir is similar to his war photography. He tends not to be overly descriptive of violence nor dwell on death—although he does grieve. There are many emotional passages relating to the losses of friends and funerals. Throughout the book, Rommel demonstrates an interest in preserving human life. He devotes energy to building and improving shelters to protect his men. He frequently spares the lives of enemy soldiers and chooses to take many POWs rather than shoot people. He shares cigarettes with POWs and treats them with dignity; in one case, he saves a POW's life and in another case intervenes to stop his own men from shooting POWs. He and his men also share water and food with wounded and dying enemy soldiers.

Love For The Troops

Rommel's writing shows his personal attachment for his troops. He frequently refers to them as "my little flock" (*"mein Häuflein"*) as if they are a flock of sheep that need to be taken care of, or a small handful of something that needs looking after. Swabians tend to use diminutives as terms of endearment. Rommel's use of these terms indicates the strong personal affection he felt for his men.

Slang and Humor

Rommel writes with an ironic sense of humor at times, often when chasing enemies or taking them by surprise. He describes these situations mischievously as though he is playing a trick on people and finds it funny. Rommel uses these opportunities not to shoot the enemies, but to demand surrender and thus takes many prisoners. He sometimes uses slang expressions or ironic terms to describe the effects of gunfire or artillery barrages.

Nature

Some of the most vivid passages in Rommel's memoir are those describing the natural world. He uses vibrant and sometimes poetic terms to describe forests, trees, geographic features and natural forces like thunder and lightning, clouds, and mountains. Some of my favorite sentences in the book are descriptions of thunder across high peaks, shadows on snow, and the view of the morning sun across mountain mists. Rommel was a talented writer and this especially comes through when his focus is nature.

Censoring Rommel

Many of Rommel's words were removed by the U.S. Army. In some cases whole sentences and entire passages were scrubbed. Many were anecdotes or descriptions that relate to human interest or which show a sympathetic side of Rommel. For example: Rommel's anecdotes of finding a wounded Frenchman by a mountain hut, caring for his horse, and a friend's funeral were removed.

Other deleted passages relate to violence, grim scenes or descriptions of soldiers' deaths. Some statements of praise for the ability of German troops were removed, as well as some bravado and sarcastic statements about enemy forces.

None of the removed passages were overly gory, obscene or contained hateful Nazi ideology or nationalist propaganda so as to have merited censorship. They were ordinary passages for a soldier's memoir—and actually quite tame compared to the writing of Ernst Jünger. However these words were still erased from history in the English language. This might have been done to shape public opinion: to make Rommel seem less sympathetic to English-speaking readers, and likely also to make him seem less intimidating as a commander.

Rommel as an officer candidate in 1910.

Young Rommel on the Eve of Battle

To form a clearer picture of Rommel's transformation from an inexperienced young solder into a future general, it helps to examine his general state of affairs before he marched away to his first war. Taking a glimpse at Rommel's character before he heard his first shots fired in battle casts greater light on how his personality changed during the conflict. What was the state of Rommel's personal life when he began his first war?

Lack of Maturity

Rommel was a member of a somewhat wild regiment of soldiers stationed in the small, scenic town of Weingarten. The garrison was located in a very rural area of south Germany. Aside from drills and training, soldiers there did not have much else to occupy their time with. Many of the rowdy young men sought to alleviate boredom through sports, drinking, and carousing.

Photos show the men of the garrison amusing themselves at various festivals and parties. In his youth, Rommel engaged in this

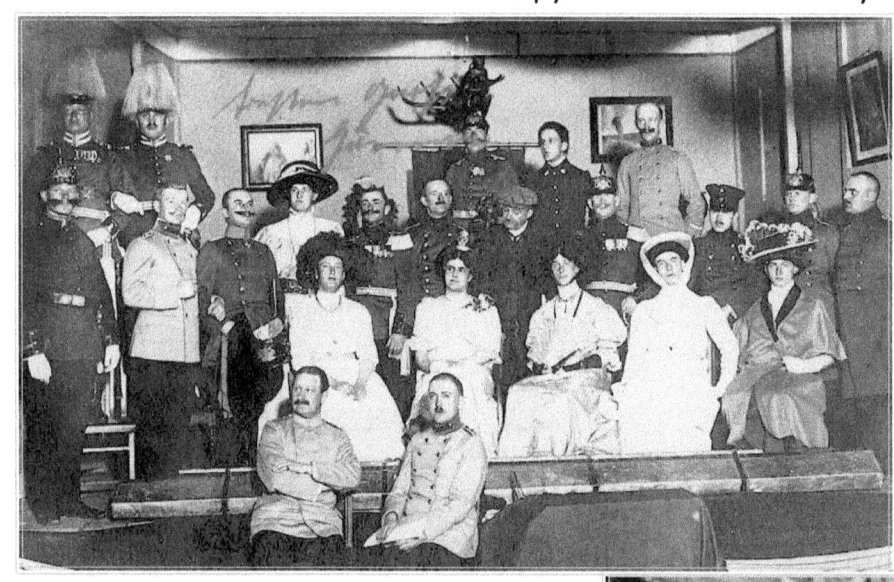

Rommel (second row from bottom, second from right) appears dressed in drag at a soldiers' carnival party in Weingarten. German carnivals (Karneval) are occasions when much drinking and overall goofiness are highly encouraged. It is a centuries-old German tradition for men to dress in drag for carnivals. It is culturally viewed as comical and is more prevalent in Germany than in most other European countries. The excessive popularity of cross-dressing among German soldiers, including during the Second World War, has been the subject of many books and articles.

tomfoolery and was photographed wearing outlandish costumes at carnivals.

Like many young people in today's era navigating their college years, Rommel struggled to break from his sheltered existence and crashed awkwardly into adulthood—he tried to be financially independent from his parents, to "fit in" with other young men at the garrison, and juggled relationship drama with girlfriends.

Girl Trouble

Confusing & Conflicting Narratives: Rommel had two known relationships with women: Lucia Maria Mollin, who he would marry, and Walburga Stemmer, an on-and-off girlfriend. Historians have struggled to make sense of the fact that Rommel appears to have been romantically involved with both women during the same time period. The confusion owes to the fact that Rommel's wife Lucie after World War II portrayed Rommel as an introvert who had no relationships with other women besides her and as a homebody with few interests apart from spending time with her and their child.

At face value, this rather idyllic version of events seemed convincing to me. However, as I studied Rommel more closely over the years, I found that elements of Lucie's portrayal of Rommel do not add up. Rommel was a very social person with many friends. He enjoyed going to dances and parties and was a snazzy dresser; photos indicate this remained consistent both in his youth and as an older man. He seems to have enjoyed carnival parties and wearing outlandish costumes. Rommel formed a veterans' group after World War I and had an active social life with his friends. He wrote letters to many relatives and seems to have thrived doing activities with groups of people, especially outdoors.

A sultry photo of Lucia (Lucie) Maria Mollin that she sent to Rommel during their courtship. Being Polish and Catholic, Lucie was an unlikely match for a German Protestant officer. Yet it appears Rommel deliberately chose girlfriends from outside his own social sphere.

Rommel's son Manfred later described his father as having great "charisma" which attracted other people to him. This charisma is evident both in Rommel's military career and in his personal life.

The most glaring inconsistency in Lucie's postwar testimony is the fact that Rommel had very much been involved with another woman, Walburga Stemmer, with whom he fathered a daughter. Certain advocates for Walburga Stemmer claim she was Rommel's "true love" and—for some reason they cannot explain—tossed aside coldly for Lucie. Their arguments as to Rommel's motives are inconsistent— they guess Lucie was rich or more respectable, or that Rommel was obligated to keep an engagement promise, or that Rommel's parents forced the break-up, or that Rommel was simply an ambitious scoundrel who perceived some advantage in marrying Lucie (hardly plausible since Lucie, a Polish Catholic orphan, was at the low end of imperial Germany's social strata).

A photo of Rommel and Lucie after they became formally engaged during World War I. They don't look particularly happy; the image seems slightly awkward. Note the lack of warmth and physical contact between the couple. This remains visible in many photos of Rommel and Lucie throughout their marriage, in contrast to other German couples from the same time period. In many photos with Rommel, Lucie appears distant or uncomfortable. Some theorize their marriage was forced. I do not believe so, but their relationship appears to have been less blissful than previously imagined. I have described apparent distance between the two in my book, "Erwin Rommel: Photographer: Vol. 4, Personal Encounters."

Compounding the overall confusion are conflicting dates. Some biographers of Walburga claim she first met Rommel in 1910, when he joined the Army garrison at Weingarten. Lucie apparently first met Rommel in 1911. However, Lucie claimed Rommel instantly fell in love with her and that they had "an understanding." Facts show Walburga became sexually intimate with Rommel in 1913 and had a child with him, a daughter named Gertrud, in December that year. I was not able to verify the date of Rommel's final breakup with Walburga. Rommel did not become engaged to Lucie until 1915—during the middle of World War I—and did not marry her until November the following year, 1916.

If Rommel and Lucie really had a whirlwind romance as she claimed, it seems to have taken a long time to develop. By contrast, Rommel's relationship with Walburga happened in a whirlwind. But was it really a romance? Some historians, struggling to make sense of events, attempt to suggest that either one of these two women was Rommel's lifelong "true love." But what if the answer was: neither?

I make no claim to be an expert on Rommel's love life—like anyone else studying him, the most I can do is speculate. Please keep in mind that the analysis and conclusions here are my personal opinions based on my own observations. I offer them to readers in an attempt to shed some light on this chapter of Rommel's life—which clearly affected him before and during World War I.

Facts & Observations: Rommel and Lucie first met in Danzig, Poland (present day Gdansk) in 1911 when Rommel was attending cadet school. Lucie was apparently living in a boarding school and studying to become a teacher. According to Lucie, they spent lots of time together. Lucie was Polish, Catholic and orphaned during the time of their courtship. In a society ruled by middle-class ethnic German Protestants, marrying a German officer would have been a social step up for Lucie—and a social step downward for Rommel. Germans generally discriminated against Poles. Ethnic tensions aside, Protestants and Catholics kept apart. Lucie was reputedly excommunicated by Catholic officials when she and Rommel eventually married. It is not known how Rommel's Protestant family reacted to news of the courtship or if they were aware of it in its early stages. In any case, Lucie was not the historically "ideal" wife for a German officer. Most of Rommel's military contemporaries married ethnic German women. Writing in 2010, Rommel's son Manfred wrote that Lucie did not correspond to Nazi ideals of being blonde or blue-eyed, and was faced with a crisis when she had to verify her "Aryan" ancestry for Third Reich authorities. Thanks to the help of her uncle who was a Polish Catholic priest, Lucie was able to supply documents and a voucher stating that her grandparents were "Aryan," but this was a difficult effort for the Rommel family.

We do not know whether or not Rommel and Lucie became sexually intimate after they first met. Lucie made a point of telling biographers after World War II that she "always" saw Rommel accompanied by a chaperone. However we do not know for certain if this is true. We do know that, while the couple did not become engaged, Lucie claimed that they had "an understanding."

Rommel wrote many letters to Lucie during and after his time at Danzig. His early letters suggest he was besotted with Lucie, although he doesn't seem to have known what to call her at first—he referred to her alternately as "Molly," "Lucia," and eventually "Lu." His writings are overconfident, moody and sometimes childish. In contrast to his strict Protestant parents, Rommel's writings from this time suggest that he, like many adherents of the German Romanticism movement, desired "free love" without involving legal or church authorities. Upon returning to Germany and moving into a new billet, Rommel allegedly wrote to Lucie on Feb. 22, 1913:

"All that I lack now is a woman to manage the whole place. Not that I'm thinking of getting married; I detest the idea of an officer's marriage with its 80,000 marks…I'll get married when I'm ready, and for that I will need neither church nor magistrate."

An excerpt from this same letter, described as 8 pages long, indicates that his relationship with Lucie had become strained by that point. Rommel complained that she no longer communicated with him regularly nor sent him pictures of herself. Evidence suggests that Lucie expressed feelings of estrangement toward Rommel.

"Why do you no longer send me pictures of yourself? You merely write, 'You have changed, and I don't know you anymore,' but I never get a picture of you. You have promised me for a long time, for God's sake!...Or are you already engaged or even married?" Rommel allegedly complained in the letter. *"I like you so much and I would at least like to see you on a photograph that I might kiss. Well, I hope that you will 'confess' to me how things are going, what you have been doing and what has happened to you."*

Seemingly around the time this letter was written in early 1913, Rommel engaged in a sexual relationship with Walburga Stemmer. I have not been able to establish as yet whether Rommel was physically intimate with Walburga before or after this letter was written. Descendants of Walburga claim that she and Rommel first met in 1910 after he first arrived at his garrison in Weingarten. I have not been able to verify this. However, it is known she and Rommel had an affair in early 1913 and that Walburga became pregnant.

What do we know about Walburga? Letters indicate that Walburga lived with her mother, who rented rooms to soldiers. Walburga appears to have been the daughter of Rommel's landlady. She seemingly complained to Rommel about money problems. In a letter to his family later on, Rommel expressed that he felt sorry for her and her mother, who were financially poor due to her father's financial irresponsibility. Rommel described Walburga's mother as *"once the wealthy daughter of a noble landlord,"* whose husband caused her to *"lose all her goods and possessions."*

The Stemmer family was ethnic German like Rommel's, but of a lower social status. Aside from renting rooms, the family allegedly earned money through seamstress work

A picture postcard that young Rommel sent to Lucie in 1911. Rommel appears on horseback on the far right wearing a dress uniform and a fancy spiked helmet. He is in the company of senior officers. He may have hoped the photo would impress Lucie during their courtship.

and by selling fruit at a market stall. By contrast, Rommel came from an established middle-class family; his father was a school principal and his maternal grandfather had been a local mayor. Again, this romantic relationship was a social step down for Rommel, but an advantageous one for Walburga. It is fair to say that, from Walburga's standpoint, a young officer of Rommel's social background would have been a "good catch."

Rommel initially kept the relationship secret. He wrote Walburga a series of passionate love letters, including passages such as: *"I'm in such good spirits, darling, so full of life, so happy, because you care for me, too,"* and flowery signatures like: *"Tender kisses and greetings, your eternally loving Erwin."*

After becoming pregnant, Walburga pressured Rommel to quit his job in the military and pursue a civil service-type career to better provide for her, according to several histories. To make a career transition in Germany has always been difficult. This is due to the fact that most Germans follow a single career path for life. Therefore Walburga's pressure on Rommel to abandon his career—after he had invested time, money, specialized training and an officer's formal education in it—was highly unusual. It also suggests that Walburga believed or at least hoped that Rommel was capable of switching careers in Germany's rigid social system.

In this photo of the young recently married couple, Rommel is wearing his Pour le Mérite medal that he received for his actions at Mount Matajur. Lucie looks away from the camera. Photos of the pair often appear stiff; unlike other couples they often look stoic and don't appear to touch each other much.

Rommel ended the relationship. The breakup allegedly took place after Rommel was sternly reprimanded by his father; I am unable to confirm this. Some scholars insist the breakup was forced. Others theorize Rommel could not socially advance if he married a fruit vendor. Rommel's friends in the military allegedly urged him to stay true to his hard-earned career and not quit over a girl. Nobody can say for certain why the relationship ended. Whatever the reason, it is clear that Rommel—headstrong, stubborn and ruled by nobody—decided to break up. I have not been able to clearly establish the date of the couple's separation. However it is safe to say that the disintegration of Rommel and Walburga's relationship occurred around the outset of World War I.

Going to war, Rommel confessed to his family about his illegitimate child, writing a sorrowful letter saying he had taken out a life insurance policy to benefit the unwed mother and child and asking his family to provide for his "orphan" in case he was killed.

Meanwhile Rommel continued writing letters to Lucie and frequently sent her postcards with pictures of himself. He was severely wounded during World War I in 1914 and had near-death experiences on a regular basis. The couple became formally engaged in 1915 and married in 1916 during the war. Lucie at some point became aware of Rommel's illegitimate daughter and allegedly decided to accept the situation. I have not been able to verify exactly when or how she found out about it.

Regardless, Rommel's initially passionate romance with Lucie appears to have cooled somewhat by the time they got engaged. Despite Rommel's outgoing personality, photos of him and Lucie as a couple appear cold, staged and often have a distinct lack of physical affection. This lack of outward affection stayed consistent in pictures throughout their marriage. Although there are lots of photos of Rommel and Lucie together as a couple, its rare to see them smiling together, looking spontaneously happy or touching each other. This is in contrast to photos of Rommel with other people; Rommel flashed plenty of smiles in photos with friends and also took a hands-on portrait hugging his former girlfriend Walburga Stemmer when she was pregnant. However there aren't many visible signs of affection to be seen between him and Lucie in most of their pictures.

The couple posed for many portraits when they were engaged and as a young married couple. That there are numerous photos may have been due to the fact that Rommel loved photography and collected as many pictures as possible of himself, relatives and friends, and even places he visited, for postcards and scrapbooks. With the exception of one candid snap of Lucie grinning seemingly as they went away on their honeymoon, the majority of pictures of the pair appear awkward. Rommel's signature to Lucie changed over time as well, from *"Greetings, kisses and thousands of hugs"* in 1913 to simply *"Yours"*— the German equivalent of *"Sincerely"* or *"Regards"* often used for casual acquaintances—for most of their married life. They remained childless for more than 10 years of marriage before their only son was born in 1928.

Strikingly, Rommel made no mention of Lucie in his book, *"Infantry Attacks!"* This omission is especially odd since they married during the middle of World War I and had also had a child together, Manfred, by the time of the book's publication. Rommel was a devoted father and adored his son. In fact, family photos show Rommel looking happier and more animated when he is interacting with Manfred than with Lucie. Despite this new and positive dimension to their family life, Rommel still did not mention his wife or their wartime courtship in his memoir.

Walburga appears to have kept tabs on Rommel for over a decade. She allegedly received regular financial support from his family. She did not marry and is said to have kept updated on what Rommel was doing, apparently becoming aware of Lucie's pregnancy in 1928. Citing information obtained from Walburga's relatives, some sources claim Walburga was upset by news of the other woman's pregnancy and hoped Lucie would miscarry Rommel's unborn child. These histories claim she committed suicide when the baby's due date drew near—allegedly killing herself in the presence of her teenage daughter by taking a drug overdose. Walburga's cause of death was officially recorded as pneumonia. Whether or not it was truly suicide, Walburga's death in October 1928 occurred only two months before the birth of Rommel's son, Manfred, in December that year. Her death would certainly have been shocking to Rommel and probably also to Lucie, and cast a shadow over Rommel's family.

While examining photos from Rommel's private collections, I found no photos of Walburga, nor postcards or letters to or from her. This is the reason why I did not include

Walburga in my *"Rommel Photographer Vol. 4 Personal Encounters,"* which examined Rommel's relationships and interactions with individuals who featured in his photo collection. It's entirely possible that Rommel—like many people might do in such a situation—got rid of materials relating to his ex-girlfriend; their relationship seems to have had an acrimonious ending. Alternatively it is possible that Rommel saved mementos and another family member could have removed them. Based on my study of Rommel, I'm inclined to believe that his feelings towards Walburga soured as he broke off the relationship; I'm not inclined to believe he saved mementos of Walburga.

Rommel acknowledged his paternity of Walburga's daughter, Gertrud (see section below). Rommel maintained a relationship with his daughter, provided her with financial support, and wrote letters to her. Judging from photos from his private albums, Rommel also spent a lot of time with Gertrud, including her on vacations and family outings.

However, despite Rommel's seemingly warm relationship with his daughter, Gertrud did not live with him after he married Lucie. Instead she lived with his sister and passed herself off as Rommel's niece. After Rommel's death, Lucie concealed the fact that Rommel had another child. Apparently Gertrud did not come forward either. The truth behind Gertrud's paternity was not revealed until after Gertrud's death in 2000.

Final Thoughts: One conclusion I arrived at is that Rommel was a rebel in his middle-class family. Despite coming from a conservative family and being named after his zealous Lutheran father, Rommel made decisions contrary to the norms of his community and likely against what his family would have wished for him.

Rommel's relationships with both women indicate he was unconcerned with social conventions. In both cases, Rommel courted a woman of lower social status with less money, from a group of society his family would not normally associate with. Both women lacked stability and male protection—Lucie's father had died, while Walburga's was irresponsible and absent. Lucie was getting by in a student boarding school and Walburga was hawking fruit to make a living. Due to the apparent weakness of these women, perhaps young Rommel wished to be "gallant" and come to their rescue. This could have been due to egotism or a sense of chivalry. Young Rommel had also led a relatively sheltered life with his middle-class family. Therefore being plunged into a rough-and-tumble world of young soldiers far from home, he may have faced loneliness or peer pressure that motivated him to seek female company. The overall impression I have is that Rommel—like many naïve young men—temporarily went "girl crazy" after leaving home and paid a heavy price for it.

It sometimes happens that women take advantage of intimate relationships with soldiers; military men often make jokes and have sayings about these types of situations. Rommel may have been "girl crazy," but we can't ignore the possibility that the girls took advantage of his situation. Walburga seemed eager to convince Rommel to solve her financial problems, to the point of pressuring him to give up his career. Lucie also demonstrated some materialistic tendencies (I have discussed some of this in Vol. 4 of the series).

In my opinion, neither relationship had a particularly happy outcome. Rommel's lack of self-control and impulsive choices left him with an illegitimate child, a bitter ex-girlfriend and what seems to have been a strained, or at least stale, marriage. Situations like this are common. It is normal for people to make imprudent relationship choices and experience romantic dramas especially when they are young. That is part of ordinary life; Rommel was no exception.

Rommel was very close to his mother and older sister Helene, shown sitting with him in this family photo.

Questions remain. Did the passionate Rommel really stay true to Lucie despite having lost interest in her once before? Did Rommel, perennially sociable and fond of parties and dances, look for companionship in affairs with other women during his life? I believe it is possible. We may never know.

For the purposes of this book, it suffices to say that young Rommel was up to his neck in relationship dramas that doubtlessly weighed on his mind as the First World War loomed.

A Death in the Family

Rommel's parents were pillars of influence in his life for different reasons. His father, Erwin Rommel Sr., was a strict school principal, math teacher and devout Lutheran. He was a demanding patriarch, who was probably frequently at odds with his namesake son who had a tendency of slacking off in school and flunking classes in juvenile acts of defiance. His mother, Helene appears to have been a gentler soul who Rommel remained close to throughout his life. He kept lots of photos of his mother and visited her frequently.

Rommel's parents were by all accounts very devoted to each other. Helene took great pride in Erwin Sr.'s work as a schoolmaster and kept a sign outside her home indicating it was "The Principal's House" for her whole life. They had a total of five children together: Helene, Erwin, Karl and Franz all survived to adulthood. One son, Manfred, an older brother to Erwin, died in childhood. Erwin would later give his own son the name Manfred after his lost sibling.

Tragedy struck when Erwin Sr. died unexpectedly during an appendix operation in late 1913. He had been the central pillar of the Rommel family. Suddenly the patriarch was gone, leaving Rommel's mother widowed. To make matters worse, war was looming on the horizon. Erwin, the eldest son, and his younger brother Karl would have to take part in the fight.

Concerns about his mother and siblings, left to care for themselves, would surely have added to Rommel's anxieties on the eve of war.

An Unexpected Child

Another issue that tormented Rommel before going off to war was the vulnerability of his illegitimate daughter, Gertrud, known by him as "Trudel," who was born in December 1913. Historical evidence indicates that Rommel neither resented his child nor ignored her. Letters show that

Rommel and his younger brother Karl (left), who also served during World War I.

Rommel was fond of the child from the outset, although he was plunged into extreme anxiety about how to take responsibility.

Before deploying in 1914, Rommel wrote an agonized letter to his older sister, Helene, to confess about the birth of Gertrud and to ask his family to provide material assistance to the child and her mother. Rommel took out a life insurance policy on himself, which he described to his sister in the following passage of this letter written before he went to war on Aug. 3, 1914:

> *"Should I...be killed, I have only one wish: to know that poor Trudel and Walburga are cared for. The 10,000 Marks of my life insurance policy will be paid to Walburga Stemmer for the upbringing of Trudel. I further request that you support Walburga with 50 Marks per month, until she is in a position to nurture and bring up Trudel without worries or for as long as you wish to support my innocent orphan child. You and Mother will surely be so kind and will fulfill for me this last wish."*

He indicated that he was skeptical of returning alive:

> *"Should I have the providence to return healthy, it is my only goal to see that Trudel is raised well and to come to the aid of Walburga in word and deed. Until this point I've done it with scarce resources; I can renounce every amusement with friends if I can do something good for them both; I care about them more than myself. I want to make amends for the wrong I have done."*

Going Off to War

Rommel went off to fight in his first war under the shadow of these various distressing situations. At the close of his August 1914 letter to his sister, he indicated feelings of depression.

He described fellow soldiers accompanying him to the train station and a spirit of loyalty to him that he felt through the ranks. "People all around have such a horrible love for me," he wrote to his sister in apparent dejection. However, he added: "It does me good."

After taking out a life insurance policy and leaving financial instructions in case of his death, Rommel proceeded engaged in extremely risky behavior on the frontlines, continuously putting himself directly in harm's way. Was it easier for Rommel to take risks in battle knowing Gertrud would receive his life insurance money if he were killed? History will never know.

Fellow veterans who survived World War I battles with Rommel later said that he acted without regard for his own life. Despite gambling his life in daring stunts, exposing himself to enemy fire and being wounded several times, Rommel survived his first war and lived to tell the tale.

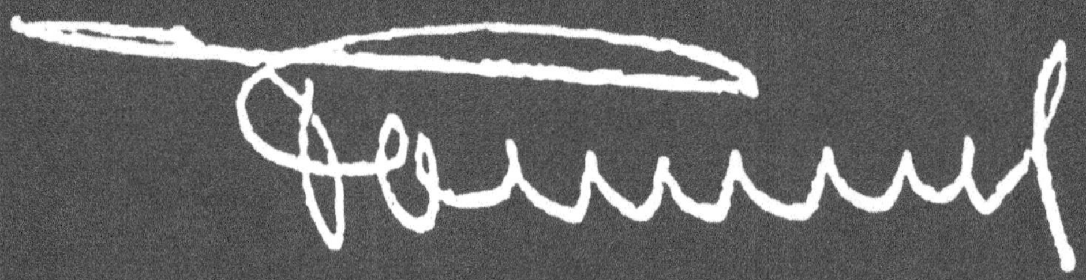

Rommel's Foreword

In this book I describe numerous battles of the World War of 1914–18 as I lived through them as a young infantry officer. Most descriptions of battles conclude with short observations, in order to draw lessons out of the corresponding combat operation.

These records made immediately after the battles should demonstrate to German youth capable of bearing arms, with what boundless self-sacrifice and valor the German soldier and especially the infantryman fought for his Germany in the 4 ½ year war[1]. They should demonstrate what enormous accomplishments the German infantry was capable of in the face of the enemy —despite inferiority in materiel and numbers – and also how supreme our German leadership was compared to that of our opponents.

Additionally the book should guarantee that experiences gained in the most difficult times of war —sometimes amid great sacrifices and deprivations – do not sink away into the past.

—The Author

---

1   It is interesting that Rommel referred to it as a 4 1/2 year war. He could perhaps have meant that the war did not end until the June 28, 1919 Treaty of Versailles, although that would be seven months after the November 1918 Armistice and 10 months after the August 1918 fourth anniversary of the war's beginning.

1
Mobile Warfare in Belgium and Northern France

"Chocolate and Bread Rolls"—*Marching Out, 1914*

> ◊ Rommel's journey begins in South Germany, where local Germans were extremely enthusiastic about the war's beginning. Rommel seems to have felt a sense of idealism and even excitement.
> ◊ The men begin to experience minor difficulties including sickening rations, stormy weather and fatigue. The enthusiasm of the young soldiers, including Rommel, fades with the looming approach of danger.
> ◊ Rommel specifically describes saying farewell to his mother and siblings. This is the only time he mentions any family members in the entire book. There is no mention of any romantic relationships—neither with Walburga Stemmer nor, more strikingly, with Lucie, who he would marry during the war and would later travel with to visit former battlefields. Moreover, the couple had been married for many years and had a son together at the time Rommel's book was published. Since this was clearly an important book for Rommel, Lucie's absence from it is odd.

Ulm, July 31, 1914—With sinister gravity, "the imminent danger of war" overshadowed the country of Germany. Everywhere were solemn, disturbed faces! Unbelievable rumors ran wild and spread rapidly. All the poster pillars were mobbed even during the gray hours before dawn. A special newspaper edition made opposite claims. In the early hours, the 4th Battery Field Artillery Regiment Nr. 49 moved through the old imperial city. The Watch on the Rhine blared loudly through the narrow streets. Windows shutters flew open; old and young enthusiastically sang and accompanied the marching soldiers.

I rode as an infantry lieutenant and platoon commander of the smartly uniformed Fuchs Battery, to which I had been attached since March 1. We did our jogging exercise in the sunny morning, completed our drills as on ordinary days, and then returned—accompanied by a crowd of about a thousand excited people—to our barracks.

During the afternoon as horses were already being purchased in the garrison courtyard, I received the revocation of my command from the regiment. Since by all appearances the situation was becoming grave, I was recalled with great urgency to my home regiment—the King William I. Infantry Regiment (6. Württ.) Number 124, back to the enlisted infantrymen of the 7th Company, who I had trained for the last two years as a recruiting officer. With the infantryman Hänle, I packed my few personal belongings

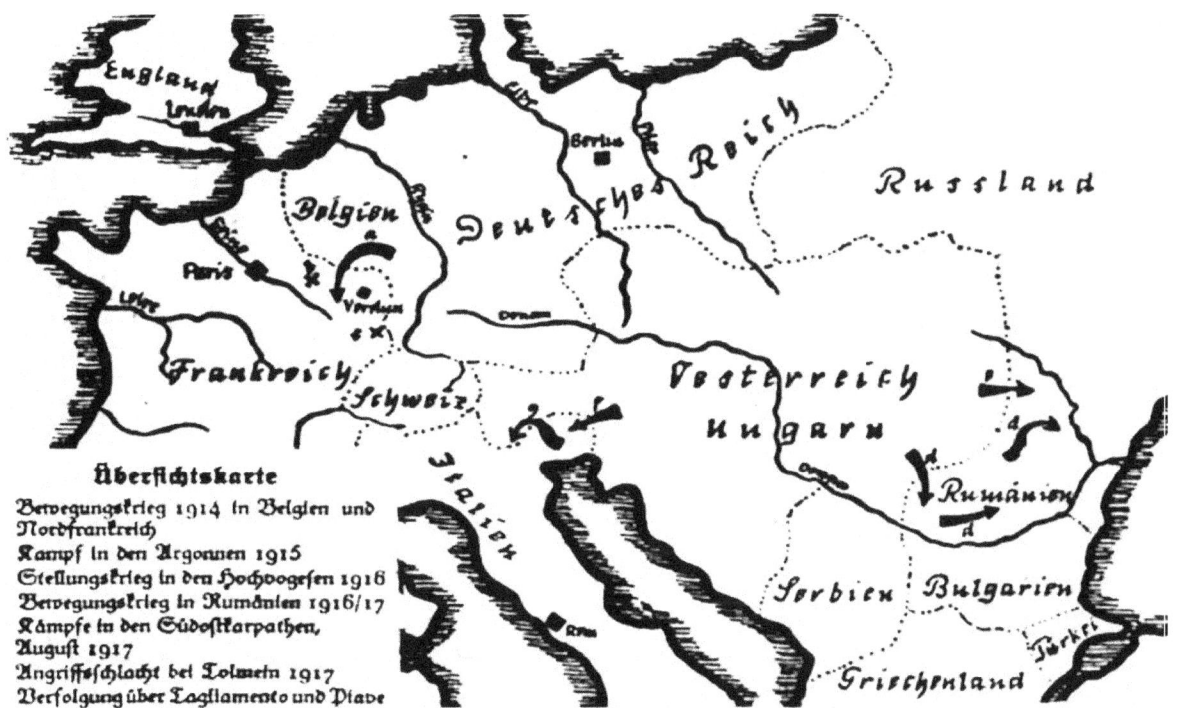

Rommel created this map of World War I Europe and marked all the locations where he saw combat along with the dates of when he was there, along with distinctions between mobile and static warfare.

with all due haste. Late in the evening we reached our garrison town Weingarten.

On Aug. 1, 1914, a flurry of intense activity dominated the regiment's barracks in the mighty old abbey buildings in Weingarten. There was an initial try-on of field gear!

I reported myself as having returned from my command and greeted the men of the 7th Company, whom I now would ostensibly accompany to the battlefield.

There was a light of cheerfulness, energetic enthusiasm and eagerness for action in all the young faces. Is there anything more splendid than to lead such soldiers against the enemy?

At 1800 hours regimental muster was called. After *Oberst* Haas punctiliously inspected the regiment, who were all dressed in field gray, he gave a rousing address. As we were dismissed, we received the command to mobilize. Now it was decided! A cheer of battle-ready, German youth rang and echoed through the old, gray abbey buildings.

The second of August 1914 was a sober Sunday! A farewell church service for the regiment was held amid streaming rays of sunlight. In the evening, the proud 6th Württemberg Regiment marched with ringing band music and snapping strides from the garrison to deployment from Ravensburg. Crowds of thousands of people accompanied. Endless streams of military columns rolled in short succession toward the threatened western border. Accompanied by unceasing cheers, the Regiment departed as night fell.

To my great distress, I was obliged to remain for another few days in the garrison to bring up reserves. I worried that in doing so I would arrive too late to the first battle.

The journey to the battlefield on Aug. 5 through the splendorous mountain valleys and meadows of our German homeland, accompanied by the exultation of all the local people, was indescribably beautiful. The troops sang song after song. At every stop, we were showered with fruit, chocolate and bread rolls.

Erwin Rommel reclines on a hillside during World War I circa 1915.

In Kornwestheim I saw my mother and siblings only for a few minutes; then the sharp whistle of the train urged our separation.

At night we crossed the Rhine. Spotlights probed the sky for enemy planes and airships. The songs faded away. The soldiers slept on benches and on the ground in sitting positions and lying down. I stood on the train, stared into the flickering fire of the engine and then looked out again into the rushing noise and whispers of the dark, humid summer night. What could the next days bring?

On Aug. 6, we arrived by evening in Königsmachern near Diedenhofen[1], happy to be released from the tight quarters of the transport train. We marched through Diedenhofen to Ruxweiler. Diedenhofen presented us with no pretty picture. Dirty streets and houses, few friendly people! That is so very different from at home in my native region of Swabia.

The march went briskly forward. As night fell, we found ourselves in a cloud-shattering lightning storm. Soon we had no dry thread of clothing left on our bodies and our rucksacks pulled down on us heavily. A fine beginning! The sound of sporadic gunshots rang from a great distance.

After a six-hour march, the column arrived unharmed in Kurzweiler at midnight. The company commander, *Oberleutnant der Reserve* Bammert, awaited us. We sheltered on straw in tightly crammed quarters.

---

1. Present-day Thionville, France.

Rommel marches in step with fellow German infantrymen during World War I. Rommel appears in the front row, third from right. His step and posture while marching resemble a famous photo of him taken during World War II (see page 79).

"Soldier Songs"— *At the Border*

In the days that followed, the company, now at full strength for war, was forged together by unrelenting practice. Aside from drill and company exercises, all forms of combat exercises were taught, most often and especially work with spades. Afterwards I spent a few rainy days as an outguard with my platoon in the vicinity of Bollingen. Nothing transpired. However, myself and some of my men were weakened due to stomach problems caused by the greasy rations from the field kitchen and freshly baked bread.

On Aug. 18, the advance to the north began en masse. I rode the second horse of my company commander. With cheerful songs, we advanced across the German-Luxembourg border. The population was friendly and brought fruit and drinks among our marching columns. We moved into billets in Budersburg.

Bright and early on Aug. 19 we passed below the cannons of the French fort at Longwy and across towards the northwest. Bivouacs were drawn up in the afternoon at Dahlem. In the evening, regimental music played amid the tents. By the light of the sinking sun, soldier songs were sung. All of us felt it: the first fight was near.

My stomach really gave me a hassle. I nourished myself only on Zwieback crackers and chocolate, but without making any recovery. I did not want to go the doctor so as not to be suspected of shirking from duty.

After a very hot march on Aug. 20 we reached Meix-le-Tige in Belgium. The 1st Battalion went to the outposts. The Second Battalion (outpost reserve) provided local security. The population was very withdrawn and cagey. A few enemy planes appeared in the cloudless sky and were unsuccessfully shot at.

"I Had To Be Tough!"—Reconnaissance against Longwy and Preparation for the First Fight

> ◊ Rommel is unexpectedly summoned by a colonel and ordered to take a patrol to scout enemy positions. His efforts to complete his mission get off to a very awkward start and continue to be clumsy as he navigates dangerous territory.
> ◊ Afterwards Rommel is caught up in a dispute between his military superiors. Assigned to messenger duty, he tramps back and forth to relay messages between officers. Rommel discovers that army life in wartime is more rigorous than he seems to have expected. Constant marching and orders drain his strength, leaving him hungry and exhausted. He recounts feeling frustrated as he misses opportunities for food and rest.
> ◊ Instead of becoming more sympathetic to his own men over time, Rommel will become almost mercilessly exacting in the demands he places on his troops as the war progresses, sometimes driving his men to the point of utter exhaustion.

The following day was supposed to be a day of rest. Yet already in the earliest morning hour, I was summoned to *Oberst* Haas with some other comrades and ordered to take a five-man reconnaissance detachment across Barancy—Gorcy to Cosnes by Longwy, located 13 kilometers away, to clarify and firmly establish where exactly and how strong the enemy was. In order to move ahead quickly, we were supposed to drive a hay wagon up to the border of our frontline positions.

A harnessed wagon already awaited us in Meix-le-Tige. However, the ghoulish nag of a Belgian horse carried us away with a loping gallop. Trying to bring the horse to a stop did not help. In the end we landed uninjured on a pile of manure. The wagon was broken in half. We resumed our journey on foot.

Fundamentally more cautious than during peacetime exercises—since one becomes very conscious of his responsibility for the lives of his men!—we traversed through ditches and alongside the street through hedgerows and grain fields toward Barancy, which was reported on the previous day to be occupied by weak enemy forces. We found the place free of enemy combatants.

Traveling off road through grain fields, we crossed the Belgian-French border, reached the southern edge of the Bois de Mousson and then descended towards Gorcy. *Leutnant* Kirn's patrol, which had been following me, camouflaged itself on the hill facing our advancement through the area. On the dusty road we recognized fresh tracks from French infantrymen and riders retreating in the direction of Cosnes. Now we became even more cautious!

In the thick underbrush a few meters parallel to the street, I continued the march forward, constantly keeping a sharp eye on the street, and reached the edge of the forest 500 meters west of Cosnes. From that standpoint I systematically surveyed the entire terrain with binoculars, especially all town boulevards. Nowhere, however, were there any French troops to be seen.

During a further advance across an open field, we came across an old woman peacefully at work. She explained to us in German that the French troops had pulled out from Cosnes and headed to Longwy an hour ago. In Cosnes and its immediate surroundings, there were presently no troops to be found. Was her statement correct? Vigilantly we stalked through grain fields and orchards to our destination. Then we moved towards Cosnes, each of us with a fixed bayonet and finger on the trigger, keeping all windows and open doorways under sharp observation.

However the residents behaved very trustworthy and affirmed what the woman in the field told us. They voluntarily brought us food and drinks. We remained suspicious at first and asked the benefactors to sample their own food and drinks before we took any refreshment. Since the local civic leader was nowhere to be found, I obtained from a church official six good-quality bicycles in exchange for receipts. These were intended to speed up the delivery of my report to the regiment.

We rode on the bikes for another 1.5 kilometers towards Longwy, where French outworks were now bombarded by heavy German artillery fire. Far and wide, enemy troops were nowhere in sight. The mission of my platoon was fulfilled.

With great spaces between each of us as we rode, we now whizzed downhill on our bikes—clutching our rifles under one arm, ready to fire—through Gorcy toward Barancy. From there I rode ahead of my men in order to give my report speedily. On the village street of Meix-le-Tige I met the regimental commander in person and gave my report.

Tired and hungry, I went off afterwards to my quarters, already rejoicing at the thought of a few hours rest. Nothing came of that. I found my battalion standing there ready to march away. The dutiful Hänle had already packed my belongings and saddled my black horse. Furthermore there was no time left to eat. The battalion moved off.

We marched to the hill located one kilometer southeast of St. Leger. The sky became overcast. Intermittent sounds of gunfire and sporadic artillery shots came from the south. It became known that part of the 1st Battalion, which still remained at the outposts near Villancourt, had already clashed with the enemy in the afternoon near Mussy-la-Ville.

At twilight the regimental staff, along with the 2nd and 3rd Battalions, moved towards Hill 312, located 2.5 kilometers south of St. Leger. The men settled down for the night under the open sky with rifles in their arms. Security was posted one kilometer to the southwest.

I had just settled down to sleep on a bundle of oats amid my platoon with every intention to immediately sleep, to regain my strength for the next day, when I was summoned to the forward area of the

Men of Rommel's regiment are stuck in a traffic jam while moving through a town circa 1914.

regiment. *Oberst* Haas asked me if I was willing to make a journey through the woods to the 1st Battalion at Villancourt. At this point, should I have protested that I was completely exhausted from my explorations around Cosnes? No. After contact with the enemy has been made, extraordinary efforts must be performed immediately.

I received the order to convey a regimental command to the 1st Battalion: that they should move along the shortest possible route to Hill 312 about 2.5 kilometers south of St. Leger. I was ordered to accompany the Battalion personally as they moved up. I went on my way accompanied by *Unteroffizier* Gölz and two men of the 7th Company. Meanwhile it grew pitch-dark. We followed the compass through the meadowland southeast of Hill 312. Slightly ahead to our right we heard the voices of our sentries calling out. Gunshots rang intermittently. Soon our path took us into a deep forest straight up a steep hill. From time to time, we stopped and listened tensely to the noises of the night. After arduous climbing and groping in the dark, we finally reached the densely wooded mountain crest west of Villancourt. The glow of fire from the burning fortress of Longwy, set alight by bombardment, lit the landscape from the southeast.

We climbed through thick bushes towards Villancourt. Suddenly a sentry called out from almost right next to us: "Halt, who goes there?" Was he German or French? We knew that the French often called out in German [as a deception]. Quickly we dove to the ground for cover.

"Watchword?"

None of us knew it. I then called out my name and rank—and was recognized.

At the edge of the forest stood the sentries of the 1st Battalion. It was not far to Villancourt. We met companies of the 1st Battalion resting in closed formation on the street towards Mussy-la-Ville. I conveyed the command of my regiment to the battalion commander, *Major* Kaufmann. However he could not carry out the order since his battalion remained under the command of the Langer Brigade. I was led to the headquarters of *General* Langer on the hill one kilometer southwest of Villancourt and there repeated my orders. *General* Langer relayed a response to my regimental commander—through me—that all the troops of his brigade were not yet assembled there and for the time being he could not spare the men of our 1st Battalion to move off [in the way my regimental commander wanted].

Utterly crestfallen at the futility of my mission and physically spent, I trod the path back to Hill 312 with my three companions. Midnight had already passed when I reached the forward area. I woke up the regimental adjutant, *Hauptmann* Wolters, and gave my report. *Oberst* Haas listened to my report. He was displeased and ordered me to walk or ride to the 53rd Brigade near St. Lager to personally inform the brigade commander there, *General* Von Moser, that the Langer Brigade refused to release our 1st Battalion.

I dutifully repeated his message. Should I have said that this demand was totally beyond my physical strength? That I had already been traveling constantly for 18 hours and was now totally exhausted? No, I had to be tough! The mission had to be accomplished. I grabbed the second horse of the company leader, fastened the saddle belt and rode off to the north past my sleeping comrades.

I met *General* Moser in a tent on a dense hill. He, also, was extremely indignant at hearing my report. He ordered me to go straight back again to Villancourt and tell *General* Langer that he needed to release my regiment's 1st Battalion—that the battalion needed to be reunited with the rest of my regiment by daybreak no matter what the circumstances.

I alternately rode and walked over rough country, through thick forests, and up and down steep hills in the darkness to Villancourt and unburdened myself of my report there. At last when I returned to my regiment on Hill 312, it was already dawn. The companies were all standing ready for battle. Coffee had already been served. The field kitchens had already gone. My loyal buddy Hänle helped me by giving me a drink from his canteen. Slowly the daylight became brighter. Thick, wet fog enveloped us. Orders were issued in the forward area.

> **Considerations:**
> - When facing the enemy, **a patrol leader** will become aware of t**he weight of his responsibility** for the lives of his men. Every **error incurs a cost in casualties**—perhaps the lives of the entire team. Therefore, extraordinarily cautious and analytical stalking of the enemy, use of camouflage, indirect approaches to streets and paths, constant surveillance of the terrain with binoculars, **thorough organization within the patrol**, and covering fire in place before crossing open terrain must be implemented. When **entering provincial towns**, advance with the patrol divided into groups on the left and right while approaching houses and have weapons ready to fire. Make quick reports back, because the information you gather becomes less valuable with every delay.
> - Peacetime **training** should include **navigation at night with an illuminated compass** in rugged, trackless, forested terrain.
> - War imposes totally extraordinary and severe challenges on the **performance ability** and **versatility** of a soldier—therefore make high demands of soldiers during peacetime! —*Rommel*

"A Horrific Sight"—Combat at Bleid

Shortly before 0500 hours, the 2nd Battalion approached Bleid in the direction of Hill 325 located about two kilometers northeast of the town.

A thick ground fog wafted across the dewy meadows. Visibility was hardly 50 meters ahead. The battalion commander, *Major* Bader, sent me to scout the way ahead to Hill 325. Since I had been active practically nonstop for 24 hours, I could hardly keep from falling out of my saddle due to extreme exhaustion.

The terrain on either side of the country road I rode down was layered with numerous hedges and enclosed pastures. Using a map and compass, I located Hill 325. After moving there, the battalion halted and spread out on the northeastern slope.

◊ Rommel has his first intense combat experience and makes several amateurish mistakes. He gets lost in fog, chases enemy soldiers through a potato field and ends up losing contact with his comrades. Rommel proves often unable to resist chasing after fleeing enemies, which often causes him to lose contact with his own forces. He was criticized during World War II for this.

◊ Rommel attacks more than 30 enemy soldiers in a street with only three comrades. Three against more than 30 is a risk few men would take. Why did he tempt disaster?

◊ He tries to smoke snipers out of hiding and causes a huge fire. Two of his best friends are killed and his regiment loses many men. Rommel's postwar "Considerations" infer he

Soon afterwards our advance security patrols on the southern and western slopes clashed sporadically with the enemy in the fog. Brief exchanges of gunfire broke out in various locations. Sporadic rifle bullets passed above our heads with whistling sizzles—a unique noise! One officer who rode barely 100 yards toward the enemy was shot at from a location almost immediately next to him. Acting speedily, several of our men managed to lame a fleeing Frenchmen wearing red uniform trousers with a gunshot to the leg and take him prisoner. Now we began to hear German commands issuing out of the fog behind us to our left. "March halfway left! Spread out!" A group of riflemen materialized—the right flank of the 1st Battalion. I was ordered by my company commander to deploy my platoon and advance in coordination with the 1st Battalion on the southeast of Bleid.

> regretted his mistakes; he expressed that, due to loss of life, violent battles in towns should be avoided.
>
> ◊ Notably, Rommel's behavior to enemies changes to become more compassionate. At first he flushes French riflemen out at bayonet point but, when he encounters French soldiers hiding later on, decides to "convince them to set aside their weapons." He and his men offer them cigarettes. Rommel later adopted similar approaches to POWs during World War II. Attempts to spare enemy lives and mercy to prisoners is a consistent theme in this book; later on, Rommel will risk his own life to save an Italian POW from drowning. Although mercy was not a Prussian military virtue and publicly derided as weakness in Hitler's Germany, Rommel was very open about his humane treatment of enemy soldiers in this book which he intended as a teaching manual.

I handed over my black steed to Hänle, exchanged my Browning for his bayonet and then incorporated my platoon into the advance. We advanced in light battle formation through potato and cabbage fields. Thick plumes of fog wafted across the fields and now fought against the brightness of the steadily rising sun. Visibility was hardly 50 to 80 meters ahead.

Suddenly a volley of bullets struck out at us from close range. We dove to the ground and lay well-camouflaged between the leaves of potato plants. The next salvo passed high over us. In vain, I searched for the enemy with binoculars. Since the enemy was clearly nearby, I mustered my platoon for a direct assault. However, the French made off quickly before we could spot them, leaving behind clear tracks through the muddy cropland. We followed them in the direction of Bleid. In the heat of battle, we lost coordination with our forces on our left.

Yet again, my platoon was fired at from the fog. As soon as we charged to attack, the other side speedily gave ground. After that, we advanced about another 800 meters without harassment from the enemy. Presently a fence topped with high hedges rose up before us from the fog. Behind to our right, the silhouette of a farm building appeared, while on our left was a grove of tall trees. The enemy footprints we followed turned uphill to our right. Were we now actually in Bleid? I ordered my platoon to stay undercover in the hedges and sent out a reconnaissance patrol to reestablish contact with our lost neighbors on the left and our own company. Until this point, our platoon had sustained no losses.

To scout the farm building ahead of us, I went ahead myself and took *Vizefeldwebel* Ostertag and two distance estimators with me. Nothing was seen or heard of the enemy. We reached the east side of the farmhouse. Here, a narrow footpath led downward to a street on the left. Across from us, a second farmhouse materialized in the fog. There was no doubt we were now on the outskirts of Bleid towards Mussy-la-Ville. Cautiously we approached the street.

I peeked around the corner of the farm building. There! Hardly 20 steps to my right, standing the middle of a street, were 15 to 20 Frenchmen, drinking coffee, chatting, holding their weapons casually slung in their arms! They did not see me.

I went back into cover behind the building. Should I call my platoon over? No! The four of us could certainly handle this situation. Hurriedly I signaled my companions about the intended surprise fire. Quietly we released the safety catches on our rifles, then sprang out from around the corner and stood there shooting at the nearby enemy. A few of the opponents remained dead or wounded on the open ground, but the vast majority scattered in all directions, reached cover in the immediate area behind flights of house steps, garden walls, and woodpiles and returned fire. Thus a very intense gunfight developed at a very short distance.

I stood by a woodpile during the attack. My opponent lay 20 meters ahead of me, well-concealed behind the front steps of a house. Only a small piece of his head was showing. We aimed at each other, both pulled the trigger almost at the same time—and both missed. The opponent's bullet whistled by a hairsbreadth past my ear. Now I had to load fast, aim quietly and quickly and hold steady. That is no easy task at a distance of 20 meters when your rifle scope is set for 400 meters, because that is never practiced in peacetime. My shot cracked. The opponent's head fell heavily onto the steps in front of him.

About 10 Frenchmen remained facing us; some of them were now in full cover. I signaled my men to charge. With a loud battle cry we rushed forward onto the village street. At this moment, in the blink of an eye, French soldiers appeared in all windows and doorways and shot at us. Against this overpowering force, there was nothing we could do. We quickly retreated and returned—all four of us, unharmed—to the hedge where I left my platoon. The platoon had already rushed out of cover to provide help. That was no longer necessary. I took the platoon back under cover. We observed that shots were increasingly being fired at us through the fog from the farm building on the opposite side of the street. The shots whizzed past high above us. I recognized—from hardly 60 meters away—that the enemy was firing not only from the second floor of the building, but also from inside the roof.

A bunch of rifle muzzles were sticking out from under the roof panels. It was impossible for the enemy to hit a target using their rifle sights this way. No wonder why the enemy bullets whirred past so high above our heads. Should I wait until they came out from cover, or immediately storm the town outskirts with my platoon? The latter option seemed right to me.

The highest concentration of enemy strength was in the farm building on the opposite side of the street. Therefore this had to be our first target. My attack plan: open fire on the enemies in the second floor and attic using one half of my platoon hiding behind the hedge, and thoroughly storm and capture the building with the second half of my platoon!

Quickly, my assault troop armed itself with a large piece of timber lying around. This made an ideal battering ram to break through doors and openings. We also made a whisk of straw which would work to smoke enemies out of their hiding places. The second half of the platoon waited ready to fire from behind the hedge, and the storm troop had completed its preparations in full camouflage. Everything was ready to go.

At the signal, half the platoon opened fire. With the other half, I led the assault across the street using the same path that I had retreated on with my men a few minutes earlier. The enemies in the building opened rapid fire, mostly aimed at the other half of my platoon hidden in the hedge. Their fire could not hit my storm troop, which had just reached protection at the wall of the house.

My men bashed through the doors with heavy blows and splintering crashes. One half of the barn door flew off its hinges. They hurled whisks of burning straw onto the threshing floor, which was covered with grain and fodder. We surrounded the farm building. Whoever wanted to get out jumped straight into our bayonets. The pale flames soon burst out through the roof ridge. The surviving enemies cast down their weapons. Our only casualty was one man slightly wounded.

Now we stormed from farm building to farm building. The second half of the platoon joined us. When we confronted opponents with immediate physical danger, they either surrendered immediately or hid themselves in the recesses of the buildings. Yet my soldiers tracked them down even in those places, and flushed troop after troop of Frenchmen out of hiding with extreme cool-headedness.

Further elements of the 2nd Battalion, mixed together with parts of the 1st Battalion, moved everywhere with driving force through the town, which was now burning in many places. Confusion broke out among the various groups. There was shooting from all sides. Casualties increased.

I stormed a church located in a side street and surrounded by a wall. Heavy fire was striking our ranks from that location. Using available cover, we sprang from house to house, nearer approaching the enemy. As we rushed forward in the assault, the enemies retreated to the west and quickly escaped our firing range through the fog.

From our left flank—the southern half of Bleid—came a sudden burst of lively fire. Our casualties multiplied. Everywhere, frightened cries for medics rang out. Wounds were dressed behind a washhouse. A gruesome sight! Most of these were severe shot wounds. A few of the men screamed in pain; others held still and stared approaching death in the eyes like heroes.

Lots of French soldiers remained in the northwest and southern parts of the town. The village behind us blazed like a furnace. Meanwhile, the sun had become master over the fog. There was nothing more to be done in Bleid. Therefore I gathered what I could, let the wounded be taken away and moved off to the northeast. I wanted to get away from the source of the fire and reestablish contact with my own company, particularly the 2nd Battalion.

The way was blocked by fire, thick asphyxiating fumes, smoldering timbers of caved-in houses, and frantic livestock. Finally—half-suffocated—we made it into the open. First we tended to the numerous wounded, then I organized our group of about 100 men and led them into the valley 300 meters northeast of Bleid. I left the platoon there, spread out and facing west, and went with the squad leaders to scout from the nearest rise in terrain.

To our right stood Hill 325, still encircled by flowing plumes of fog. Neither friend nor foe was recognizable in the fields of tall grain south of it. From the edge of a golden, freshly tilled wheat field, the red trousers of a French rifle formation gleamed out at us—about 800 meters in front of us to the right, opposite a deep ravine. (This was the 7th Company of the French 101st Infantry Regiment.)

In the lower terrain to our left, an increasingly violent fight churned for possession of the burning town of Bleid. Where were our company and our 2nd Battalion? Were they with the groups still left in the town, or with the forces that still lay further to the rear? What was I supposed to do?

Since I did not want my platoon and myself to remain inactive, I decided to attack the enemies across from me, who also happened to be located directly in the advance route of our 2nd Battalion. The camouflaged deployment of our platoon behind the hill, the move into position and our opening fire was executed with the same silence and precision as during peacetime exercises on training grounds.

Soon the groups were echeloned partially among the leafy potato plants, partially behind sheaves of oats. They were well-camouflaged and fired steady, well-aimed shots—as we had been trained during our excellent peacetime schooling. The enemy vigorously opened fire as soon as the foremost part of the platoon moved into position. Yet the enemy's shot groupings struck far too high. Merely a few shots struck the area in front of and next to us. We soon became accustomed to it. After a gunfight spanning one quarter of an hour, the lone casualty my platoon had to report was a bullet-riddled mess kit. German rifle formations appeared on Hill 325 from 800 meters behind on our right. Cooperation with our forces was now assured. Now my platoon could directly attack without worries.

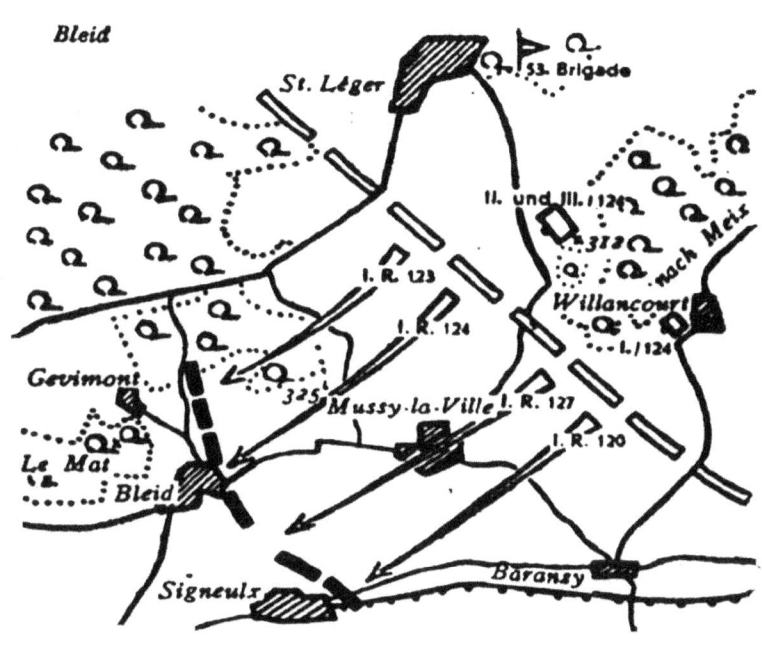

Rommel's map of his first fight at Bleid in 1914.

We sprang forward in groups with synchronized mutual fire support. This maneuver was practiced quite often in peacetime. We crossed a basin that enemy shots could not reach. Soon I had my whole platoon assembled on the same hill in a blind spot to the enemy. Thanks to our enemy's poor performance at shooting, my platoon until this point had suffered no setbacks in the offensive combat.

With fixed bayonets, we stalked within storming distance of the enemy position on the rising slope. During this maneuver, we were not bothered by enemy fire since it passed high over our heads and was mostly aimed at the sections of the platoon still further back in the rear. Suddenly the enemy ceased firing completely. Did the enemy plan to assault us directly from the hilltop?

We charged quickly in an assault, but found the enemy position completely empty aside from a few dead bodies left behind. Traces of the retreating enemies led in a westerly direction through the rows of tall grain, which rose as high as a person's head.

Yet again I found myself and my platoon far ahead of our own forces. I wanted to wait for further action from our comrades on our right. My platoon occupied the territory we had taken. Then I went with the leader of the first platoon section, a *Feldwebel* from the 6th Company, and *Unteroffizier* Bentele to scout west through the tall grain field.

We wanted to determine where exactly the enemy had gone. The platoon remained in contact. Without opposition from enemy forces, we reached the road to Gevimont—Bleid about 400 meters north of the town of Bleid. The road ascended north through a path hewn into the hillside. On both sides of the road were tall groups of bushes. These bushes severely obscured our views to the north and west.

We looked out from one of these bushes above the road, casting a very sharp eye in all directions. Bizarrely, there was absolutely nothing to be seen of the retreating enemy. Suddenly, Bentele signaled with his arm to the

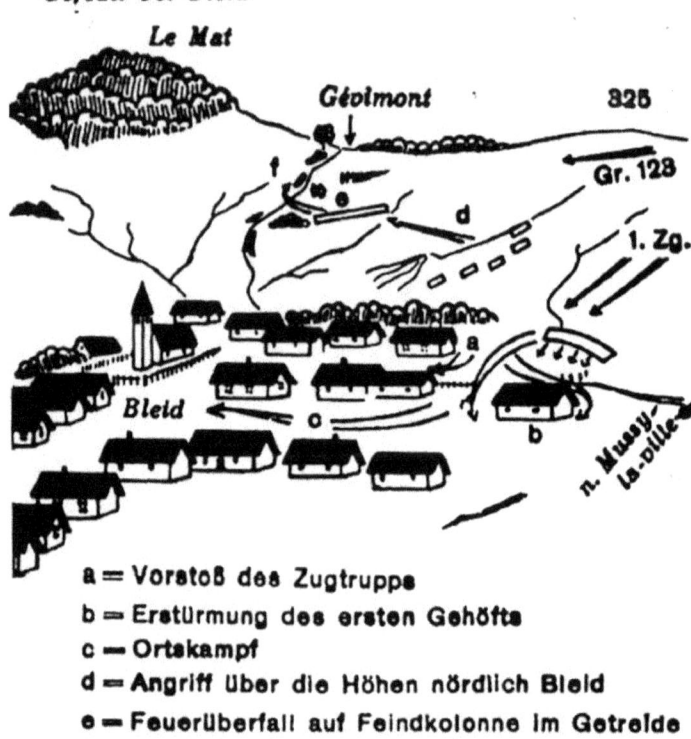

A more detailed map of the fight at Bleid. A) Forward thrust of the platoon; B) Storming of the first farmhouse; C) battle in the town; D) attack across the hills north of Bleid; E) surprise fire on the enemy column in the wheat field; F) storming the thickets on the route to Bleid-Gevimont.

right. Hardly 150 meters away from us, the grain stalks were moving. Through the grain, we glimpsed the twinkle of bare metal mess kits glinting in the sunlight from the top of French soldiers' loaded backpacks. The enemy over there was making a withdrawal, ducking to avoid the fire from our forces on Hill 325 which sprayed across the highest part of the western ridge.

An estimated 100 Frenchmen were coming straight at us in rows. None of the enemy lifted a head above the level of the tall grain stalks. (These were members of the 6th Company of the French 101st Infantry Regiment. They had been attacked on the slope west of Hill 325 by members of the German 123rd Grenadier Regiment and were now retreating to the southwest.)

Should I call the rest of my platoon over at once? No! They would do much better in their place as tactical support for the three of us. The penetrating force of rifle ammunition was plain to my eyes: [a single shot could take out] two to three men at this distance!

I quickly shot from a standing position at the front of the French column. The enemy soldiers ducked for a few seconds under the grain, yet resumed their march in the same manner and direction afterwards. Not a single Frenchman lifted his head out of the grain to look about for new enemies approaching them so suddenly and from such a close distance. Now, all three of us began shooting.

Again, the French column vanished for a short time, then split up into smaller sections hurrying west to reach the road between Bleid and Gevimont. We spread rapid fire among the ranks of the fleeing enemies. Astonishingly, we received no return fire, although we were standing upright and must have been very easy for the enemy to see. Fleeing Frenchmen now ran across the left side of the street across from the bushes where we stood. They were very easy to shoot at a 10-meter distance through a gap in the bushes. We divided our fire at different groups of enemies. Dozens of Frenchmen were put out of action by our three rifles.

Now our 123rd Grenadier Regiment advanced across the hillside to our right. I waved my platoon forward and stormed both sides of the Gevimont—Bleid road heading north.

We were very astonished to find Frenchmen hiding in all the bushes along the roadside. It took a lot of lengthy coaxing on our part to convince them to set aside their weapons and come out of hiding. They had been convinced that the Germans would slit the throats of all prisoners. Out of the bushes and grain fields we drew out well more than 50 men. Among them were two French officers—a captain and a lieutenant who was lightly wounded in one arm. (These were members of the 6th and 7th companies of the French 101st Infantry Regiment.) My men offered cigarettes to the prisoners, at which they visibly became more trusting.

The grenadiers also reached the road to Gevimont—Bleid from the hill to our right. Gunfire struck the area between our forces, aimed from the direction of the high forest of Le Mat, 1500 meters northwest of Bleid. Rapidly I moved my platoon into the cover of the ravine above us, planning to launch an attack against the enemies in Le Mat from here.

Suddenly, however, everything went black in front of my eyes. I lost consciousness. The exertions of the past day and night, the fighting for Bleid and on the northern hill, and not least of all the ravaged condition of my stomach had completely sapped the strength from me. I must have lain unconscious for a very long time. When I returned to my senses, *Unteroffizier* Bentele was tending to me. Sporadic French grenades and shrapnel struck in the surrounding area. Members of our own infantry were heading out of the Le Mat forest back towards Hill 325. A retreat?

I assumed command of a portion of these soldiers, occupied the hillside along the Gevimont-Bleid road and ordered them to dig in. The men in the ranks told me that they had suffered most severe casualties in the Le Mat forest, and had lost their commanding officer. Afterwards they retreated on orders from higher authority. Above all else, they had been badly shelled by French artillery.

A quarter of an hour later, the buglers blared the "Regimental summons!" and "Call to assembly!" From all sides, members of the regiment streamed in groups to the area west of Bleid. I also went there with my own little handful of men. Various companies arrived one after another. Their ranks were very thinned. In this first battle, the regiment had lost ¼ of its serving officers and 1/7 of its manpower in dead and wounded.

For me, it was especially painful to find out that two of my very best friends had been killed. Sadly it was impossible for me to render them final burial services.

After our groups were reorganized, our battalion moved off immediately to march on Gomery. Our path took us through the south section of Bleid.

Bleid presented a horrific sight. In between the smoking ruins lay dead soldiers and civilians, along with cattle that had also been shot. Our troops were informed that the enemy facing the 5th Army had been beaten across the entire frontline and had retreated. Joy over the first victory, however, was very subdued due to our grief for our fallen comrades.

We marched south. The march often faltered. Enemy columns moved in the distance. Batteries of our 49th Artillery Regiment moved forward and into position to the right of our marching path. As soon as their first shots fell, the enemy columns in the distance disappeared.

Night fell. Dead tired, we finally reached the village of Ruette. This place was already overcrowded with our own troops. We camped out in the open. Straw [for bedding] was no longer available to us. The troops were much too tired to conduct a long search for it. The moist, chilly earth of the farmland prevented sleep from being refreshing. Before morning it became cold—we all froze to the point of shivering. During the last half of the night, stomach problems kept me and other fellow sufferers constantly up and moving. At last, dawn broke. Thick fog spread across the fields again.

Considerations:

- It is difficult to maintain **contact** with other troops in a **fog.** In the foggy battle near Bleid, contact with our own forces was torn apart shortly after our clash with the enemy. It was not possible for us to immediately reestablish it. Advancing in a fog with a compass must be practiced since much is being done nowadays with artificial fog.
- During **a clash in fog against an enemy** of the same combat strength, whichever side is capable of firing the most rounds within the blink of an eye during this collision will win. Therefore carry machine guns ready to fire while advancing. **Urban battles**,

as happened during the battle of Bleid, often unfold across distances of only a few meters. Have hand grenades and machine pistols ready. Let covering fire from machine guns, automatic weapons and rifles take effect before you launch a direct infantry assault. Combat in urban areas tends to incur heavy casualties—therefore it should be avoided whenever possible. Pin down the enemy in urban areas with gunfire or blind them using artificial fog and attack them outside city limits.
- **Tall grain** offers good concealment—however, bare metal objects such as bayonets or mess kits can betray the location of troops. French troops failed to secure the city limits of Bleid, just as they failed to observe the battlefield during their retreat through the grain field and the ensuing skirmish.
- The German soldiers had, after the first exchange of gunfire, **a feeling of strong superiority** over the French infantrymen. —*Rommel*

"Cries of the Wounded"—At the Meuse, Engagements at Mont and in the Forest of Doulcon

◊ Rommel's interactions with enemy soldiers and his own men continue to develop. He changes his approach to battle, attempting to shield his men from danger using stealth and surprise.

◊ He describes the death throes of a wounded infantryman with great pity. Despite being caught in crossfire, Rommel risks his life to try and help this individual.

◊ Rommel has an eerie experience surrounded by the cries of wounded men. Instead of ignoring them, or carrying out "mercy" killings, Rommel helps all—even Frenchmen who fought him hours before. His behavior is extraordinary given Prussian military ideals then; emotional detachment and indeed ruthlessness were considered virtuous. In many cases, highly respected German and Austrian commanders have viewed coldness as toughness and mercy as weakness. In Hitler's Third Reich, German soldiers were being urged to cruelty and ruthlessness like never before. If Rommel had wished to court Nazi favor, he would have excluded his story of compassion on the dying Frenchmen or perhaps invented a brutal fiction worthy of Hitler's "Master Race." His anecdote of mercy to foreign foes did not correspond to Nazi ideology or rhetoric.

◊ Rommel's actions reveal his optimism. Other soldiers in his position might have decided, "Oh well, let's save our rations—these men are going to die anyway." Many

> soldiers at that time gave into bitterness. In the writings of Erich Maria Remarque, nothing really mattered. Others like Manfred von Richthofen thought of war as a "game" and that fallen enemies were just unlucky "sportsmen." Rommel is not so cynical. It seems his personal example influences his men. Despite total physical exhaustion, one of Rommel's men later volunteers to guide medics to retrieve the wounded. This young infantryman is the same Rommel who British soldiers would face many years later during World War II—who gave beer and cigarettes to British POWs like Capt. Roy Wooldridge and treated wartime enemies humanely.

In the days following the fight at Longwy, the enemy was pursued in a southwestern, and later western direction. In the Chiers-Othain area, it came to brief but heavy fighting. During this, the French artillery covered the retreat of their infantry through strong and extremely versatile fire, and in some cases sacrificed themselves. During the night of Aug. 28–29, the 7th Company of Infantry Regiment 124 formed an outpost south of Jametz. All positions and outguards dug themselves in.

On Aug. 29, the advance to the Meuse continued. During the rest west of Jametz, the 13th Engineers, who found themselves far ahead in the marching column, were ambushed by strong enemy forces who attacked from the nearby forest. Bitter close-quarter combat ensued; the engineers tackled the enemy using shovels and axes. There were heavy losses on both sides. The 123rd Grenadier Regiment and the 3rd Battalion of Infantry Regiment 124 intervened in the action. The fight ended with the capture of the governor of Montmedy fortress and 2,000 men who occupied it, who had wanted to cut a way through to Verdun. We passed right by the bloody battle scene.

Eastward of Murveaux, the French received us with shrapnel shot from the west bank of the Meuse, which ultimately did little damage. Their aim was set at a very high angle. In the glowing heat of the sun, we marched at about noon to Dun on the Meuse. The French artillery fire became stronger. In a forest 1,500 meters east of Dun, the battalion halted and deployed. The companies arranged themselves into columns amid the many branches of the leafy forest. Just afterwards, heavy French artillery fell scattering into that section of the forest.

We distinctly heard from across a great distance the sound of shells launching, then the grenades spinning towards us. Several seconds later, they growled through the forest roof of leaves and then burst with a massive crash, sometimes against trees, sometimes deep in the earth. Splinters wailed with gurgling tones through the air; clumps of dirt and branches pattered down on us below. At every impact, we huddled together and pressed ourselves flat to the ground. The incessant danger unnerved us! The battalion remained in that same position until evening. Our losses were astoundingly few.

Ahead of us at the edge of the forest—900 meters southeast of Dun—the 4th Battery of Field Artillery Regiment 49, with whom I had served a month earlier, was in a heavy firefight in a partially camouflaged position. They did not prevail against the numerical force and caliber of the far-off artillery of the French, and suffered losses of manpower and equipment. In the early twilight the 2nd Battalion withdrew toward Murveaux. We spent the night beneath the open sky. My stomach growled because I had absolutely nothing to eat that day aside from a handful of wheat kernels. There was no bread.

On Aug. 30, our battlefield worship service which took place before noon near Murveaux was brought to a speedy end by a French artillery barrage. The artillery duel over the Meuse intensified. To our great joy, 21 cm Mörser howitzer batteries with wheels were brought forward into position and soon sent their enormous shots whooshing towards the enemy. We spent the night of Aug. 31 in extremely crammed quarters in Murveaux. The next morning the 2nd Battalion went across Milly towards Sassey, and there crossed the Meuse via a pontoon bridge laid by some engineers, and began the march towards Mont-devant-Sassey as the advance guard of the 53rd Brigade. Shortly after the advance into this region, searches of each farm building resulted in 26 men of the active French Infantry Regiment 124 (the same number as our regiment) being brought out of the cellars.

On the southwestern route from Mont, the front of our infantry column was targeted by fierce fire from the wooded heights that dominated the landscape west of Mont. Shortly afterward our own artillery began shooting towards Mont from the summit southwest of Sassey and caused casualties. This happened due to the report of a mounted reconnaissance patrol, who had been shot at from Mont half an hour before. It took a very long time until the mistake was clarified and the battery adjusted its fire.

A platoon of the 7th Company was now deployed to attack the enemy on the hills west of Mont. The attack however was halted by strong enemy fire. Additionally the assault of another platoon had no success. Situated faraway in a dominating position, the enemy wreaked severe losses on attacking forces as they climbed the very steep slopes before they could fire a shot. After the attacks failed, the 7th Company retreated and received the command to immediately go help the heavily besieged Infantry Regiment 127 in the Doulcon Forest, located two kilometers south of Mont. The company went southeast through the village of Mont and climbed Hill 297 in loose row formations alongside hedges—thus vanishing from the sight of the enemy west of Mont. The company had hardly arrived in the forest and had closed ranks into a marching column when French shrapnel fire forced us to the ground. We found cover behind trees and in hollows in the ground. There was nothing to be seen or heard of Infantry Regiment 127.

At the order of the company commander, I went ahead with a pair of men to the southern edge of the Doulcon Forest to establish contact with the regiment. Encountering various bursts of shrapnel fire, we reached the southern edge of the woods. Our own troops, however, we found nowhere. Below us to the left, in the Meuse valley, Dun was under heavy French artillery fire. According to the sounds of the shots, French artillery was positioned behind all hills west of the Meuse near Dun. Our infantry and that of the enemy was nowhere to be seen.

After my return the company moved westward along a forest path. On the edges of a clearing about 100 meters wide, our marching column rested with security posted on all sides. Now the company commander dispatched reconnaissance patrols in different directions to establish the location of Infantry Regiment 127. They were hardly gone—and the company had only rested about five minutes—when the entire clearing fell under massive French shrapnel fire. Coming in abrupt bursts like rain during a thunderstorm, the lead shots clattered down upon us. We huddled behind trees and quickly removed our backpacks for cover. To spring sideways or forward was impossible due to the concentration of the impacts. Although the barrage lasted for several minutes, there were no casualties. The backpacks absorbed some shots, and only one man's uniform tassel was torn in shreds. We all found it puzzling how the French artillery had discerned our exact location in the middle of the forest, and how it was possible for them to direct their fire at us in such a short time. Was it just a coincidence?

Presently one of our patrols returned with a badly wounded man of Infantry Regiment 127. This man told us that his regiment had retreated hours ago. Nothing remained ahead of us in the forest but dead and wounded men. Whole battalions of Frenchmen had marched past him about two hours ago through the forest in a northerly direction, and were almost certainly still concealed in the woods.

To remain totally alone as the only company in the huge forest was hardly an attractive prospect under these circumstances. Should we, also, retreat? Our deliberations about this ended when an infantry battalion appeared behind us along the same route we had taken to arrive. After a short conversation with the battalion commander, the 7th Company accompanied them as an advance guard on their march west through the forest. My platoon and I became the infantry point men.

Barely five minutes into our march, fierce gunfire roared from the forest halfway to our right, accompanied by loud battle cries. The distance of the battle scene from us is difficult to estimate—I guess about one kilometer. Now, we headed towards the uproar of battle by marching along a narrow footpath. On both sides of us stood very dense thickets between tall, leafy trees.

Black shapes lay on the narrow, straight path several hundred meters ahead of us.

A German soldier with a fixed bayonet leads a column of French prisoners-of-war off into captivity.

Frenchmen? Bullets from there now began whistling past our ears. We went into the bushes on either side of us for cover. Our company spread out along both sides of the path. Meanwhile the enemy ahead of us shot in sweeping salvos. The enemy's cone of fire—mostly ricochets—clattered through the thickets. Slowly we moved, without shooting, into crawling positions and closer towards the unseen enemy with short lunges. As soon as we believed the enemy to be about 150 meters apart from us, we also opened fire through the thick bushes at our invisible foe, yet kept advancing. As platoon leader, I only had a few of my riflemen next to me; the others were all hidden from sight.

The woods ahead of us lit up. Judging by the noise of the fire, we were hardly 100 meters from the enemy. I launched an assault with my platoon. We reached a clearing. Rapid enemy fire struck at us from the opposite side. It appeared that the clearing, only about 50 meters wide, was completely overgrown with blackberry bushes in such a way that a breakthrough was unthinkable in this location.

The frenzied fire of the enemy forced us to duck to the ground. Now the firefight along the edge of the clearing began on our side. However, despite the close distance, not the slightest trace of the enemy was visible in the dense greenery. Now the other two platoons of our company swarmed into our line. We lay in thick firing lines with only 2 to 3 steps between us. "Keep firing and dig in!" came the command.

A few meters to my right, *Oberleutnant* Bammert lay near a strong oak tree. Under this heavy enemy fire, to move sideways or even backwards seemed totally impossible. Fortunately most of the enemy fire was aimed too high. Nevertheless, casualties soon began to occur. While a portion of our men kept up a lengthy gunfight, the others dug in while pressed flat against the earth. Due to the forest roots, this work was exceedingly problematic. Shot-up pieces of branches and leaves rustled incessantly down upon us.

Then—suddenly, what's that? Were we now getting shot at from behind? Shots from behind hit the ground next to me. Dirt sprayed into my face. My neighbor on my left screamed aloud and writhed in pain! A shot from behind had ripped through his whole body from his heel all the way to his shoulder. "Help, medic, I'm bleeding out!" roared the poor guy in total desperation, frenzied with pain, covered with streaming blood.

The fire from ahead increased. I crawled over to the wounded man, but there was no more help for him. With his face distorted with pain, his hands clenched into the earth, he grew stiller and stiller. Then a shudder passed through his entire body. The life of this brave trooper flew away.

The gunfire from ahead and behind, to which we were exposed without cover, drove us nearly insane. By all appearances, remnants of the battalion following behind us were joining the fight as they entered the area in which we were being fired at. Amid the thick forest brush, it was impossible to clarify the confusion. This elicited rapid fire from those opposite us.

Bam! A shot struck through the center of my shovel blade right as I was digging. Shortly after that, *Oberleutnant* Bammert took a shot to the thigh. I took command of the company.

A direct assault got underway from our right: beating drums, bugle calls, battle shouts, intermingled with the slow tack-tack-tack of French machine guns. We breathed somewhat easier. I ordered the 7th Company also to storm in assault around the clearing from the left. The troopers, glad to emerge from this dreadful situation, swarmed forward with the utmost determination. Those opposite us, however, now preferred not to wait for our onslaught. Only a few shots were fired at us during our assault. When we reached the opposite side of the clearing, the enemy had vanished into the thicket.

We hunted after them through the thick underbrush. I wished to pursue them initially with my company as far as the southern edge of the Doulcon Forest. Perhaps there one could again reach the enemy with gunfire across an open landscape. Believing that the whole company was following hot on my heels, I rushed ahead as quickly as possible with the foremost groups. We reached the southern edge of the Doulcon Forest, however, without running into the enemy. Ahead of us to the south lay Ferme de la Briere, located on the nearest rising terrain of a broad meadow. Behind that same rise in terrain, a little halfway to our right, was a French battery, firing salvo after salvo into the Meuse valley near Dun. Astonishingly, there was nothing to be seen of enemy infantry. Judging from appearances, the enemy in the forest had shrunk away toward the west.

Meanwhile it became apparent that contact within the company had been totally shattered. Altogether I had 12 men with me. Coming from the left, a patrol of the Infantry Regiment 127 bumped into me and revealed that their regiment would shortly launch an attack left out of the forest in the direction of Ferme de la Briere. Soon, skirmishing lines from the left would advance. Should I wait for company to catch up? Or should I take my few men seize the French battery behind the hill nearby?

I decided on the latter option as I expected the company would soon catch

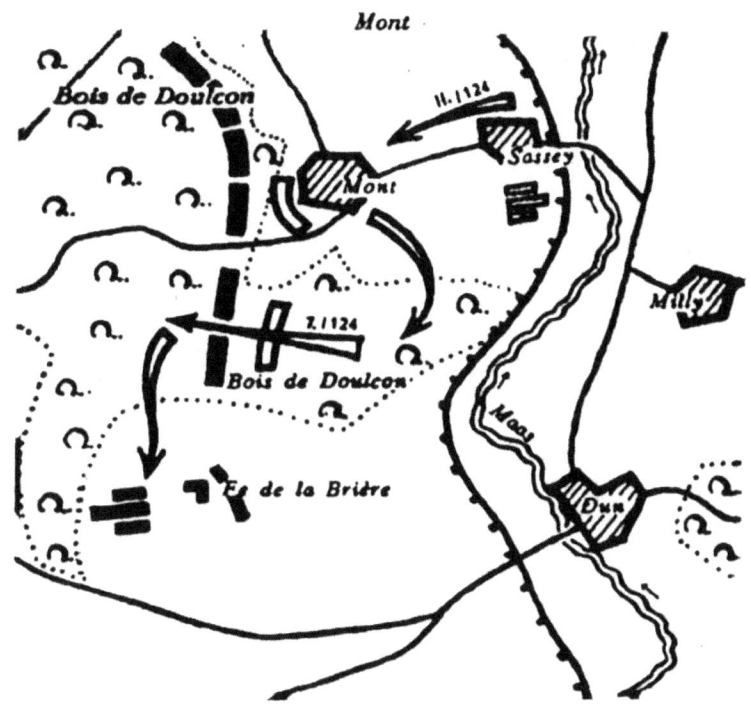

Rommel's map of actions at Mont and the Doulcon Forest.

up. We reached the area in lunges and began to climb about 600 meters west of Ferme la Briere in the direction of the French battery. Judging from the clash of the shots they were hardly about 100 meters away from us. Beyond on our left, the foremost elements of Infantry Regiment 127 pressed forward into Ferme la Briere.

Night began to fall. Suddenly, rifle fire from Ferme la Briere was directed at us. Undoubtedly, the men of the 127th mistook us for Frenchmen. The fire grew stronger and drove us to the ground. We tried to clarify the mistake by waving with our helmets and handkerchiefs. All in vain! Nowhere in our nearest surroundings was any place to take cover. Shots burst thickly all around us into the grass. We squeezed our bodies against the earth, yielding quietly to the hard fate of being fired at by our own troops for a second time within mere hours. Seconds became eternities. My brave men—mostly all reservists—moaned aloud when shots hit close to them. Lying totally still, we longed for the approaching darkness, which we hoped would rescue us. Then finally, finally—the firing subsided. In order not to bring it down on us again, we remained lying motionless for a while longer.

Afterwards we crawled back across the ground behind us. Our disengagement succeeded. To my joy, not one of my 12 men was wounded. It was too late to attack the French battery—also I had lost the will to do it. The moon emitted scarce light through light clouds. We withdrew to the Doulcon Forest—the afternoon's battleground. Nothing was to be seen or heard of our own company. Later I found out that, shortly after our charge in the forest, a technical sergeant assembled the whole company and led them in retreat on the shortest road back to the Mont after a trooper reported that I was killed during the assault.

Treading through the Doulcon Forest, we heard faint wailing and cries of the wounded coming from right next to us. The sounds echoed chillingly through the quiet night. "Comrade! Comrade!" a weak voice called out from the nearest bush. A young man of the 127th with a shot through the breast lay freezing on a large stone. The poor guy choked, as we attempted to help him, that he did not want to die. We wrapped him up in a coat and a shelter-half, arranged him in a better lying position and gave him drinks from our canteens.

Now cries from wounded men rose up from all sides. One called out heartbreakingly for his mother. Another prayed aloud. Others wailed in pain. French sounds were mixed in to all this: *"Des blesses, camarades!"*

This din of suffering and dying people was utterly disturbing to listen to. We helped friend and foe without discrimination, as much as we were able to, giving our last piece of bread and the last drops from our canteens. In the darkness, it was impossible to carry the severely wounded cross-country through the dense forest thickets without stretchers; they would have died of excruciating pain during the journey.

Exhausted—we had eaten hardly anything since morning—we reached Mont shortly before midnight. The locality had suffered much during the afternoon's battle. Countless houses were shot up. Dead horses lay in the streets. In one building I came across a medical company. I gave their leader directions to the place in the Doulcoun Forest where we had encountered the wounded and arranged for their rescue. One of my men volunteered to lead the medics as a guide. Afterwards I turned my attention to finding lodgings for the night. No trace of my own battalion was to be found.

Light was still burning behind the shutters of a window. We entered the house and found a dozen women and young girls, who initially were terrified of us. With a few French words, I asked for some food and quarters for the night for my little flock of men. Both were granted, and soon we were sleeping on clean mattresses. I woke at daybreak. We took up the search for our 2nd Battalion and found them encamped east of Mont.

Our appearance was met with great amazement. All of us were long ago written off as dead. *Oberleutnant* Aichholz took command of the 7th Company. That evening we were billeted in Mont. After our French host heavy-heartedly conceded two bottles of wine to Hänle and myself, I slept magnificently in a proper bed. Unfortunately, I was reminded for many days of this luxurious bedstead by bug bites.

Considerations:

- **The ambush against the engineer company resting en masse** teaches us that, even when massed in a large group, every unit must secure itself during rest periods. This is especially true in terrain that is not easy to observe and when one must reckon with fast enemy troops.
- In the forest east of Dun, the 7th Company, **formed in a column**, lay for a long time **in the path of heavy French artillery fire**. If a single shell had struck within the column, one to two groups of soldiers would have been annihilated at once. Increased spreading out of troops and entrenchment of single men must be required today, due to the increased potency of weapons. Begin digging in **before** the first enemy bombardment begins! It's much better to do too much shoveling work than too little! This work saves blood. As the example in Mont shows, detailed reconnaissance of locations you are walking through during an advance into enemy territory is advisable. Maybe the 26 French soldiers were shirkers avoiding duty; it could also have been they had orders to shoot German troops if it came to a fight so close west of Mont.
- **A mounted patrolman's report**, in which he stated he was fired at a half hour previously from Mont, **provoked our own artillery fire** at Mont, into the area where our own Infantry Regiment 124 was moving into. This caused casualties. It is extraordinarily important to maintain communication between artillery and infantry. Artillery must continuously observe the battlefield.
- The surprise fire of French artillery in the Doulcon Forest demonstrates that **to march or to stop in a marching column** within firing range of enemy artillery is wrong. With the powerful effect of artillery today, the most severe casualties would occur.
- The **fight in the Doulcon Forest** shows so perfectly the **difficulty** of forest fighting. A person sees nothing of the enemy. Shots clatter with loud bangs against the trees and branches. Countless ricochets zip through the air. It's hard to say which direction the enemy is shooting from. It's difficult to maintain a sense of direction and communication. The leader in the foremost lines only has influence over his immediate surroundings, and control of the remaining troops easily slips from his hand. Digging in under fire is extremely complicated due to forest roots. A situation will grow desperate in the frontline if —as in the Doulcon Forest—rearward elements of one's own troops open fire, because the frontline will be caught in a crossfire. Whether advancing or in a forest fight, it is equally advisable to have many machine guns in the foremost lines as much as possible. Shooting with machine guns while moving is appropriate for both surprise collisions with the enemy and storming assaults. —*Rommel*

"Under My Command"—*Combat at Gesnes*

◊ As if in a spell of destiny foreshadowing his later life, a motley assortment of leaderless men rally to Rommel and place themselves under his command. Young Rommel finds this very exciting. He organizes his miniature army in a good location and hopes to lead it into battle—but when his superiors find out, an "older officer" takes command immediately. Rommel is left crestfallen.

◊ Rommel's care for his horse stands out in this passage. Some German officers viewed animals as mere beasts and simply the means to an end. In fact, the German army during World War I were so callous to their horses that horses died at an unprecedented rate, resulting in a shortage of horses in Germany, which contributed to the famine known as the "Turnip Winter." Rommel's writing reveals a consistent affection for horses and he took special measures to protect his horse from danger. One could argue that Rommel cared for the animal merely for practical reasons. However, Rommel was very fond of animals and demonstrated a tendency throughout his life to collect pets. His writings suggest he had personal affection for this horse. He described placing his horse away from enemy fire. He also expressed concern about giving the animal "a good place" to sleep and added that the horse "urgently needed care after the strenuous days and the cold night."

In the wee morning hours of Sept. 2, 1914, the battalion moved to Villers-devant-Dun. There it had a short rest. Afterwards the battalion moved up to the regiment and marched in the glowing heat of the sun through Andeville, Remonville to Landres. Now the Meuse lay behind us, the enemy had yielded. The mood of the troops was exceptionally good despite the strains and fighting of the last few days. The music played as we were performing maneuvers.

Faraway in the south, in the direction of Verdun, one could see the twinkling flashes of shots and bursting shells. We marched west. Heat and dust made the march strenuous. After noon near Landres we made a sudden turn southwest. The Infantry Regiment 124 hurried along bad paths through rambling forest terrain to go to the assistance of the severely threatened 11th Reserve Division. In the forest 1.5 kilometers northwest of Gesnes, French artillery met us with a hail of shrapnel. The battalion halted.

I was sent forward to scout a way in the direction of Gesnes that was concealed from the enemy fire. With an NCO, I journeyed through thick underbrush to the southern edge of the wood. There, heavy shrapnel fire spread along the edge of the forest drove us into cover for several minutes. We went to the left and found a tolerably camouflaged path.

On our return journey, we found the battalion no longer in its former place. Hänle alone waited with my black horse and reported that the column had marched away to the right. At the forest edge ahead, dozens of enemy shells and shrapnel had touched down. Were they aiming at our battalion? To find out, I rode ahead along the scouted path to Gesnes. Hänle and the NCO came with me. Beyond the edge of the forest, however, there was nothing to be seen of the battalion for far and wide. Had they already advanced over the hill to Gesnes?

A leaderless company of the 11th Reserve Division asked me to assume command. Soon, three additional companies followed suit, who likewise had lost all of their officers. I led my very impressive-looking force in developed formation out from beyond the forest edge, which was an especially dangerous area due to enemy artillery fire, and headed in the direction of Gesnes. On a slope 1,200 meters northwest of Gesnes, I let the groups organize and arrange themselves into new units. These troops made a totally hardy, battle-worthy impression.

The crest of the hill in front of us was under fierce fire from French rifles, machine guns and artillery. It appeared that my own troops were in the battle ahead. As my new assembly of men organized themselves, I rode forward. I tied my black horse to a bush located close behind our own firing lines—but still under the protective cover of the slope.

Ahead I found parts of the 1st Battalion of Infantry Regiment 124 mixed with grenadiers of the Grenadier Regiment 123 engaged in a fierce firefight with opponents on the hills to the south and southwest of Gesnes. Our attack had been brought to a standstill at this place amid strong fire from enemy infantry and artillery. The troops had lightly dug in.

There was hardly anything to be seen of enemy infantry even using binoculars. French artillery of various different calibers made things from this spot onwards hard to endure. Nobody had seen anything of the 2nd Battalion. Was it still bogged down in the forest behind? I galloped back.

On the way, I came across the commander of Grenadier Regiment 123 and reported to him the situation on the hill and the location of the battalion that had placed itself under my command. At once this battalion was taken away from me, and an older officer of Regiment 123 was ordered to take command of it. I was very saddened about this.

I began searching for the 2nd Battalion of Infantry Regiment 124 along the edge of the forest to the west, often endangered by volleys of French shrapnel. No trace (of them) was to be found anywhere. Now I rode back to the frontline at the hill 1200 meters northwest of Gesnes and assembled the elements of Infantry Regiment 124 that I found there. Soon I had about 100 men gathered together.

Now the French batteries began driving rapid fire from their cannons. All around us, it crashed and splintered for several minutes, then another enemy battery ceased firing after the other one, and finally they all went silent. Twilight fell. Also the gunfire was hushed for a time, flaring up again only here and there.

Until late at night, I searched the hills west of Gesnes for the 2nd Battalion of Infantry Regiment 124. In vain! Afterwards I returned to the group I had assembled. The men were tired, droopy and hungry. They'd had nothing to eat since the very early morning. Unfortunately I, also, had nothing to give them. Field kitchens were certainly not coming through the woods of Gesnes. We tried to satiate ourselves with wheat seeds. I ordered a few hours of rest. At the first sign of dawn I wanted to move west in the direction of Exermont, thinking the regiment would surely be found there. The night passed without disturbances. Before morning, it [the weather] became tangibly crisp. My sick stomach proved to be a totally reliable alarm clock. Coffee was absent.

With the first light of dawn began the chattering of French rifles and machine guns along a wide front. We moved in the direction of Exermont. In a hollow two kilometers northeast of Exermont, I came across the regiment's battleground, where the 2nd Battalion of Infantry Regiment 124 was dug in close by as a reserve.

Because the adjutant of the 2nd Battalion had met with misfortune during a nighttime horseback ride, I was obligated to step in for him. Also here there was nothing to eat. Again I stifled my hunger with wheat seeds.

From time to time, the infantry fire ahead of us picked up. Enemy artillery went quiet. At about 0900 hours I rode with the battalion commander to reconnoiter ahead. The 1st and 2nd Battalions firmly held the high ground between Exermont and Gesnes.

We came up past the bodies of fallen officers, *Hauptmann* Reinhardt and *Oberleutnant der Reserve* Hollman, who had been killed two days previously. Storming forward at the head of their companies, they had become victims of French gunfire. Also, the evening's fight had torn noticeable gaps in the ranks. Our own frontline was dug in. There was hardly anything to be seen of the enemy across from us near Tronsol Farm.

After returning from the reconnaissance, I was sent away to seek out the battalion field kitchens and lead them over to us. Since marching away from Mont, or perhaps more accurately, since our short rest near Villers-devant-Dun —for about 30 hours! — we'd gotten absolutely no food provisions.

Nobody could say where the field kitchens might be. First I rode hither and thither through the woods of Gesnes and the woods of Romagne, then towards Romagne. At last I came across vehicles of the 11th Reserve Division, filled with troops. No trace was to be found of our own battalion's field kitchens. I then rode to Gesnes. Perhaps days previously, before the beginning of the battle of Exermont, the field kitchens had been ordered to head for Gesnes.

As improbable as it seemed, I had a feeling that I would find the kitchens right there near our own frontline. Gesnes was empty. No friend, no foe! In the valley between both frontlines I galloped to Exermont. The firefight on the hills on both sides had completely stopped. One kilometer southwest of Gesnes I ran into the fully assembled fighting force of the 2nd Battalion, standing ahead of the regiment's frontline. Shortly afterwards, patrols came forward and reported that the regiment was advancing in one quarter of an hour. Under these circumstances I could let the kitchens stay wherever they were.

The heights around Tronsol Farm were reached without further fighting. Leaving behind only a few dead, the enemy had withdrawn south. The regiment encamped in tents encircling the farm. My black horse got a good place in one of the farm stalls. He urgently needed care after the strenuous days and the cold night.

"A Hailstorm"—Pursuit through the Argonne, Fighting Near Pretz

> ◊ Rommel finally finds his feet in combat. Bomb blasts don't bother him as much. Hardened to the stresses and horrors of war, he thinks more clearly in action. In a dramatic scene, he gallops on horseback through artillery fire and rides zigzag to avoid shrapnel. The future Field Marshal also takes up a role as a machine gunner. He is conscious of casualties and driven by a desire to end fights quickly.
> ◊ His self-confidence brings conflict with his superior, Haas. Rommel and his men are stranded in the open with orders to remain there as a supporting force fails to appear. Haas stalls. Feeling left in the lurch, Rommel confronts Haas about it. Rommel has often been accused of being ambitious, but offending the colonel could have negatively impacted his career. Evidently Haas does not take kindly to the challenge and later accuses Rommel of friendly fire. The allegation is unfounded—Rommel produces witnesses and evidence, but Haas does not withdraw the accusation.
> ◊ Rommel is defensive of his honor. From a cultural standpoint, Germans can be sensitive when they feel a threat against their personal "honor"— a sense of outward self-esteem comparable to the Japanese concept of "*mentsu*" (face).
> ◊ German military history is rife with stories of backbiting. Later during World War II, Rommel clashed with staff at Hitler's headquarters and Hermann Goering, to whom he referred in his memoirs as his "bitterest enemy." The early clash marks Rommel's entry into the world of battlefield politics.

On Sept. 4, our path took us on dusty streets under an oppressive heat through Eglisfontaine, Epinonville, Verry, Cheppy, and Varennes toward Boureuilles. Along the entire distance, one saw signs of the over-hasty French retreat—cast-off rifles, backpacks, and motionless vehicles. On and next to the street lay many dead horses severely swollen in the heat. The march was extraordinarily overtaxing.

Late in the evening we first reached Boureulles and made camp in tents. During the night, my sick stomach robbed me of urgently needed sleep.

Days afterward we marched—again in the glowing heat of the sun—through the Argonne, through Clermont—Les Ilettes towards Briceaux, which we reached late in the evening. Far and wide there was no enemy. The French rearguards had moved out an hour before our arrival. The French stronghold of

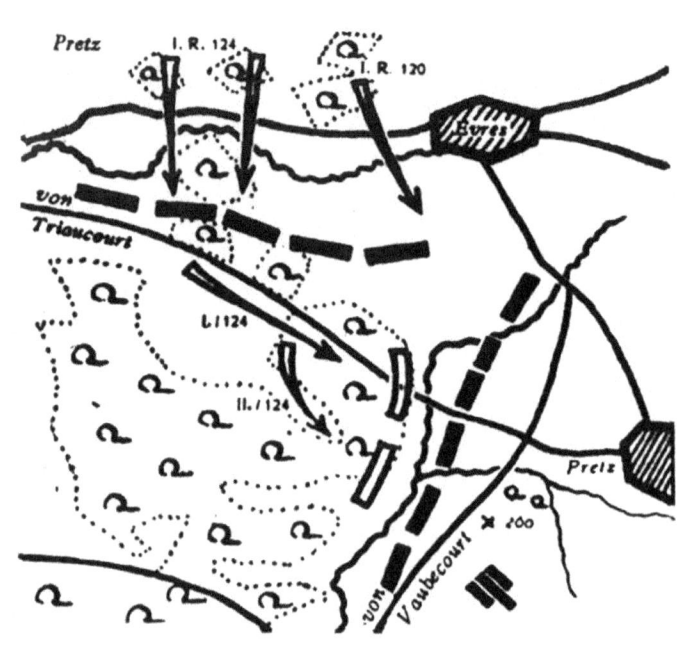

Rommel's map of the fighting near Pretz.

Verdun now lay 28 kilometers northeast of us. Quarters in Briceaux were good. Nobody made special demands. A mattress and a little something to eat was enough. *Hauptmann* Ullerich took over command of the 2nd Battalion. Mounted reconnaissance patrols sent out at daybreak on Sept. 6 were shot from the woods just south of Briceaux. At about 0900 hours the regiment deployed from Briceaux in formation to the southwest. The foremost sections clashed with the enemy near Longues Woods. The 1st Battalion attacked and speedily won the road between Triaucourt and Pretz. Countless Frenchmen were taken prisoner.

Afterwards the 1st Battalion moved along the road towards Pretz and the 2nd Battalion followed. On both sides of the road stood a thick, tall forest. Over to the left, a violent fight unfolded. Coming out from the southern edge of the woods, the 1st Battalion collided with a strong enemy again. A ferocious fight arose at a distance of barely 100 meters. Throughout it, French artillery yet again made things difficult to endure. Aside from the fact that they had vast munitions supplies at their disposal, they were extraordinarily mobile with their fire. Also the forest on both sides of the road from Triaucourt to Pretz, in which the 2nd Battalion was resting, was soon made unsafe by French shellfire.

Around noon, the 2nd Battalion received the order to go ahead to the southwest edge of the forest two kilometers west of Pretz, launch an attack to the right of the 1st Battalion and capture Hill 260.

The battalion marched away under security. *Leutnant* Kirn led the point, and I rode next to him. Without running into the enemy, we reached Hill 241. The journey took us through high gorse bushes which almost completely overgrew the narrow path. After 100 meters we had to emerge from the forest. Suddenly, a strong French reconnaissance patrol bobbed up close ahead of us. An exchange of bullets played out at distances of only a few meters. Then the French retreated. Casualties on our side were nothing to be counted.

Afterwards it became apparent that our contact with our battalion behind us was broken off. It had to be reestablished. The point men halted. I rode back. I found the battalion lying sideways in the forest. I reported the clash of our infantry point and the withdrawal of the enemy. Then the march to Hill 241 resumed. However, after a few hundred meters, French shrapnel fire forced the battalion to the ground.

For long minutes, bullets and splinters rattled down on us like a hailstorm. The raging shellfire made any movement impossible. The troops found some shelter in the smallest hollows, behind trees and raised backpacks. Yet casualties began to happen.

As the strength of the firing diminished, I galloped left through the forest to establish contact with the 1st Battalion. Yet the woods proved to be too marshy. Without having achieved anything, I had to turn around again. Then I stalked ahead on the eastern border of the forest on foot. Here I was shot at by the enemy many times, who occupied the knoll about 300 meters east of the forest edge.

At last I found the 3rd Company. They were holding back from attacking until the 2nd Battalion intervened. This attack [from the 2nd Battalion] was prepared immediately after my return. As the 6th and 8th Companies in the foremost lines broke out of the woods in a surprise storm attack towards Hill 260, the French infantry evacuated their positions without resistance and retreated. Also the French artillery, which had been so very active all day long, now made itself scarce. We took and occupied Hill 260 and gave pursuing fire to the retreating enemy. Soon, the looming nightfall brought an end to the fight. Reconnaissance patrols were sent out. The companies entrenched themselves. Ahead of us to the right, mountains of ammunition cartridges lay in an abandoned battery position.

To make a report to the regiment, and to fetch the battalion field kitchens, I rode back where we had come from. Since our march from Briceaux, the troops yet again had gotten no sustenance. *Oberst* Haas spoke openly of the 2nd Battalion's achievement with great esteem. I encountered the field kitchens along the road from Pretz to Triaucourt. At about 9 p.m. they reached the battalion. The hungry troopers were nourished.

No telephone link then existed with the regimental command post. After midnight, orders for the next day came. Our reconnaissance patrols came and went. There was little time left for resting, although the enemy did not bother us.

"Death Reaped Harvest"—*Attack on the Defuy Woods*

During the night, reconnaissance patrols firmly established that the enemy had established a new position in the Defuy Woods, located three kilometers away. The Regiment commanded the 2nd Battalion to march across the road between Vaubecourt and Pretz at 6 a.m. and take the Defuy Woods. Sections of Grenadier Regiment 123 were supposed to advance on the right of the 2nd Battalion.

In the time that followed, the battalion marched with the 6th and 7th Companies in the frontlines (which were about 600 meters across). The 5th and 8th Companies were in the second advancing line behind the left exposed flank, all poised to attack. Without being shot at, we came within about 1,500 meters of the woods. The left flank approached the northeast corner of the Defuy Woods. I rode between the 6th and 7th Companies. Our neighbors to the right (the Grenadiers) had not fallen into battle formation.

Then came the regimental order from behind: "2nd Battalion, do not advance further! Stay where you are!" I first passed the order on to the companies, then galloped back to the regimental command post on Hill 260 to inquire about the reason and duration of the halt. *Oberst* Haas wanted to hold off on the attack until the grenadiers on the right came up into position. How long this would take was unforeseeable.

In the meantime, the French artillery came into action. Their fire predominately hit sections of our frontline, who were still standing there without cover, and the reserve companies that were still exposed. From the northern edge of the Defuy Woods, which dominated the landscape far and wide, the French artillery observers possessed exceptional visibility conditions.

I strove forward in a fast gallop with an order for my battalion that the frontlines were to dig in under the good camouflage of cabbage and potato fields. During my ride back, a French battery zeroed its sights on me with shrapnel. Galloping sideways and in zigzags, I nevertheless easily evaded their shots.

The French artillery fire became stronger minute by minute. Medium calibers growled at us. A shell burst amid the 5th Company, which was halted in marching position, and wiped out two groups at once. In contrast, the well-camouflaged and quickly deeply entrenched companies in the front lines suffered little under this fire. A battery of Artillery Regiment 49, which came into the attack from the direction of Hill 260, were severely harassed by the heavy French shells.

In a section of the road two kilometers northeast of Vaubecourt, the battalion and regimental command posts were situation close together. It didn't take long before additional French batteries brought this area under heavy fire. No wonder! The numerous couriers and mounted officers, the straight lines of men who came marching from there, and the various observation posts had given away the command post's location.

Shell after shell shrieked over. Splinters, dirt clumps, and rocks flew around our heads. Hours passed under this fire. Our attack was not continued.

Tired and wilted, I lay in a road ditch and attempted to catch up on my lost sleep. The shell explosions in my immediate area no longer agitated me now. We had already become emotionally dulled to it. During the course of the day, most of the trees lining the road all around us were ripped to shreds, but there were few casualties in my own area.

As the day faded, orders came to carry out the attack on the Defuy Woods. This released us from inactivity and brooding. The 3rd Battalion was supposed to advance from the right of the 2nd Battalion, and so was the Grenadier Regiment 123, still further to the right. At this exact same time, the French artillery fire dwindled noticeably.

Soon, also, the last gun fell silent. I rode ahead and got the battalion moving. Astonishingly, we were met by neither French artillery nor infantry fire. Had the enemy cleared out yet again? The frontline—a line of skirmishers with about

Rommel strides briskly along the northern coast of France as a Field Marshal in 1944. His stride is much the same as when he marched as a young infantryman shown on page 49.

four steps between each of them—ascended towards an open crossing in the terrain 500 meters northwest of the Defuy Woods near the forest edge. The Grenadiers were now on the right, and the 3rd Battalion at the same level with them. A few hundred meters behind the frontline, the reserves followed: the 1st Battalion and Machine Gun Company.

I rode close behind the lines of the 7th Company, which found itself in the left flank. It was already growing dark. Just as we approached within 150 meters of the forest, the French hit us with rapid small-arms fire. We did not expect it. In just a few seconds, a violent firefight was underway. Our extremely exposed reserve companies swarmed into the frontline. Now, men were lying pressed together without cover. Toward the rear, sections of the regiment sought cover in the earth from the raging enemy fire.

There, platoons of the Machine Gun Company—located in positions close to combat vehicles hardly 100 meters from our own frontline—opened continuous automatic fire on the forest edge. Roaring screams came from our frontline: "The Machine Gun Company is shooting us!" This all unfolded in just a few seconds. I had been sitting on my horse in the left flank of the battalion. At this point, I rushed over in a long gallop to the machine guns and ordered them to stop. I gave my horse to the next best man and took a machine gun platoon from the 7th Company's left flank forward with me.

Rommel's more detailed drawing of action near Pretz. A) Fight near Pretz on Sept. 6. B) Attack held on Sept. 7. C) Storm attack on the Defuy Woods on the evening of Sept. 7.

There lay the brave riflemen in a furious firefight. Clattering shots rang all around. Death reaped harvest. After my machine gun platoon opened fire, we ran to storm the enemy. Our neighbors on the right joined us. In this brief moment, nobody felt exhaustion or weakness. All our muscles were tight, and our fighting spirit was wilder and more determined than anything else. Finally, we saw hope of grappling with the enemy. Gunfire tore gaps among our tightly packed ranks, yet did not stop our stampede. Battle cries from a thousand throats echoed from the edge of the forest—there, victory beckoned. We broke into the forest at last with a shout. Yet again, the enemy cleared out of his position in the last moment and withdrew from close quarter combat into dense underbrush. It was difficult to penetrate.

Hadn't the regiment ordered us to attack by forcing our way through the forest until its southern edge? That would probably take a long time. Then why shouldn't I just go left around the outside of the forest to pursue the enemy and cut them off? Quickly I decided. Two groups and the machine gun platoon accompanied me.

Wheezing, we made our way uphill around the forest edge. There were no thickets to hinder us here. It was possible for the retreating enemy to withdraw back into the woods just as quickly as we came forward. Completely exhausted, we reached the eastern corner of the Defuy Woods. Across several hundred meters, one could sweep the southern forest edge with automatic fire. There was still light for shooting.

Hurrying feverishly, we made the machine guns ready to fire. The shooters nested themselves near the forest edge in an enormous bush located to the extreme east of the corner of the forest. In the blink of an eye it could be possible for the enemy to emerge from the forest. Back in the forest to our right, we could hear German signals. Minutes passed. No enemy let himself be seen. Light for shooting dimmed more and more. Over to our left, countless farms in Rembercourt were burning. The flaring tongues of fire lit the sky.

No sound, no trace of the enemy! I was already experiencing the pang of a guilty conscience that I had taken the machine gun platoon along with me without permission of the regimental commander [Haas]. Since there was no longer the prospect of a fight, I released the platoon back to its own company. They had hardly gone off when a rifleman showed me in the glow of the fires of burning Rembercourt, the silhouette of a marching column at about 100–150 meters distance across the coverless hilltop. The French! Their kepis and bayonets were clearly recognizable through the binoculars. Doubtlessly the enemy was moving off in close order. I regretted having sent away the machine gun platoon just minutes too early, but that could no longer be changed.

On my command, 16 rifles opened rapid fire on the nearby enemy. Contrary to our expectation, the Frenchmen did not immediately run away, but instead rushed at us in a storm yelling, *"En avant!"* According to the sound of their yell, there must have been about one of two companies of them. We fired as many shots as would come out of our rifles. Nearer and nearer came the enemy. One of my men voluntarily cleared out of his position and was brought back forward by me. It seemed our fire drove the attackers to the ground. In the meadow grass ahead, the enemies hardly raised their heads from their position in the fiery glow of burning Rembercourt. Their front lines must have been now about 30 to 40 meters in front of us. I was resolved to withdraw in the face of this superior strength before it came to a bayonet fight.

It did not come to that. Our rapid fire had taken away the enemy's desire to storm us. The battle cry, *"En avant!"* fell silent. Only five packhorses of French machine gun troops and two machine guns followed through with the advance up until the forest edge and were later captured at this location. In front of us, things became still. Judging by appearances the enemy had withdrawn towards Rembercourt. A reconnaissance patrol brought in about a dozen prisoners from the area close ahead of our frontline and positively identified about 30 dead and wounded Frenchmen.

Now where was the 2nd Battalion staying? Apparently it had not pushed through the Defuy Woods as it had been commanded to do. In order to establish contact, I went with a pair of men—who took the captured Frenchmen and their pack animals along with us at the same time—back to the northeast corner of the Defuy Woods. I left the remaining parts of both my groups behind in their position.

On the way, I came across the regimental commander. *Oberst* Haas was not in the least pleased about my report of what had happened along the forest edge. He expressed his unequivocal conviction that myself and the people with me had not shot at Frenchmen, but instead at elements of our own Grenadier Regiment. Not even the [French] prisoners or the packhorses bearing their machine guns convinced him otherwise.

Considerations:

- The attack against the Defuy Woods on Sept. 7, 1914 had to be carried out for three kilometers **across open terrain which offered little cover**. A regimental command **halted the attack**, because our neighbors on the right had not arrived. At the same time, the French artillery began a strong and sudden surprise fire.
- The developed elements of the 2nd Battalion rapidly found cover out of sight in the potato fields and dug deep holes with their spades to protect themselves from the fire. They incurred no losses despite the strong artillery fire that lasted the entire day.
- In contrast, the closed formation of a reserve company led to painful losses from enemy direct hits. This **teaches** again that **no massed formation within the zone of enemy artillery fire** is permissible, and furthermore demonstrates anew the **importance of spades**.
- The **regimental command post** and the 2nd Battalion were sited close together in a road pass. Numerous couriers and riders streaming to this spot from all directions betrayed the location to the enemy, and the enemy laid down heavy fire on it. So, again here, **no grouping in a mass**!
- **Demand** that **couriers**, etc. **do not all head streaming to the command post while within sight of the enemy!** The enemy must not be allowed to recognize the command post. Therefore, in addition, do not choose any obvious hills as the command post.
- When we continued the attack in the **evening**, the **French artillery** did not fire. They had already **moved out of their position**, presumably to save themselves beforehand from falling into enemy hands during a German night attack. The **French infantry** allowed the **German attack** to approach up until about 150 meters without shooting, launched continuous fire for a few minutes, then broke off the fight within

> the protection of the forest and cover of falling darkness and escaped. On the German side, **heavy casualties** were sustained during this short fight. On Sept. 7, 1914, Infantry Regiment 124 lost five officers and 240 men dead and wounded.
> - In the flurry of battle, elements of the **Machine Gun Company shot across** from ground positions past thickly massed infantry lines on a rising slope 350 meters ahead of them in order to get the enemy located 500 meters away in the forest. In doing so they **endangered** our **frontline** most severely.
> - During the **advance** to the Defuy Woods, our **especially needed distribution in depth** decreased gradually because we no longer expected to face any strong resistance by nightfall. Reserves and the Machine Gun Company had moved up much too close to the frontlines. This came back to bite us during the sudden surprise fire of the enemy.
> - It often happens **in such situations** that individual **soldiers lose their nerve** and want to get to safety. The leader must energetically confront these ones—even, in certain circumstances, with his gun. —*Rommel*

"An Absolute Hell"—At the Defuy Woods

> ◊ Rommel reflects on receiving letters from home, noting they seem to come from a "a different world." As usual, he avoids focusing on himself and uses plural pronouns (i.e. "we") to emphasize that other soldiers shared this experience.
>
> ◊ Rommel's creative solutions to problems are not always appreciated. When he decides on a daring scheme to beat back the enemy, a subordinate refuses to obey. Rommel takes charge anyway and proves successful, although he faces a complaint at headquarters. This would happen again later during World War II.

The regiment commanded that the 3rd Battalion approach the southern edge of the Defuy Woods with its left flank at the eastern corner and that the 2nd Battalion should accordingly emplace itself for defense on the left side outside the forest. The 1st Battalion approached north of the Defuy Woods in the 2nd line, and the regimental command post was located on the 1st Battalion's left flank. The area assigned to the 2nd Battalion was a sprawling, barren and coverless summit, which in no way pleased us. It was to be expected that the positions on this summit would be particularly exposed to French artillery fire. We would much rather have taken over the forest area allotted to the 3rd Battalion.

To keep our losses as low as possible from the expected enemy artillery fire, our most recent experiences suggested only one solution: deep trenches. The company commanders – three companies being led by young lieutenants – were warmly encouraged to carry out the digging with all their energy and with no regard to the tiredness of the troops. Before midnight the main portion of the work still had yet to be achieved. We then slipped in a few hours rest, and took up the work again on the positions before daybreak. We were aiming for a trench depth of 160 centimeters.

Soon the whole battalion was in feverish activity. The powerful enemy artillery fire of past days had instilled every trooper, hauntingly and before his very eyes, with an understanding of the importance of spadework. Even the battalion staff—consisting of the battalion commander, adjutant and four couriers—dug themselves into a 6-meter-long trench behind the center of the forward-right emplacement of the 8th Company.

The work was extremely and unusually tiresome. The ground showed itself to be hard as stone. Hardly anything could be achieved with our short shovels and there were very few pickaxes on hand. Despite the most extreme strain on all our strength, we only managed to penetrate the earth's surface slowly…infinitely slowly. In the meantime it was 2230 hours. The troops had gotten no sustenance since 0500 hours. The battalion commander sent me away in the direction of Pretz in order to fetch the field kitchen.

At about midnight I returned with them to the battalion. For the first time since the war started, our mail had arrived. In several hours of work, the companies had managed to dig about 50 centimeters into the earth. This result gave rise to serious concern. Before daybreak, it was essential the positions had to be much deeper—otherwise heavy losses caused by enemy artillery would be unavoidable.

Now at about midnight the troops were totally exhausted. They needed first to be nourished and then rest for at least a few hours. The field kitchen came to the companies. Food was handed out. The *Feldwebel* distributed the mail. In narrow foxholes, one could see the letters from home by the glow of a candle, which had been mailed to us weeks before. They sounded as if they came from a different world. For the first time, one became aware of how long it had been since we had been away from home. The time seemed like years ago—and yet it had only been a few eventful weeks.

Then we ate the thick soup of the field kitchen and afterwards grappled with pickaxes and shovels again. The battalion staff were the first to lay down to rest in the early morning hours after they dug their trench about 1 meter deep. My raw hands hurt. My dead-tired body could feel the hard surface of the ground as little as the coldness of the September morning, which slowly dawned.

At about this time, the companies got back to work. In the gray before dawn, a howitzer platoon of Artillery Regiment 49 established itself on the eastern edge of the Defuy Woods in a fissure in the terrain between the 3rd and 2nd Battalions, about

A photo of Rommel's men, begrimed and with shovels handy, next to their dugouts on a hilltop.

30 meters behind our frontline in a half-defiladed position. Gun emplacements were dug in since concealment for the gun squad was no longer sufficient.

In the first morning hours of September 8 we were in no way ready to do battle. Enemy defensive installations were visible through binoculars across the valley from Hills 267 and 297 (west and northeast of Rembercourt). We had visual communication with our neighbors on the left (Infantry Regiment 120) located on Hill 285, one kilometer northeast of the Defuy Woods. Gaps of about 500 meters existed between both flanks, but could be grazed with fire. A machine gun company established itself within the 2nd Battalion's area. The 8th and 5th Companies were situated in the frontline. The 6th Company was located behind the right flank and the 7th Company was echeloned and inserted behind the left flank.

The battalion commander inspected the positions with me. Overall, diligent digging work was still being done. In certain positions, a depth of 130 cm had been reached. At about 0600 hours, French artillery of all calibers opened fire in a sudden burst. The magnitude of this shelling far overshadowed anything we had experienced up to this point. Salvo after salvo came howling at us. All around us, shots cracked and splintered. The earth rocked and shuddered. Most of the shells were time fuzes which exploded in the air close above the hillside, and a portion of others as impact fuzes.

Cowering together closely, we lay at the bottom of our still rather pathetic shelters.

There was practically no protection from the splinters of the time fuzes. They burst, then came down straight or crookedly from above, even reaching the bottom of the trenches. Ceaselessly, a rainstorm of earth and stones pattered down on us, with jagged shell fragments screaming through the air in between. The fire lasted for hours with the same intensity.

Once a shell struck right above our trench and rolled over the side, landing at the bottom of the trench right in the middle of us. Fortunately it was a dud. Feverishly, we all worked with shovels, pickaxes, knives and our hands to dig ourselves deeper below the earth. My body trembled thousands of times during the course of the day as shells shattered to pieces nearby.

At about noon, the enemy fire lessened somewhat. Now for the first time it was possible to send couriers out to the companies. Ahead of us, everything was in order and the French artillery had not shown itself anywhere. The losses were happily much fewer than we had feared (2%–3%).

Soon the enemy increased fire back to a powerful level. They must have had a monstrous abundance of ammunition available to shoot. In contrast to this, our artillery was nearly completely quiet on this day. Ours was noticeably suffering from a lack of ammunition. Additionally, the southern edge of the Defuy Woods and the positions of the 8th and 5th Companies provided the only possible vantage points to observe enemy territory.

The French artillery fire lasted the whole afternoon with only short pauses. By that time our trenches were about 140 cm deep. In some areas, the men had hollowed foxhole niches into the front of their trench walls. Not even the splinters of the incendiary shells could reach them in there. With about 50 centimeters of solid, hard ground as cover, these foxhole crevices also protected against impact fuzes.

The day waned. The enemy now increased his fire to a monstrous degree of force. They shot anything from their guns that could be fired. The hill became an absolute hell. Thick black smoke of medium calibers wafted over our positions. Impact after impact burst and sprinkled from the hill. Was this the preparation for a French infantry attack? They might as well come—we had been waiting for it the whole day!

Just as suddenly as it began, the French artillery ceased fire. No infantry attack came. We crawled out of our dirt holes. I went to the front of the four companies. Overall our losses were happily few (16 men in the whole battalion) and the men, despite this difficult test of nerves, were in the best spirits. The hard [digging] work of the past night and during the enemy fire had paid off well.

The last rays of the sinking, blood-red sun illuminated the battlefield. There, on the right along the forest edge, stood both guns of Artillery Regiment 49. The gun crews lay dead or wounded beneath them. The platoon had been concealed in such a way that it could not shoot. Just as bad was the sight of the 3rd Battalion in the forest to the right. In this area, the strong network of forest roots had made trench work next to impossible. Ultimately, flanking French artillery fire—whose potency had become even greater due to numerous shell bursts that shattered trees—had assailed the companies severely.

I presented myself to the regiment to receive orders. *Oberst* Haas was deeply shaken by the severe losses of his 3rd Battalion. It needed to be recalled out of the forest. The 2nd Battalion received the order to hold the summit east of the Defuy Woods by itself—without immediately depending on [troops] on the right and left. "The 124th Regiment will die in its position!" *Oberst* Haas said in conclusion of his orders.

After returning to the battalion, the right flank of the 8th Company was withdrawn. The 6th Company took over the front on the eastern side of the Defuy Woods and dug itself in anew. Also the remaining parts of the battalion worked zealously on further fortification of the position. The field kitchens arrived shortly before midnight. Again they brought the mail with them. As on the previous night, the troops rested a few hours on the bare soil of the field. Straw was far and wide unavailable to obtain.

The next day the French artillery shelling began at the same time as on Sept. 8. In the good, deep positions we did not care very much about it. An intermittent telephone connection existed with the regiment. Shells constantly ripped the [phone] cables. I stayed for a long time in the 5th Company's position and, together with *Unteroffizier* Bentele (of the 7th Company), observed the enemy's disposition. The French artillery, for the most part, stood in coverless firing positions and also the French infantry behaved totally without caution.

We sent the results of our observations with sketches from the battalion to the regiment with the request to order observation post officers of our own artillery to the frontlines of the 2nd Battalion. Ahead to our left, at a distance of about 500 meters, lay the left flank of Infantry Regiment 120 on the southern slope of Hill 285. Across from them, the French had nested themselves along the railway line. In the sector 600 meters west of the [railway] Station Vaux Marie was a dense mass of French reserves. We could hit their flank with machine guns from the hilltop just left behind our positions. Were we not allowed to do this?

I suggested it to the leader of machine gun platoon in the sector of the battalion. He had reservations—didn't want to do it. Without further ado I then took command of the platoon. I made it perfectly clear that the French artillery would retaliate for our operation, and we alone had to work just as fast in order to get away in one piece. A few minutes later, our platoon's full-automatic fire caused dire confusion and extremely heavy losses among our densely grouped enemies. Our objective was reached. Hurriedly I broke the attack off and rushed speedily with the guns into cover to our right. The enemy's retaliating fire soon ensued and hit nothing. The operation was completed with no losses on our side.

Meanwhile, the leader of the machine gun platoon had complained to the regiment [headquarters] about my independent action. I was ordered to report to headquarters. After I reported about the progress of the operation, the matter was concluded.

During the course of the day, multiple artillery observers came into the battalion's sector. The enemy artillery emplacements were shown to them. Only their batteries controlled the stocks of our ammunition, which was so lacking that the French artillery remained undisturbed by our weak fire. Only a heavy battery reduced enemy batteries positioned near Rembercourt to silence.

That evening—as in previous days—came the French evening lullaby[2] with a colossal expense of ammunition. Then the enemy fire again went totally silent. As far as we could observe, the French artillery withdrew from their far outlying forward positions.

We worked to create more splinter-proof cover for our positions. Woodcutting commandos moved into the Defuy Woods. The day's losses were, happily, even less than the day before. Only the 6th Company had suffered from flanking fire.

The field kitchens arrived at about 2200 hours. The 7th Company's technical sergeant brought a bottle of red wine and a bundle of straw. Shortly before midnight, I lay down to rest close behind the battalion command post.

Considerations:

- It **proved very costly** for the 3rd Battalion to set up defenses during the night of Sept. 7–8 **close to the southern edge of the vast Defuy Woods.** They sustained the heaviest casualties in this position and had to be withdrawn from action on the evening of Sept. 8. Due to the hardness of the ground and the numerous forest root systems, it was very difficult to penetrate the earth's surface. The French artillery fire thus had a dominating effect on the **superficially dug-in troops** in the woods and along the forest edge.
- So many shells that would have flown past and detonated far to the rear had they been fired at a bare hill exploded on contact with trees at the forest edge—right where the mass of our troops were. The forest edge was a real bullet trap.
- It was **very easy** *for the French artillery* **to shoot at the forest edge**. Nowadays the effects of artillery are even more powerful—the casualties in the same situation today would be even heavier. In contrast, the **2nd Battalion's spadework** on the barren hill paid off well.

Despite hours-long artillery barrages under the strongest output of ammunition,

---

2. *Abendsegen:* the name of a German children's bedtime song from the 19th century "Hansel & Gretel" opera. Rommel used it here sarcastically.

the casualties remained bearable. However, the **shells with time fuzes** in particular were extremely **unpleasant** because portions of their splinters struck vertically into our trenches.
- The rock-hard earth in the 2nd Battalion's sector made digging in difficult to accomplish. With the utmost pressure and through **personal example**, leaders of all grades had to keep their exhausted and hungry men digging in with the utmost exertion during the night of Sept. 7–8.
- The **ammunition expenditure of the French artillery** was considerable on Sept. 7–9. They could afford to use it lavishly because their large reserve storage was close nearby. By contrast, on the German side, ammunition became quite scarce, so that our infantry could only draw [artillery] support in few positions during battle.
- Formation in defense is today much different than in 1914. Back then, there was a single front line, the rest in the second line. Today a battalion's position consists of combat outposts and the main battlefield, in which forces are distributed in depth. Within a space of some one to two kilometers wide and equally deep, dozens of small strongpoints are distributed [today]: rifles, light machine guns, standard machine gun nests, mortars and antitank guns, which mutually support each other. Thus the enemy fire will be broken into fragments, and our own fire can become more concentrated from rear areas. In certain locations, evacuations in the face of superior fire is possible and yet still the defense can be continued, if the enemy's position on the main battlefield is penetrated. In case of a breakthrough, the enemy would then face a long and most difficult path ahead.—*Rommel*

"Torment"—Night Attack of Sept. 9–10, 1914

I must have slept soundly on my straw bed. A very loud uproar of battle, coming from ahead to my left, startled me awake on the hill at about midnight. It was raining in streams. I was already soaked to the skin. On the left, light signals flickered through the dark rainy night. Gunshots clattered ceaselessly. According to the courier, the battalion commander had been at regimental headquarters for a long time.

The din of the firefight came

◊ Rommel confesses to almost accidentally shooting his own men. He admits being terrified of shells, soiling his trousers, frantically digging with silverware and suffering from nightmares. Grim honesty was incompatible with the prideful ideals of the Nazi regime. At the time Rommel's book was published, Nazi books on war promoted bombastic nationalistic illusions—for example, the superiority of the Fatherland, loyal braided maidens, or cold superhuman alarmingly closer. Were the French attacking at night? To find out what was actually going on, I rushed forward in the direction of the noise, accompanied by a courier.

Suddenly, about 50 to 80 meters ahead of me, I saw silhouetted figures rise up and rush closer. Two marching columns! They must have been Frenchmen. Surely they had broken through the gap between Infantry Regiment 124 and Infantry Regiment 120 and wanted to catch the 2nd Battalion in the flank and rear. The ghostly mass pressed nearer and nearer. What was I to do?

I hurried right towards 6th Company, quickly briefed the leader, *Hauptmann* Graf von Rambaldi, and asked him to place a platoon at my disposal. With this [platoon], I deployed against the enemy to attack. As their columns became more apparent in the glow of distant light signals, I ordered the men to take position and disengage their rifle safety latches [to prepare to fire]. I was not completely sure if these were actually Frenchmen, so I called out from a distance of 50 meters before I would give the order to fire.

My call was answered—it was our 7th Company. Their leader, a young *Leutnant*, had withdrawn the company from their echeloned position in the second line behind the battalion's left flank. He wanted to dig them in at a new place about 400 meters farther back. He justified this plan by saying that a fight could very well develop at this moment, and that the 7th Company was still in the second line.

In a manner that was hardly friendly, I educated him about his incorrect outlook. Later on, it sent chills up my spine to think that I had come within a hairsbreadth of shooting at my own recruits.

Shortly afterwards, the battalion commander returned from regimental headquarters and brought back a regimental command for a night attack. Our battalion—in the frontline regiment—was given the order to storm and seize Hill 287 about 500 meters north of Rembercourt. Our neighboring regiments—Grenadier Regiment 123 on the right and Infantry Regiment 120 on the left—were supposed to attack at the same time.

> desires to die in blazes of glory. As Rommel was preparing to publish *"Infantry Attacks!"*, Nazi propagandist Joseph Goebbels published a jingoistic World War I-themed novel called *"Michael,"* promoted as an epic literary achievement. By contrast, *"All Quiet on the Western Front,"* by Erich Maria Remarque contained descriptions of soldiers' human suffering, but was savaged by the Nazi government as unpatriotic and burned in 1933. Remarque's sister was beheaded for her brother's writings. Rommel, a military instructor, must have been aware of perceptions of war and soldiers in Nazi "pop culture" and could have pandered to it by reinventing himself as a "perfect" German hero, but was instead brutally honest about his war experiences at the risk of seeming "weak."

The exact time of the attack was not yet firmly determined. However, the battalion was supposed to prepare at once. The command was a solution releasing from the hell of French shellfire—the goal of the attack, however, was not very ambitious. We would have been especially glad to additionally attack and seize the French artillery positions on the hills around Rembercourt[3].

In the pouring rain and pitch-black night, the battalion prepared itself for the attack within the area to the left of its current sector. Guns were loaded and bayonets fixed. The motto was: "Victory or Death.[4]" Among our neighbors to the left, fighting seemed to have been going on already for a long time. The gunfire ceased there after intermittent pauses, in order to flare up again in a different location.

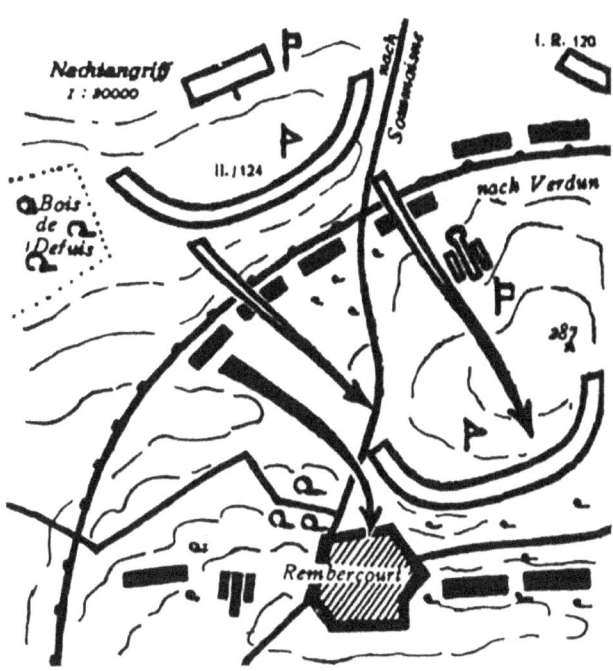

Rommel's map of the night attack of Sept. 9, 1914.

The 1st Battalion had closed ranks. The regimental commander was with the 2nd Battalion. We knew the enemy was in position along the railway line and south of there in areas of the road between Sommaisne—Rembercourt. The troops waited eagerly to begin the attack. They had been soaked to the skin and freezing for quite a long time. At last, at about 0300 hours, came the order to advance.

In a dense mass, the battalion hurtled forward with titanic momentum uphill towards the enemy at the railway line, overran the enemy, seized the sectors on the Sommaisne—Rembercourt road and stormed Hill 287 in one fell swoop. Wherever our opponents fought back, they were finished off with bayonets as the battalion stormed around them en masse right and left. With all four companies in the front line, the battalion occupied Hill 287. Since there was still no contact with [our forces on the] right and left, the flanks were withdrawn. The troops were mixed and it took a long time to organize them. Morning was already dawning. The rain dwindled.

The companies worked feverishly to protect themselves from French artillery fire, which was expected soon. The [digging] work in the wet, clayed earth was however very

---

3. i.e. .in retaliation for the shellfire.
4. "*Sieg oder Tod*": a phrase Rommel would hear again from Hitler in North Africa during World War II.

exhausting. The shovel blades kept getting caked with thick, greasy coatings of clay and needed to be cleaned. Presently, in the gray morning hours, the contours of the hills surrounding Rembercourt became distinctly recognizable. They towered over our new position. Suddenly through the twilight, as we dug in to secure the combat command posts, we recognized dense masses of Frenchmen around the low-lying northern entrance to Rembercourt. Our sentries sounded alarms.

Up until this time, I had been on the right flank of the battalion with the 6th Company (under the command of *Hauptmann* Graf von Rambaldi). It was clear to see that the closed French formations were marching into Rembercourt from the northwest. The 6th Company and parts of the 7th Company opened fire. At a distance of about 300 to 400 meters, a very fierce fight quickly developed. Some of the Frenchmen attempted to secure the upward-sloping town street of Rembercourt under their control. Most of them shot at us. Due to excitement, the majority of our men waged the gun battle in standing positions – finally, we had French infantry in our sights! After about one quarter of an hour, the enemy fire decreased noticeably. Ahead of us, at the northern exit from Rembercourt, lay a great number of dead and wounded. However, our own ranks had also thinned due to their careless behavior. This early morning battle cost just as many casualties as the night attack.

We intensely regretted that we had not been allowed to storm the Rembercourt region and the hills on both sides. Despite all experiences, the battle courage of the troopers remained unbroken. Indeed, they were spoiling for a fight with the enemy infantry, who had proved themselves to be inferior in all battles up to this point. The companies resumed digging in after breaking off the firefight with the enemy near Rembercourt. They had barely dug 30 centimeters into the earth when the French artillery came into action in the usual manner. Their rapid fire hindered further work in the open.

Until this point, the battalion staff had found no time to create the least bit of cover for themselves. The battalion's sortie on Hill 287 and the fight with the enemy at the northern exit from Rembercourt had unfortunately kept them all moving. Now a French battery, located in an open position on a hill just west of Rembercourt, fired shot after shot between us from barely 1,000 meters away. Thankfully many duds fell in the rain-soaked earth. We sought shelter from the enemy shells in a plow furrow and covered ourselves with bundles of oat sheaves to conceal ourselves from enemy spotters in some way.

Soon it started raining again in streams. The plow ditch filled with water. French shells struck close by our oat bundles. Attempting to dig with shovels while lying down proved impossible. One's shovel blade instantly turned into a big lump of clay. Also, we ourselves were soon literally coated from head to toe with sticky clay slime and froze in the wet stuff to the point of chattering. In addition to all this, my sick stomach made itself aggravatingly noticeable to everyone and drove me into the nearest shell hole to relieve myself every half hour.

It became known that our neighbors' attack had not been as successful. The 2nd Battalion was far ahead of the front of the division. At about 1000 hours a howitzer battery of Artillery Regiment 49, positioned close behind the battalion sector, tried to help us. Yet they did not hinder the far-off enemy fire and instead drew it even more unwantedly towards us. As in previous days, the French infantry barely showed themselves and only harassed us a little with gunfire.

Slowly, endlessly slowly, the day stretched on in all this torment. In peacetime, a person would never have believed that such emotional stress could be possible. We had the yearning wish to somehow be delivered from this torture. We would have preferred to go on the attack again. Throughout the whole day, the French artillery did not pause firing. Salvo after salvo flew at our positions on Hill 287 until the evening. Yet again came the "evening lullaby" —then we had survived it. The French batteries showed off, flaunting themselves and then drawing off in retreat, as far as we were able to observe. They wanted to get themselves to safety at nighttime. Our casualties on Sept. 10 were quite severe: 4 officers and 40 men were dead, 4 officers and 160 men were wounded, and 8 men were missing.

After the night attack, the French fortress of Verdun was almost encircled. Only an area of hardly 14 kilometers across separated the 10th Division, located eastward near Ft. Troyon on the southern front of Verdun, from the divisions of XIII and XIV Corps attacking from the west. The only train connection point in the Meuse valley leading to Verdun was under German fire.

Night fell. The battalion began fortifying the positions. Then around midnight came the field kitchen. Hänle, feeling concerned for me, brought me dry clothes, underwear and a blanket. I preferred not to eat. My sick stomach had plagued me too much all day long. Should I report sick? As long as my legs could carry me and I could perform my duty, that was out of the question. I slept for a few hours in my new dry things, although I had gruesomely horrifying nightmares. In the early morning hours I set about work with a pick and shovel.

On Sept. 11, the French artillery acted as it had during the previous days. Our companies were now deeply entrenched in the clay earth and our casualties were few. However, the ceaseless downpour of rain, together with the very cool temperature, made our stay in our positions uncomfortable. Again the field kitchen came at about midnight.

Considerations:
- It happens very easily **during night battles** that **one's own troops on opposite sides will shoot at each other.** The 2nd Battalion barely managed to avoid this.
- The **night attack** from Sept. 9 –10 led the 2nd Battalion some 1,000 meters ahead of the front of the division. With **minimal casualties on our side** the established goal was reached. Additional further advances would presumably have met with no serious resistance. The rain was advantageous to the attack. **Heavier casualties** first occurred **during the firefight** with densely massed French retreating towards Rembercourt and **while digging in exposed to French artillery**. Things would have been even worse if the troops had not already dug 30 centimeters into the ground by the time the French artillery opened fire. Therefore—**before daybreak, spade work**!
- **Support from our own artillery** was extremely scarce on Sept. 10 and 11 **due to ammunition shortages.** French batteries could direct rapid fire on our ranks from open positions without being punished.
- Throughout all these days of fighting, the **field kitchens** could only come **at night** behind the company positions due to the intense enemy fire. During the day they were many kilometers behind the frontline. One quickly became accustomed to this type of nourishment. —*Rommel*

"Blacking Out"—Countermarch through the Argonne

On Sept. 12 at about 0200 hours —while it was still pitch dark—I was summoned to the regiment to receive orders. In a trench with makeshift camouflage of doors and boards, located hardly 100 meters behind the 2nd Battalion, *Oberst* Haas issued orders by the glow of a candle: "Clear out of your position before dawn, begin a retreat to Triaucourt. The 2nd Battalion will be the rearguard. Hold the hills one kilometer southwest of Sommaisne with two companies until 1000 hours and then follow the regiment."

On one hand, we were

◊ Rommel has developed an aggressive streak and is reluctant to retreat, although he describes his position as a "witch's cauldron." He prefers action and is tenacious about holding territory he takes, which becomes more apparent as the narrative progresses.

◊ Utterly exhausted, he experienced a blackout, falling into "death-like sleep" and remembering nothing later. This happened to him again in North Africa during World War II; Rommel pushed himself to his limits but would suffer from blackouts.

◊ In a prelude to his World War II photography, Rommel witnesses a town hall burning from a distance and refers to it as an "eerily beautiful wholeheartedly happy to get out of this witch's cauldron, on the other hand we could not grasp mentally why we were supposed to retreat. Pressure from the enemy was nonexistent from our point of view. It was a shame that the fortress of Verdun—located 32 kilometers behind us to the left and no longer having a railway connection to central France at

> sight." He would later take many wartime photos of explosions, smoke, action, vast desert landscapes and battle scenes that emphasized contrasts between light and darkness, in addition to other themes I have identified before in my *"Erwin Rommel: Photographer"* book series.

its disposal—would now be allowed to breathe freely again. Now the high command which oversaw the overall situation must have had their reasons for this measure. Maybe we were more urgently needed in a different position.

Before daybreak the 2nd Battalion disengaged from the enemy unnoticed. Our clothing, thickly coated with lime crust, and the shabby fighting conditions made our march extraordinarily strenuous. Two companies stayed behind on the hill two kilometers north of Rembercourt for several hours as the rearguard. At dawn, as on previous days, the French artillery opened heavy fire on our now vacant positions. We all found this especially delightful. The jokesters in our companies once again had lots of fodder for humor.

We assembled in the forest west of Pretz, then the 2nd Battalion moved to outposts by Triaucourt. *Hauptmann* Ullerich rode forward with me to reconnoiter the positions. Again it began to rain in streams. I was happy to be sitting on a horse again. The 5th and 7th Companies were inserted as security for the outpost sectors. The remainder of the battalion became forward reserves in Triaucourt.

After I rode back to the position of the outpost companies in the afternoon, I returned to staff quarters and fell into a death-like sleep. Neither shouting at me nor shaking me availed my battalion commander in his attempts to wake me up so that I could draft a handwritten report with sketches about the disposition of the outposts. When I got a tongue-lashing about this on the morning of Sept. 13, I had no idea that anyone had tried to wake me up.

We began the further retreat march on Sept. 13 as early as 0600 hours in regimental formation. Passing Briceaux, our path was supposed to go through the Argonne. The sun beamed brightly for the first time in a long while. The streets and paths had no firm foundation due to excessive use by supply groups during the rain.

Entering the Argonne 1.5 kilometers north of Briceaux, the retreat faltered to a halt. Substantial sections of artillery and columns were stuck in mud. Every single team of horses had to be guided through with a lead and helped by groups of men. It was lucky that the enemy was not closing in on us and did not shoot at us with their long-ranged artillery as we entered the Argonne. After about three hours of waiting and helping others, the march resumed. Marching on the rain-soaked earth of the forest road behind the constantly halting artillery was extremely stressful for the troops. The infantrymen continually had to grapple with the artillery wheels to move them forward. At about evening we arrived in Les Ilettes. There we had a short meal break. Then the march resumed north through the Argonne. Throughout the 12-hour march already behind us and the difficult paths, the energies of the troops were nearly depleted. Nevertheless we kept marching further into the dark night with a goal that still seemed very far off.

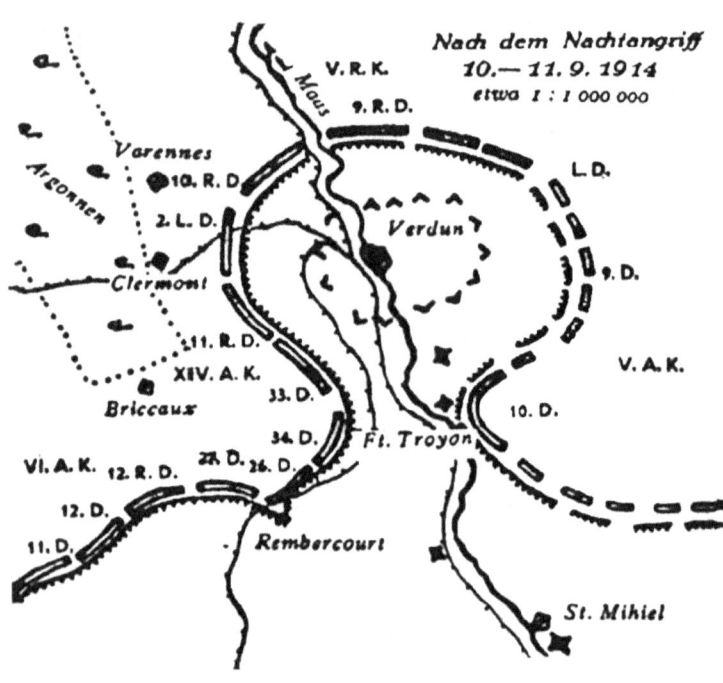

Rommel's illustration of troop positions after the night attack.

Men steadily began blacking out due to exhaustion. At every standstill in the march, men fell where they stood due to fatigue and were fast asleep in the blink of an eye. Then we continued onward, and every exhausted individual had to be shaken awake. We marched...stopped...and marched. I also slipped off my horse very often during the march because I constantly nodded asleep.

After midnight we drew closer to Varennes. The town hall was burning—an eerily beautiful sight. I received the order to ride out ahead to scout for quarters in Montblainville. This small place yielded only a few sleeping areas and no straw.

Early on Sept. 14 at about 0630 hours the silent column of the exhausted regiment staggered through the dark town streets. We obtained quarters very quickly. A few minutes later Montblainville was still again as if in a state of extinction. Everyone slept. Nobody felt the hardness of our sleeping quarters. On the same day *Major* Salzmann took command of the battalion. Afternoon we marched onward to Eglisfontaine. There we found cramped, dirty shelter. The battalion staff huddled in a tiny room, which was filled with bugs, but in any case better than being out in the open under an incoming downpour of rain. My stomach by this time was in deplorable condition day and night. Often I fainted.

Throughout the following days and nights, the French artillery laid fire on all locations behind the front, including Eglisfontaine. We entrenched ourselves in an area near the town. On Sept. 18 we marched for Sommerance in order to rest there for several days. I obtained a room with a bed. I now hoped to get my stomach problems under control to some degree. Bathing, shaving and fresh clothing felt like receiving special acts of generosity.

In the first night (Sept. 18–19) we heard the alarm at 0400 hours to depart for Fleville. There the battalion stood for three hours in streaming rain as a corps reserve, then was sent back to our quarters afterwards. On Sept. 20 we had a real day of rest. The troops organized their weapons and equipment.

> **Considerations:**
> - **Breaking off the fight** occurred during the night from Sept. 11–12 unnoticed by the enemy. Also on September 13 the enemy did not pursue us, which could have been very unpleasant indeed for us as we entered the Argonne. During the retreating march on Sept. 13 the troops, on outpost duty the previous night, were expected to march 45 kilometers. This march was **made complicated** due to **frequent delaying stops** and **demands for help** from columns who had become stuck. The battalion was on the move without pause for more than 24 hours. —*Rommel*

"Monstrous Bitterness"—Assault at Montblainville, Storming of the Bouzon Forest

There was another alarm on the afternoon of Sept. 21. We retreated to Apremont. There the battalion received the order to relieve a battalion of Infantry Regiment 125 after nightfall in the foremost line on a summit 1.5 kilometers west of Montblainville. The description of the position we would take over was hardly attractive: "Forward

◊ Young Rommel witnesses an atrocity. As a medical team is about to rescue one of his wounded comrades, the man is mercilessly gunned down by group of French soldiers right before the medics reach him. Rommel describes being overcome with "monstrous" bitterness. He says German soldiers later took revenge on the French troops in what he describes as

slope position, within full view of the enemy, wet trenches, lots of small arms and artillery fire, therefore casualties are occurring daily. Movement to the rear is only possible at night."

In the pitch-black night, again in a downpour of rain, our path took us cross-country through a soaked grain field to the front, under the leadership of a group of guides from the troops we were supposed to relieve. Before midnight we took over. The positions we took over—small trenches, 50 centimeters deep, not connected with each other—were filled to the brim with water. The defenders were lying wrapped in their coats and shelter-halves, clutching bayonets in their arms, far behind the position. It was said that the enemy was barely 100 meters across from us.

The troops quickly adjusted to this situation. They scooped the water out of the trenches using cooking utensils and diligently began deepening and fortifying the position. They had learned to appreciate the value of trench positions at the Defuy Woods. Progress on the work went quickly in the wet earth. After

"carnage" or "butchery" *(ein Gemetzel)*. Unlike his other passages, Rommel provides no details about where he was personally or his individual actions leading up to and during the brutal retaliatory attack. Instead, Rommel disappears from the narrative and shifts the focus of the story onto the actions of his group. Two possibilities arise: either Rommel lost his self-control and indulged in a murderous desire for revenge with the others, or witnessed his comrades engaging in cruel actions and failed to prevent these things from happening. Situations like this are common in battle. Soldiers can lose their self-control particularly when overcome by intense stress or tragedy, such as when they are severely injured or when a comrade dies. I once spoke with a U.S. World War II veteran who told me he was unable to stop his friends from shooting an unarmed enemy soldier who surrendered; this veteran became emotional due to feelings of guilt. Rommel could have been unwilling to describe his actions at this battle, possibly out of guilt.

◊ Brutality would have made a more popular story in Hitler's Germany—how steely young German troopers avenged their fallen comrade in a bloody, Wagnerian act of vengeance. Yet, Rommel is silent here. We do not know the details.

◊ Rommel's passage describing the cruel death of his comrade at the hands of the French and the Germans' desire for—and acts of—retribution were completely removed from the U.S. Army's 1944 English translation of this book.

a few hours, the majority of the trenches were connected with each other. Now the battalion could face the coming day with reassurance.

On Sept. 22 the sun finally shone again. In the first morning hours, everything in the battalion sector was very peaceful. The enemy was positioned to the right on the eastern edge of the Argonne at a corner of the forest some 400 to 500 meters across from us. Directly in front of us, on the road between Montblainville and Servon, nothing was to be seen of the enemy. However, slightly to the left of us a forested section of that same road was occupied by the enemy. Despite the relatively short distance, it was possible to move outside of the trenches without being shot at. Ripe plums on the trees close by our positions were quickly taken in these circumstances.

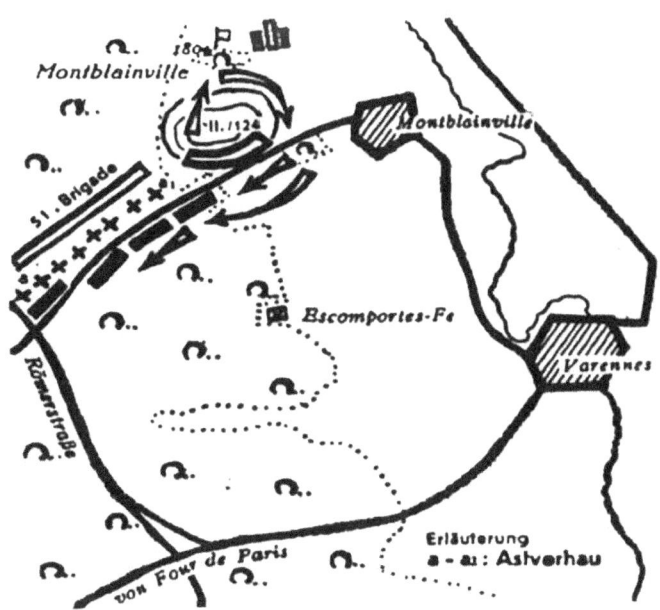

Rommel's map of action around Montblainville. A and A1 (near the tiny crosses on the upper left) mark the sites of abatis barriers.

At about 0900 hours a French gun battery began shooting at our newly built-up positions. Casualties that occurred from this were very minor thanks to our work during the night. The enemy fire stopped after half an hour. Later from time to time brief artillery harassing fire was directed at our position. The French infantry had made themselves totally scarce until about noon. A reconnaissance patrol was therefore sent out in order to establish whether the enemy was located in the woods ahead of us and where they were.

At a distance of about 50 meters from the edge of the forest, the reconnaissance patrol received heavy fire and had to withdraw, leaving behind a severely wounded man. We took part in the firefight from our position to support the reconnaissance patrol. After it was over, French soldiers and our medics both approached the wounded man, who was lying very close to French lines. From all outward appearances, the wounded man was going to be recovered. In order to make it possible, we completely stopped firing on the enemy. However, as soon as the French reached our wounded man, they shot this defenseless person—firing all together.

All of us were overcome with monstrous bitterness about this malicious deed. We opened fire immediately. We desired to charge in a storm into the forest to avenge our comrade. During the course of the day, an opportunity for this would eventually come to us.

In the afternoon the field kitchens came into the hollow 800 meters north of our position. Despite presently very heavy French harassing fire with grenades, shrapnel and gunfire, the troops including those in the frontline were able to receive nourishment from meal runners. At about 1500 hours I presented myself at the regimental command post located at Area Point 180 some 1.5 kilometers northwest of Montblainville. I was briefed about the situation and received the 2nd Battalion's order to attack.

In the Bouzon forest, along the Montblainville—Servon road, a strong enemy presence was situated in a position behind an abatis[5]. All frontal attacks against this position made until this point by the battalions of the 51st Brigade, located ahead of us to the right, had failed. On our left—to the east of the Argonne—Infantry Regiment 122, strengthened by the 1st Battalion of Infantry Regiment 124, made an attack through Montblainville against the mountain ridge located one kilometer south of the town and was making good progress.

Now, as night fell, the 2nd Battalion was supposed to attack the flanks of the enemy—who was located in a forest, behind an abatis on the Montblainville—Servon road—and roll up to the west. It was a grand mission, but also a difficult one! Advancing to my battalion I studied the terrain with precision and contemplated how to best lead the attack. A forward storm charge from our current positions to secure our first goal of the Servon—Montblainville road was inadvisable. This attack would not come to the enemy as a surprise. We would be flanked from the forest and cost severe casualties before even reaching the road. Also, this attack would not damage the enemy's flanks.

After I transmitted the regimental command, I proposed the following suggestion to the battalion commander: First, to clear out the positions on the summit 1.5 kilometers west of Montblainville and to assemble the battalion on the northern hillside of this mountain, which offered more cover. Then we would go into the hollow very close to the eastern side of our previous position—distributed in depth – and seize the small forest 600 meters west of Montblainville.

This small forest had in fact recently been shelled by our artillery with terrific force and thus from all outer appearances had been completely evacuated by the enemy. Due to the terrain behind it, it was possible for our movement to go unseen by the enemy, who until that point was straight across from us. Upon reaching the small forest, the battalion would then be located south of the road, facing the west and sited opposite the eastern edge of the Argonne forest, and would from there be able to prepare for the attack. This attack would then be required to meet the flanks of the enemy on the Servon—Montblainville road. Provided that we cleared out of our current positions right away, this attack could still take place, even in the twilight.

---

5. A fortification consisting of bundled tree branches.

A painting called "Storming of the Hill 285," by Georg Schöbel (1915). The painting shows German infantrymen attacking the French in their trenches in the Argonne in July 1915. Trench warfare was fierce and intimately brutal; men often stabbed and clubbed their opponents to death.

This suggestion was put into action. Springing away one by one across the hillside, platoon after platoon cleared out of their positions on the southern slope. Very energetic fire from enemy artillery caused a few light injuries. Soon the whole battalion was together on the northern hillside. The enemy kept shooting at the vacant positions. Then we went into single file—battalion staff leading the way—amid the grooves in the terrain to the small forest. Our erstwhile opponents still fired continuously at our abandoned positions, now to our right. Apparently they did not notice our movement.

Without encountering the enemy, we reached the small forest. A trench for kneeling riflemen stretched from its northern edge. Inside lay numerous pieces of equipment including knapsacks, canteens and some firearms. The occupying force had seemingly rushed to evacuate the small forest due to the German artillery fire during the afternoon. We took position facing west and prepared to attack the enemy on the forest edge. It still seemed as though the enemy there had noticed nothing of our presence—or at least we got no fire from over there.

The forest edge we were about to attack was 400 meters away from us and the terrain sloped upward slightly. South 500 meters, the road led straight along the forest edge to a hollow, which seemed especially suitable for our encroachment. Here, the fully camouflaged 5th Company moved up single file until they were about 100 meters from the forest edge. During this time, the 7th and 8th Companies reported they had finished moving into position between the road and the hollow. The 6th Company, in reserve, closed the gap left by the 5th Company. The individual companies were quickly briefed about the plan of attack and their tasks. By heavy echeloning from the left going forward, we hoped to surround the enemy on the road.

It was already raining heavily by the time *Major* Salzmann gave the signal to move. Soundlessly the battalion moved closer to the forest edge. Soon, the foremost rifleman of the 5th Company were slinking forward, diving from bush to bush in the hollow as the 7th and 8th Companies approached the forest edge across 250 meters from under tall trees. No enemy movement stirred. Their entire attention seemed focused on the abandoned positions north of the road.

A view of Montblainville from the south. A) French abatis (near crosses); B) (hilltop) position of the 2nd Battalion; C) (on the right below the hill) the launch point for the attack.

The 5th Company moved ahead in the same direction through the undergrowth. Soon the whole 5th Company with the battalion staff submerged in the forest of timber. The 7th Company charged the street and the enemy from 80 meters distance. A violent battle rapidly unfolded there. It didn't last long. The 5th Company and the battalion staff turned right, the 8th Company and the left wing of the 7th Company turned half-right—and presently the whole battalion gave forward storm signals and rushed with a powerful battle cry at the enemy.

The enemy's abatis availed him nothing. The astonishing and massive charge into his flanks and rear had a strong impact. Panic seized the enemy reserves and the defenders of the abatis. Those of them who did not perish by our troopers' bullets, bayonets and blows of rifle butts fled head over heels to the west.

Savage rage for the dastardly shooting of our severely wounded comrade before noon was wreaked. Only the night brought an end to the butchery. Fifty Frenchmen, multiple machine guns and 10 ammunition wagons of a battery fell into our hands—plus many mess tins over open fires filled with warm, ready French evening meals. We lamented our losses: *Leutnant* Paret and three men dead, one officer and 10 men wounded.

Our storming attack developed yet further. Due to the complete panic that broke out on the [enemy] right flank, an entire French brigade was swept along with it. Scrambling, they abandoned their strong position behind obstacles. Across from them, the 51st Württemberg Brigade fell on the intersection of the Montblainville—Servon with the Roman road during the course of the night and large numbers of deserters fell into their hands. The battalion encamped on the site of the battle. On damp earth and without straw, wrapped only in our coats, we were quite freezing in the cool September night. Our horses, however, finally got to eat their fill again on captured oats.

At daybreak on Sept. 23 I had to accompany *Oberst* Haas on a reconnaissance ride to the Roman road. Afterwards the 2nd Battalion received the command to advance along the eastern edge of the Argonne toward the south until [reaching] the farm of Les Escomportes. While I was detained among the regimental staff, the battalion—contrary to the regimental order—marched off on a diagonal route through the forest. During an attempt to reach the eastern edge of the Argonne parallel to the Les Escomportes farm, I established that the French were still sitting there with machine guns. I finally ran into the battalion in the afternoon. Before noon it had advanced into the forest without reaching Les Escomportes farm, instead reaching the hill one kilometer south of it and driving away a few enemies they found posted there.

Just as I arrived, shells struck in the space where the battalion was located. Yet again it baffled me how the French artillery had obtained information about the exact standing location of the battalion in the middle of the forest and how they managed to shoot with such precision.

Hungry and tried, the infantrymen lay under trees and in French-built shelters made of tree branches. They had once again gotten nothing to eat since they went on the march in the early morning. I rode back in order to fetch the field kitchens, which should have been in stationary position at Apremont. I found them one kilometer north of Montblainville. However, in the deep mud, the utterly exhausted horses could not bring the kitchens forward to the battalion.

The kitchens remained stuck 400 meters east of Escomportes farm. Between midnight and 0300 hours the companies went in groups to fetch their food. Meanwhile the command arrived from the regiment that the 2nd Battalion must reach Escomportes farm by 0500 hours. Under these circumstances, our sleep was very brief.

Considerations:
- **Relief of a frontline battalion at night.** A guide detachment leads the way forward. The takeover has to proceed soundlessly, otherwise the alerted enemy will react with

fire and cause the most casualties. Again the **2nd Battalion needed to do heavy spadework** before daybreak and therefore withstood the enemy artillery fire with **minimal casualties**.
- **Battle analysis.** It is advisable to keep strong covering fire ready for reconnaissance of the kind that took place before noon on Sept. 22. Casualties can thus be prevented. In some circumstances it's advisable to provide a light machine gun patrol as fire support and thereby guard forward rushes from point to point.
- **Clearing a forward slope position** by day was possible for the 2nd Battalion on Sept. 22 with few losses despite heavy enemy fire from only 500 to 600 meters distance. The troopers jumped one-by-one back over the hill. Also, **with the effectiveness of today's weapons**, a similar disengagement would in my opinion be possible. Certainly the enemy must be held down via artillery and heavy infantry weapons (which did not happen back then). Furthermore, **deploying artificial fog** would make such a movement easier.
- The **clash of the 2nd Battalion** on the evening of Sept. 22 **into the flank and rear of the strongly entrenched enemy in the Argonne forest** led to penetrating success with minimal casualties on our side. Due to the favorable terrain, a company was able to echelon forward far on the left during the advance which produced very good effects during the clash with enemy troops. The right flank of the enemy in the forest was overrun and the panic of those in flight seized an entire French brigade and prompted them to give up their position.
- The night of Sept. 23–24 showed just how **difficult it can be to nourish** troops during mobile warfare. —*Rommel*

"Hit by a Bullet"—Forest Battle along the Roman Road

◊ Rommel discovers a wounded Frenchman and expresses empathy for him; his men help the Frenchman. This anecdote of a peaceful interaction was also omitted from the 1944 English version of the text.

◊ Rommel is reluctant to dwell on his own pain after being wounded. Instead he focuses on other soldiers—on not wanting to leave his men, on other soldiers suffering in the ambulance and in the hospital. He claims not to have felt much pain and, admitting the ambulance ride was rough, says it "caused pain" without directly referencing himself. This is typical of Rommel. He tends to minimize his own discomforts and instead focus on others. When describing discomforts, he often attempts to do so in a group context, using unique features in the German language to "depersonalize" the experience and remove references to himself as an individual. This indicates that Rommel, like many

> other combat veterans, was very conscious of other soldiers' sacrifices and did not allow himself to complain about his own pains even in writing.
> ◊ The bullet Rommel took damaged his leg muscles. Doctors did not believe he would walk again. However, he made a full recovery. For a long while he had to walk with a cane and thigh brace and was left with a large scar.

As ordered, the 2nd Battalion arrived on Sept. 24 at 0500 hours at Les Escomportes farm, stopped and rested. In a narrow, dark chamber, Col. Haas gave Battalion Salzmann the order to advance diagonally through the Argonne approaching the crossroads between the Four-de-Paris—Varennes roads, and to take and hold the Roman road.

I forgot my exhaustion, sick stomach and the feeling of total fatigue. The new task tensed my strength for action. As the battalion began to march, the sun rose from the morning fog in a blood-red fireball. The verse automatically came to mind: "*Morgenrot*..."[6] Without a path, our route went crosswise through dense underbrush towards the crossroads. Keeping direction with my compass, I marched on foot at the head of the single-file column. Sometimes impenetrable bushes forced us to turn.

We young officers of Infantry Regiment 124 had, during the previous few years of peacetime, been trained often in navigating alone by compass at night crossways through the vast forests surrounding the garrison. That now paid off. After an hour-long march we reached the Roman road. We were still one kilometer away from our established goal. Under security on the march we moved ahead to the south. The battalion staff were riding far behind the head of the column.

At a junction of six intersecting forest footpaths, lying beside a ramshackle mountain hut, lay a severely wounded Frenchman. He stammered and shivered from coldness and fear. According to his utterances, the poor guy[7] had been lying on that spot since the battle near Montblainville. His comrades had abandoned him during their retreat. Our medics tried to help the man and bound his wounds.

At this time, a mounted reconnaissance patrol returned from the Four-de-Paris—Varenne road and reported that the enemy had entrenched ahead near the street. Hence caution!—The 6th and 5th Companies were sent ahead, each moving on a different path, towards the street. Security preceded the companies. The tall trees became less dense, yet the undergrowth remained as thick as before. As the battalion leader remained in the area of the mountain hut with the 8th and 7th Companies, I accompanied the point of the 6th Company's the column, advancing on the right.

---

6. A German folk song, "Morgenrot" (Morning Red) by Wilhelm Hauff. The verse Rommel refers to is: "Morning red, morning red, will your glow soon leave me dead? Soon the trumpets will blast, Then my life will end at last, My own and many comrades'!" The song's cheerful tune makes it more horrifying.

7. Lit. "*armer Kerl*."

Along the side of our path lay numerous dead Frenchmen. Suddenly the earth shook from the hooves of galloping horses. Was it friend or foe? The winding road provided only about 70 meters of visibility. Quickly, the point of our column took position in the bushes on the left and right. Within the next blink of an eye, a herd of riderless horses galloped around the corner, stopped as they saw us, and broke away running to the right.

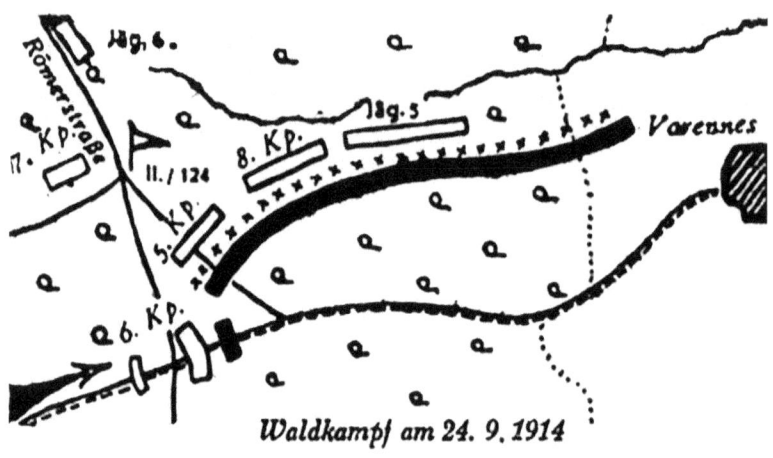

Rommel's map of the forest battle where he was wounded.

Without any further unforeseen incidents, the 6th Company then reached the great street as, over on our left near the 5th Company, an energetic firefight developed. I galloped back to the battalion staff and gave my report. At the same time, the 5th Company reported that, 500 meters south of the mountain hut, they had clashed with the enemy who were located behind abatis obstacles, and that the enemy was growing stronger. The company could not advance further and asked urgently for support. Shortly after that, two officers of the company were carried back, severely wounded. The din of battle ahead near the 5th Company grew louder. Also gunshots were now audible near the 6th Company. Countless shots clattered through the forest. Were there snipers in the trees? Nothing was certain.

Major Salzmann deployed the 8th Company to the left of the 5th Company. Both companies were supposed to attack the enemy together and hurl them back across the Four-de-Paris—Varennes road. This attack also failed about three quarters of an hour later due to truly severe casualties. According to the statements of numerous wounded men, the enemy was located in very strong positions behind abatis fortifications, through which [the enemy] fired machine guns. In the meantime the leader of 6th Company, Graf von Rambaldi, returned with a minor wound and reported that his company on the Four-de-Paris—Varennes road was right across from an enemy of equal strength located about 200 meters east. Also the woods to the west of the company were not clear of the enemy. At that point I went forward to the 6th Company to get a look at the situation there for myself. Accompanied by a strong reconnaissance patrol far south of the Four-de-Paris—Varennes road, I pressed forward 50 meters east of the all-around defense position of the 6th Company towards the enemy.

I exchanged shots with the enemy over a long period and gained the impression from it that there was only a strong sentry position facing me. After returning to the battalion I suggested to advance with the 7th Company and *Jäger* Battalion 6 towards 6th Company and attack on both sides of the street in the direction of Varennes. That way we would catch the enemy across from 8th Company, 5th Company and *Jäger* Battalion 5 in the flank. Before this had yet been decided, a dispatch rider delivered the regimental order that the 2nd Battalion of Infantry Regiment 124, in unison with the subordinate *Jäger* Battalions 5 and 6, needed to clear out the enemy from the areas surrounding the road to Varennes. At the same time, the 6th Company reported that groups of Frenchmen in close order were marching in approach from the direction of Four-de-Paris. So, it was high time to give them some fresh air to the east! As fast as humanly possible, we now began preparing to attack. *Jäger* Battalion 6 approached south of the road, and was supposed to advance with its left flank parallel to the street. The 7th Company was emplaced north of the street. The 6th Company was supposed to attack on the left of the 7th Company, but would leave heavy security on the road towards Four-de-Paris.

When all components reported their final preparations ready, the attack began. The battalion staff followed 7th Company. Just as soon as we passed the first 100 meters, rapid fire from a seemingly strong enemy forced us to take cover.

One could hardly see 20 meters ahead through the thick undergrowth. Not the slightest trace was to be seen of the enemy himself. The companies now also opened fire and moved closer to the invisible enemy by crawling one-by-one and advancing in small lunges. Due to the ear-splitting crashing and banging [of gunfire] in the forest, it was not possible to even guess the enemy's distance from us. The firing increased. Our own attack came to a standstill.

In order to drag the 7th Company forward again, *Major* Salzman and I went to the frontline. I took a gun and an ammunition magazine from a wounded man and took over command of about two groups. It wasn't possible to oversee more in this forest. Several times, we hurled ourselves through the bushes with a war cry at the enemy who we supposed to be in our immediate area. We never succeeded in catching him. But constantly his rapid fire forced us to the ground for cover. Our casualties mounted minute by minute as cries for medics rose all around us. Pressed flat on the ground or taking cover behind thick Argonne oaks, we let the enemy's rain of fire pass over us, in order to immediately take more ground closer to the enemy at the next pause. By now it was difficult to get my own men to move forward. We only gained ground very slowly. The sounds of battle indicated that our neighbors nearby had advanced to a position parallel to us.

Once again I led a storm at the enemy in the bushes ahead of us. A little herd of my former recruits dashed with me through the undergrowth. Again, the enemy shot like crazy. Then—finally!—I saw five Frenchmen hardly 20 paces in front of me. They shot freehandedly from standing positions. In a flash, my rifle was pressed to my cheek. Two Frenchmen, one standing behind the other, collapsed as my shot cracked. Now I only had three enemies facing me.

Apparently my men had retreated back to cover. They could not help me. I shot again. The shot missed. Rapidly I ripped the chamber open. It was empty. There was no time to load in the face of my nearby enemies. There was no cover available in the immediate area. To flee in retreat was out of the question. The only opportunity I saw was in my bayonet. I had been, in peacetime, an enthusiastic bayonet fighter and had acquired quite some skill at it. So now, despite that I stood alone against three opponents, I had complete trust in the weapon and in my capability. But, as I rushed forward, the enemies fired. Hit by a bullet, I tumbled over and now lay only a few steps away from the enemies' feet. A ricochet had ripped my left thigh. Blood sprayed out of a wound as big as a fist. Every second I expected a shot or a deathblow. With my right hand I pressed the wound down and at the same time tried to wriggle myself behind an oak tree. For minutes I lay between the two fronts. Finally my men broke through the bushes again with a war cry. The enemy retreated.

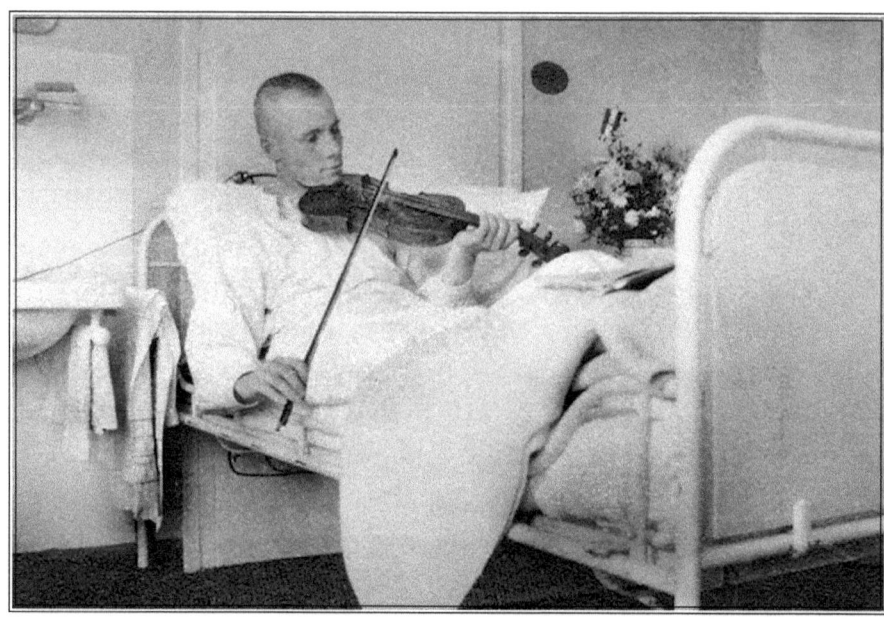

A photo of Rommel during his recovery in Stuttgart's Karl-Olga-Krankenhaus hospital. Rommel continued to play violin later in life.

Gefreiter Rausch and the new recruit Rutschmann, who had been in the military for just one year, cared for me. A coat belt was used to tie the blood vessel, then the wound was plugged with compression bandages. Afterwards they carried me in a shelter-half through the bushes back to the rear by the mountain hut.

From ahead, the report came that the enemy had been driven out of his fortified positions and out of the forest. The enemy had left prisoners in our hands. However our own losses were awfully hard: the 2nd Battalion alone had suffered 30 dead, among them two officers, and 81 wounded, among them four officers. However the battalion had, for the third time in three days—as would later be recorded in our regimental history—fulfilled its task brilliantly.

The parting from all of these brave men was hard for me. As the sun set, two men carried me in a shelter-half, fixed to a pole on either side, five kilometers behind the lines until Montblainville. I hardly had any pain. However, my consciousness kept fading, likely due to the heavy blood loss.

At night I woke up in a barn in Montblainville as *Stabsarzt*[8] Dr. Schnizer worked over me. My loyal friend Hänle had brought him. I was bandaged again and ultimately loaded into a horse-drawn ambulance. Three fellow sufferers lay moaning next to me. Our trotting journey took us cross-country to a field hospital. The road had been torn up by shells. The rocking of the vehicle caused severe pain. By the time we were unloaded at about midnight, one of those lying next to me was already dead.

The field hospital was overcrowded. The wounded lay in rows, wrapped in blankets, on the road. Two doctors worked feverishly. Eventually I was sought out and then was brought into a hall to lie on straw. As day broke, an ambulance was ordered to take me to a rear area hospital in Stenay. The Iron Cross, Second Class, reached me there a few days later. After surviving an operation, I was taken in the middle of October by a courtesy car back to the German homeland.[9]

Considerations:

- The enemy on the Four-de-Paris—Varennes road made it difficult for the 2nd Battalion to complete its objective. Ultimately three battalions were deployed to **attack in the woods**, and they only managed to drive the enemy into the dense forest **through heavy blood sacrifices**.
- Even at the beginning of the forest battle there were severe losses. Three officers fell, among others. **Whether French tree snipers** were at work here is difficult to say because none were spotted or brought down.
- It proved **very hard** in this costly fight **to bring the men forward**. The **personal example** of the leader is only effective **in a dense forest** from within his immediate surroundings.
- In **a man-to-man fight** between opponents of equal strength, whoever has one more round in his gun barrel will win. —*Rommel*

---

8. This rank denotes a staff surgeon with an officer's commission.
9. Rommel consistently uses the word *"Heimat"* to refer to Germany, rather than the preferred Nazi terms *"Reich"* or *"Vaterland."* I will discuss this word choice in greater detail later on.

2

Battles in the Argonne

1915

"Whiz-Bangs"—The Company Sector in the Charlotte Valley

> ◊ Here 23-year-old Rommel flourishes as a company commander. His distinctive leadership style develops. Instead of writing that his men will "follow" him, Rommel uses the term *"mitgehen,"* meaning to "go with" or "accompany." This emphasizes his belief that he and his soldiers were equals. Throughout his life, Rommel believed a leader should share the burdens of the men serving under his command and work closely with frontline soldiers to solve problems, regardless of hierarchical or administrative distinctions. This often caused misunderstandings among Rommel's staff and upset the rigid bureaucracy of Germany's military. Rommel's hands-on mindset brought him into conflict with Hitler and the Nazi High Command during World War II, who took a detached, compartmentalized approach to issues facing troops in the field.
> ◊ Here we can see Rommel taking pride in physical labor and engineering achievements. He describes construction work in favorable terms and is always ready to build things. He also expresses admiration for enemy fortifications he considers well-built. Other commanders might not have wanted to be involved with building work, but young Rommel never hesitates to get his hands dirty. His men respond positively. As an older man, Rommel during World War II engaged in the same sort of behavior—he was always ready to get in the mud alongside his troops and help fix things.
> ◊ Rommel is very preoccupied with using his engineering skills to prevent men from getting injured. He continuously seeks new ways to improve fortifications and shelters, and shield other men from danger.

Shortly before Christmas I was released from the hospital. The wound was not yet healed. It impaired me a lot while walking. Service operations at a replacement training battalion I was with were not much to my liking. So I went with the next troop transport back into the field.

In the middle of January 1915 I came upon my regiment, located in position in the western part of the Argonne. The deep, boggy path from Binarville to the regimental command post gave a foretaste of conditions in the Argonne forest.

I was given command of 9th Company, whose command had recently been made vacant. A long, extremely narrow path of cut logs for pedestrian traffic [over the marshy ground] led outward from the regimental command to the front. Sporadic bullets occasionally crashed through the winter forest. Now and then grenades came hissing towards me and forced me to dive flat on the ground quickly. I was glad of the cover offered by communications trenches, which were as deep as a man's height, and eagerly

Rommel (farthest on right) enjoys a lighthearted moment with other German officers, circa 1915.

acquiesced to the deep, gummy mud paste of these trenches. By the time I reached the company, you could no longer tell at all by looking at my clothing that I was arriving fresh from home.

I took command of about 200 bearded warriors and an estimated 400 meter-wide company sector in the front lines. The French welcomed me by shooting harassing fire with their "Whiz Bangs"[1]—a gun battery. Our own position was situated in a continuous trench at the frontline with numerous traverses, through which multiple communications trenches led to the rear. Because there was a barbed wire shortage, there was no barrier towards the front. The position had been built with great scarcity. Due to underground water, part of it stood at only one meter deep in solid earth. Also, due to this subterranean water, the shelters—most of which were supposed to hold from eight to 10 men—jutted outwards with their roofs protruding up over the solid earth and acted as bullet traps. In many cases they were made of thin oak branches and only protected against shell splinters.

Even during the first hours that I was active as company leader, a shell struck a fully occupied shelter and severely wounded nine men. I ordered, that from that point onwards, the imperiled shelters should be evacuated immediately in every instance of fire and cover should be shared in rifle trenches. At least here in these trenches it was not possible for an entire group to be hit by a single shell at once.

---

1. As in English, the German word imitates the sound of these shells—they are known as *"Bumratsch."*

At the same time, I took measures to ensure that during the coming nights all protective roofing would be strengthened to such an extent that it would hold against field artillery shells. The strong oak trees located close to the positions also proved to be especially dangerous. Shells that shattered against them sent splinters flying even into the very bottom of the trenches. I ordered for some of these trees to be felled.

My life with the company during the following days was very fast-paced. For me, a 23-year-old officer, there was no duty more wonderful than to be a company commander. Through circumspection, clear instructions, through unceasing care for the men who trust in him, through austerity for himself and through shared living under equal demanding conditions, a leader can earn the confidence of his subordinates in a very short time. This trust can be so great that the troops will go with him[2] through thick and thin.

Every day brought an abundance of work. There were so many shortages: of boards, nails, hinges, roofing paper, wire and tools. The table and bed of my underground "company commander" dugout—which was about 140 centimeters high, and shared with a platoon leader—were made of beech sticks bound together with wire and string. The walls were not braced, and water incessantly trickled down from them. During damp weather, water also came down continually through the roof, which was made of two layers of oak logs with a thin earth covering. The dugout had to be bailed out every four hours, otherwise it would be completely submerged. Fire could only be made at night. During the daytime we froze severely in the wet winter weather.

Through the dense underbrush ahead on our front, there was nothing to be seen of the enemy position across from us. In contrast to us, the area the French possessed, in which their frontline was located, was such that they did not need to retrieve material to fortify their positions; they had brought this material, already prepared, from their own country. Furthermore they did not have much to worry about in terms of harassing fire, because the German artillery was very short of munitions, and also because the French positions were well-camouflaged in a thick forest.

The terrain ahead of our position in the direction of the enemy dipped slightly from 100 to 150 meters. We estimated the enemy position to be at about 300 meters distance from the opposite end of the dip ahead of us. From that position, the enemy sprayed small arms and machine gun fire at the terrain behind our trenches at different times of the day and night. Thus the enemy severely hindered our work [when we were] exposed from cover.

Most especially unsettling for us was the fire of the "Whiz-Bangs." With these, shell fragmentation and impact would occur at the same time. Whoever was surprised by a

---

2. This choice of the German verb *mitgehen* shows that Rommel wants his soldiers to "go with" rather than blindly "follow" or "obey" him. It's an unusual word choice for a German military leader and says a great deal about Rommel's philosophy.

"Whiz-Bang" while outside the trenches had to dive to the ground with lightning speed exactly where he stood to present a smaller target for the sharp-edged splinters that sprinkled from all sides.

Rain and snow alternated during the latter half of January 1915. From Jan. 23–26 the company moved into a reserve position located 150 meters behind the frontline. The shelters were even worse there, the enemy fire was more vexatious, and daily casualties were no fewer than before. The company was put to work with the Labor Service—transporting materiel, building shellproof shelters, cleaning communications trenches, constructing log pathways.

When we moved forward into the old position again, we were wholeheartedly glad and went about our work with new zeal. The mood and the solidarity of the troops were exemplary. Officers and men earnestly wished to endure all hardships in order to protect our homeland from an enemy invasion and to end the war in victory.

On Jan. 27, myself and a few others reconnoitered one of the trenches on the

Rommel smiles as he wears his Iron Cross and wound badge after recovering from his injury in 1915. Rommel had to walk with a cane for a long time after his release from hospital; the cane is visible propped next to his left boot.

left side of my company sector which led towards the enemy. We were actually situated in a previously French position, which our regiment had taken by storm on Dec. 31, 1914. Creeping forward carefully after dismantling the abatis we found in the trench, we came across a dead Frenchman lying about 40 meters ahead, who had probably been lying unburied between the fronts since the time of the aforementioned December 1914 assault. From here onwards, we discovered to the left of the trench a small French soldier cemetery and at the end of the trench—about 100 meters ahead of our company position—in the deepest part of the terrain dip, an abandoned French medic station, which had exceptionally good roofing and could comfortably shelter 20 men.

During this reconnaissance we came across no living enemies. During this, the enemy was firing, shooting the usual harassing fire with small arms and machine guns from a position that, judging from the echo of the shots, must have been located about 100 to 150 meters across the dip in the terrain. Nothing was visible of the enemy position through the thick underbrush.

I decided to rebuild the old French medic shelter as a forward base and get the work started that same afternoon. We heard echoes of the Frenchmen's conversations coming from the point beyond us. It was unadvisable to send reconnaissance patrols to probe any farther ahead. The crackling of dry branches would quickly betray their presence to the vigilant enemy and they would probably get shot before they could see anything of the heavily camouflaged enemy position in the underbrush.

"Bitter Fighting"—Storming on Jan. 29, 1915

> ◊ The fledgling soldier Rommel exhibits the daring he would become famous for in World War II. He decides to occupy a French position to use it against its creators. His brilliant plan backfires when higher-ups don't provide reinforcements. He and his men are trapped in an enemy hornet's nest. Rommel and his troops narrowly escape. Although his actions were undoubtedly brave, his independent decision-making here is questionable. He shows fierce determination in bailing his troops out of the predicament, however, and was later awarded the Iron Cross 1st Class for his efforts.
> ◊ An odd incident arises when Rommel and his men discover items "of a female nature" in a French dugout. Rommel's description of the Germans' bewildered reaction indicates the items gave rise to some speculation and were probably more than just a random "souvenir." The language seems to suggest there was possibly a woman present among the Frenchmen or that an enemy soldier had been cross-dressing.
> ◊ Faced with a life or death situation, Rommel threatens to shoot one of his men for not obeying his orders. This is the first and last time that Rommel recounts an instance of insubordination.

To pin down as many enemy forces as possible in the Argonne, small-scale operations were ordered for all regiments of the 27th Division for Jan. 29, 1915.

In our regiment, a sector of the 2nd Battalion (right) planned a "storm" troop assault preceded by detonating a French sap. During this, the enemy was to be pinned down with artillery fire ahead of 10th Company (forward right) and the 9th Company (forward left), in the middle of the regiment's sector by the 3rd Battalion. A howitzer battery of

Artillery Regiment 49 had already adjusted to the precise target on the 27th and 28th of January. Because 10th Company would foreseeably transfer position, the 9th Company was not to advance, but instead to shoot or capture enemies fleeing from the flanks.

Jan. 29 dawned as a cold winter day. The ground was frozen. At the start of the operation I stood with three groups at embrasures within a newly constructed strongpoint, 100 meters ahead of our company's actual position. Our own shells passed roaring above us. A few shattered on the trees or struck behind us. Then came the explosion of the French sap to our right. The earth quaked. It hailed dirt clods, branches and rocks. Small arms fire began. Hand grenades cracked. A single Frenchman sprang up from the right, coming at our position, and was shot.

A few minutes later the adjutant of the 2nd Battalion arrived and reported that the storm assault was going exceptionally well and asked on behalf of the battalion commander whether or not we in the 9th Company wanted to join in the advance. Of course we wanted to! Now we could get out of these horrible trenches and this eternal ducking for cover!

I did not think it was advisable to go climbing from the position straight into the open with the whole company. The enemy had set sights very well on our position with machine guns and artillery, and could most likely see us from observing through the trees. Our advance would surely be noticed very quickly, and fired upon. Therefore I had my men go creeping out through a tunneled sap on the right flank of the company position. After this, by creeping forward in a skirmishing line, we assembled for action on the left. After about 15 minutes the company was positioned ready to attack about 80 meters ahead of our previous position on the hillside sloping towards the enemy. Cautiously we crept closer towards the enemy through the bare branches of the underbrush.

However, before we reached the dip in the ground ahead of us, the enemy on the other side opened fire with small arms and machine guns. Every movement came to a standstill. All around us, bullets slammed into the frozen ground. Cover on the hillside was scarce to be found. Thick oaks could only provide cover for a few riflemen. Furthermore, there was not the slightest trace to be seen of the enemy through binoculars. It was very clear to me that staying there longer would result in stronger enemy fire, which even if imprecise was bound to cause severe casualties in my company, positioned without any opportunities to take cover on the hill.

I wracked my brains for how I could bring my little flock of men out of this wretched predicament without harm. In such moments, the responsibility for the welfare and woe of subordinates weighs heavily on the leader. Suddenly I came to the decision to lunge ahead in sections, with our goal to reach the terrain depression 50 meters ahead. There we would at least have some more cover than we could get waiting here on the hillside. As soon as I decided this, I heard the storming signal resound from the distance on my right. My bugler was lying next to me. I ordered him to blare the same signal.

Moving in unison as one man, the courageous 9th Company rose up despite the gunfire that with undiminished intensity continued to oppose us. With a loud battle cry, they stormed forward. We crossed the dip in the terrain and reached the French wire obstacle.

Now we saw the enemy across from us—from a man-to-man distance—evacuating their strong position immediately. Our yell and the flashing of our bayonets came as a bone-rattling scare to them. Their red trousers lit off through the bushes and the tails of their blue-gray tunics went flying.

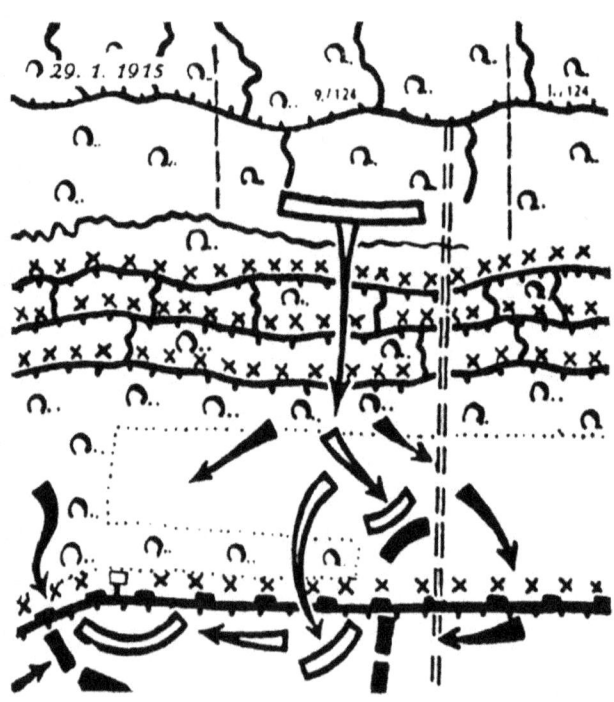

Rommel's map of the action on Jan. 29, 1915.

We hurried to chase them. Why should we worry about the machine guns and rifles they had left behind in the position? Ahead of us, the enemy fled head over heels through thickets. Staying close on their heels, the 9th Company overran, in a single dash, two additional well-fortified French positions which were furnished with strong wire obstacles—their [French] occupants burst out in a hurry [to escape]. Because of the enemy did not shoot during this rush [to get away], we suffered no casualties.

We reached a summit. The forest grew lighter. The enemy fled before us in dense masses. Shooting, we kept on their heels. Sections of the company were still taking prisoners from a dugout while those in the lead reached the forest edge 500 meters west of Fontaine-aux-Charmes. Now we were 700 meters south of our line of departure. From this point onward, the terrain sloped again. The fleeing enemies rapidly disappeared into the low-lying scrubland. Contact with right and left was nonexistent.

On both sides to the rear, a bitter fight was still raging. I occupied the forest edge 500 meters west of Fontaine-aux-Charmes, organized the companies and sought to establish contact with our right and left.

To our unanimous bewilderment[3] one trooper brought out from a French dugout

---
3. Literally, *"zu unserem allgemeinen Hallo."*

various different objects that beyond all doubt were of a female nature. These had been left behind in their immediate flight.

Shortly afterwards a reserve company arrived. I transferred my concern about making contact with left and right over to them, and pressed on ahead with the 9th Company in a southwesterly direction. Our path down the slope took us through scrubland. The timber forest here had in a large part been cut down. Soon we crossed a basin. Safer than ahead at the front, I followed with the company in single file.

Suddenly fire from half left drove us to the ground. Amid the bushes there was nothing to be seen of the enemy. In order not to become bogged down in the pursuit, I withdrew the company in a westerly direction from the enemy zone of fire and hiked upwards through a light timber forest south again.

At the uppermost edge of the forest, we suddenly found ourselves faced by a wire obstacle of an enormous magnitude we had never before seen—it was 80 by 100 meters deep and stretched away on either side as far as the eye could reach. The French had cleared the entire forest here. Beyond the obstacle, situated on a slightly rising slope, three men of my company—among them the extremely young volunteer soldier Matt—stood and waved. So this strong position was not occupied by the enemy at this spot. To take it and hold it until the reserves arrived seemed to me a worthwhile and important undertaking.

I attempted to spring through the obstacle through a narrow lane that was still open—but drew such strong fire from half left that I had to jump away. The enemy at a distance of about 300 to 400 meters could hardly see me now, because the obstacle was very thickly netted. Ricochets banged around my ears.

Crawling on all fours, I reached the enemy position. Beforehand I had ordered the company to crawl behind me one by one. However my foremost platoon leader had not found the courage—and thus his platoon, and therefore the entire company, stayed motionless outside of the obstacle. All my calling and waving availed nothing.

It was impossible to hold this fortress-like, reinforced position with three men. The company HAD TO[4] come up. I rushed west and found another path through a very shallow ditch back through the obstacle. I crawled back to the company and informed my foremost platoon leader that I would shoot him if he did not immediately carry out my order. That helped.

All of us crawled through the obstacle into the enemy position despite small arms fire that increasingly sprayed the area around us from the left. By inserting the company in a half-circle formation I secured our possession of the position we had taken. The company dug in.

4. Rommel's emphasis. This is the only instance in the book where he writes in all capital letters. It's a very dramatic touch.

The fortification, "Central," in which we found ourselves, was an exemplary construction. It was part of the main French position running diagonally through the Argonne. Strong blockhouses stood with about 50 meters space between them. Inside them were machine gun positions, from which the vast obstacle could be defended frontally as well as from the flanks. The breastworks, linked together by the individual blockhouses, were situated so high above the natural ground that the obstacle in its entirety could be protected with sweeping fire from their embrasures. Behind an earth embankment about two-meters tall lay deep covered trenches. Between the protective wire and

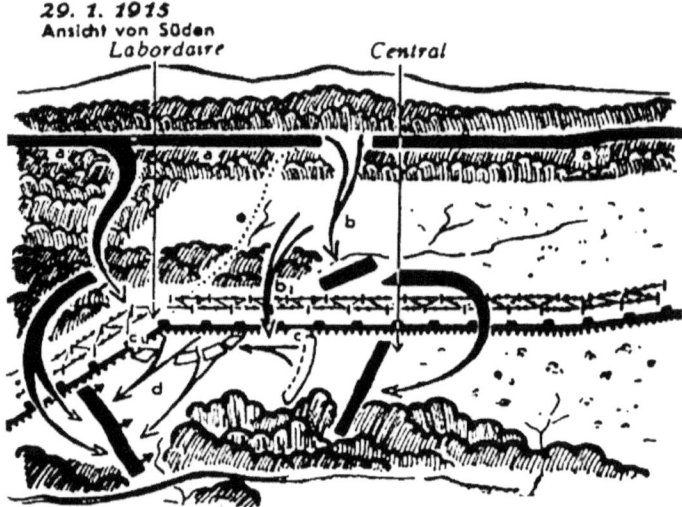

a: 3 französische Stellungen.
b: 9. Kp. verfolgt nach Durchbrechen der 3. franz. Stellung und dringt in Central ein.
c: 9. Kp. hält Teile von Central und Labordaire.
d: Sturmangriff vor Abbrechen des Gefechts.
e: Rückweg

A view of Central from the south. A) Three French positions; B) 9th Company's pursuit and entry into Central after the breakthrough; C) 9th Company holds elements of Central and Labordaire; D) storm attack before breaking the fight off; E) retreat.

the position, five-meter wide trenches were drawn from blockhouse to blockhouse; these were partially filled with water and were now frozen over. A forest path led off 10 meters behind the position. The high elevation of the fortification made concealed movement with vehicles on this path possible.

Now we came under heavy rifle fire harassing from the left. The compound over to our right did not appear to be occupied at all. At about 0900 hours I sent my battalion the following written message: "9th Company is embedded in the fortress-like, heavily constructed French earthwork, 1.5 kilometers south of the position where we broke out by storm, and the company holds sections of the works running diagonally through the forest. Requesting the most urgent support, machine gun ammunition and hand grenades."

In the meantime the troopers tried strenuously, to no avail, to overcome the solidly frozen ground with shovels. Only by using the few picks and pickaxes we had could any progress be made in this endeavor. We had been in the fortification for about 30 minutes when a sentry sent out to our left reported that over to the left (eastwards, at about

500 meters distance), the enemy was retreating in closed columns through the protective wire. With a platoon, I took the enemy under fire. Sections of the enemy sought cover. Other elements located north of the wire turned further east and most likely headed for the concealed [forest] path leading inside the works, because soon after our surprise fire we were attacked from the east in this direction.

The company had made hardly any progress at digging in. In the meantime I had found an especially convenient position for holding a bridgehead, located over to our right about 200 meters away at a sharp angle created by the enemy works there (Labordaire). Thus the company retreated to that location at a lightning pace. Hurriedly we created cover for ourselves from the enemy fire using tree trunks lying around and held the enemy at bay with intense gunfire. Therefore the enemy in the east did not risk advancing closer than 300 meters towards us and by all outward appearances also began digging in. Presently the firefight became quieter before dying away to a full stop.

I had captured four blockhouses within the bridgehead position. The company occupied the area in half-circle formation. I kept a platoon of 50 men with me under cover between the protective wire and the position as a reserve. Also leading to this spot was a narrow, zigzagged lane through the wire.

Time stretched by. We anxiously awaited the arrival of reinforcements and ammunition. Suddenly a report came from the right, that French infantry were retreating through the wire entanglement scarcely 50 meters ahead of our position. The platoon leader asked if he should fire. What was left for us to do? If we did not fire, the enemy could then cross unhindered through the obstacle to occupy the position next to us. Yet in a few moments it would certainly come to a fight. If we were to fire, the French could then shift out to the west and successfully penetrate our position using the nearby passage. It was also possible that enemy elements could cut across our only connection to our regiment. Then we would be fully encircled. I gave the order to fire.

Rapid fire struck out at the nearby enemy from above the high parapet of the French fortifications. Soon this developed into a bitter fight. The French scrapped very bravely here. The majority of our new opponents—we were dealing with about a whole battalion—shifted fortunately towards the west, crossed the wire obstacle about 300 meters away and then pressed against us from the west on a broad front. The ring around 9th Company was closing.

Only a small passage led north through the wire obstacle towards the battalion. It also lay alone in the path of grazing fire from enemy positions to the west and east. However, in our position, the enemy was approaching alarmingly near to us. Combat was rapidly consuming our fighting supplies. Hand grenades and ammunition were getting scarce. The reserve platoon had to hand over the majority of their equipment. Brave troopers brought it up to the line of fire. To make the ammunition last as long as possible, I ordered intermittent pauses in firing. Closer and closer crawled the enemy from the west towards us. What would happen if the ammunition was all used up? Yet still I hoped for help from our battalion. Minutes stretched on like eternities! Now bitter fighting raged around the blockhouse furthest to our right. Our last hand grenades were sacrificed here. A few minutes later—it must have been about 1030 hours —a French storm troop managed to take the blockhouse. From its embrasures, rifle and machine gun fire struck elements of the company from behind. Almost simultaneously, a communication about our situation reached me with the command from our battalion, called out to me by a messenger above the barbed wire: "Battalion in position 800 meters north, and dug in. Company Rommel come back. Reinforcements not possible." Again, our front line called for ammunition. Our entire ammunition supply would not last another 10 minutes.

Now a decision! To break off the fight and spring away in retreat through the narrow passage in the barbed wire through the crossfire at close-quarters from west and east would mean at least 50% casualties, if the whole company was not gunned down all at once. Then, should we give ourselves up as prisoners after firing our last bullet? Never! It remained possible to attack the enemy in the west, which posed the strongest menace to the company, and then retreat afterwards. That could, and must, save us! It was true that this enemy was vastly superior to us in numbers, but until this point, no French infantry had been able to stand fast against the storm attack of my troopers. If the enemy was first pushed back into the west, there would likely be the opportunity to retreat across the wire obstacle. That would leave only the fire of the distant enemy in the east to deal with. But everything had to be done very quickly, before the enemy in the west recovered from the shock.

My attack plan was communicated rapidly to the reserve platoon and also our front line through a messenger. Everyone knew how difficult the situation was and were resolved to give their utmost. Then I rushed forward to the right with the reserve platoon. The blockhouse was retaken, the [French] front line was swept away, and a forceful battle cry reverberated through the forest. Now the red color of French trousers lit up spaces in the bushes ahead of us. Shots cracked—then the enemy started running. Now, the moment had come for us to break off the fight. The French fled westward, we hurried in retreat towards the east.

We went off at a fast run through the broad wire expanse, single file. Soon enemy fire from the east struck between us. It was good that this enemy could only aim at single lunging men from a side distance of about 300 meters. Only seldom did a shot strike its target. Soon, the majority of the company had reached safety, just as enemy fire also commenced from the west. Except for five severely wounded men, who could not be rescued, the company returned and reached the battalion position without becoming involved in further fighting. The battalion had in the meantime entrenched itself in a dense forest just south of the three captured French positions. My company was inserted on the left flank. Connection with the left was absent since the battalion could not throw back the enemy positioned across from it. Contact with the right flank of the 1st Battalion was held through individual groups. The company dug in at between 80 and 100 meters distance from the edge of the forest. Due to the frozen ground it was tiresome work.

Throughout the whole attack the French artillery had laid heavy fire on only our existing positions and the terrain in the rear. During the fight, they were probably not informed about the situation of the skirmish in the forest. Now however they began revenge fire at a great expense of ammunition. The foremost edge of the forest bore an especially heavy brunt of the fire. Our work to dig in was greatly disrupted. Writing on a message blank, I reported the course of the attack before noon and the fight within the French fortifications "Central" and "Labordaire" with a sketch attached.

Following a sharp increase of their artillery fire, the enemy made a counterattack at twilight on Jan. 29. French troops stormed in thick masses through the underbrush at our new position. Bugle signals and commands drove them forward. Then our rapid fire burst into their ranks. They were toppling over, seeking cover, shooting! Here and there little groups of them dared to come closer to us—in vain! Their attempts to storm us withered under heavy casualties by our defensive fire. Numerous dead and wounded lay close ahead of our front. When total darkness came, the French withdrew to the forest edge 100 meters ahead of us and dug themselves in there. The infantry fire was silenced and we renewed our efforts to dig in because our positions were barely 50 centimeters deep. However before we dug conspicuously deeper, French shells burst between us. All around us everything flickered, burst open and cracked. Razor-edged splinters of American steel shells howled through the winter night and snapped strong trees like matches. Our positions provided no sufficient cover from the fire, which now continued the entire night with only brief pauses. Wrapped up in coats, shelter-halves and blankets, we lay freezing in the paltry trenches and jolted all together at every new impact over and over. During the course of the night, 12 men of the company fell victim to enemy shells. That is more than the storm attack before noon had cost. It was impossible [for the kitchens] to bring food to the front that night.

As dawn broke, the enemy artillery activity tapered off. At once we went about our work to deepen our positions. This would take a lot of time but again we didn't have much. At 0800 hours, renewed heavy artillery fire forced our work to stop. Shortly thereafter French infantry attacked. Again these attacks were repelled without any trouble. Additional attacks that followed throughout the course of the day met with the same fate. Around midafternoon we had dug ourselves into the ground so deeply that we no longer had much regard for the strong French artillery fire. Because the communication trenches to the rear no longer existed, warm food came to the front only after the fall of darkness.

> **Considerations:**
> - The superiority of the German infantry demonstrated itself during the storm attack on Jan. 29, 1915. The fact that the French infantry, armed with machine guns and holding three positions situated in a row and fortified with barbed wire, lost their nerves and ran off, is made even more incomprehensible by the fact that the attack of our 9th Company was no longer a surprise. The enemy had recognized our preparation for an attack and had thus started to fire on us. Also that we successfully broke through the encirclement in Labordaire with an attack on an enemy vastly superior in numbers bears witness to the fighting quality of our troops.
> - It was unfortunately not possible for the battalion and regiment to **exploit the 9th Company's success**. Reserves were only available in few groups during the attack of three battalions in our front line. The fight in Labordaire became difficult through the scarcity of ammunition and hand grenades. The situation became especially perilous when the enemy took the blockhouse on our right flank. At the same time the battalion's command to retreat came, the ammunition was threatening to run out and the path to retreat through the strong enemy resistance and fire was becoming blocked. Any other decision from my point of view would cause severe losses, if not lead to our total extermination. Above all it was impossible to wait for darkness, because our last bullet would have been shot well before 1100 hours. The **breaking off of the fight in Labordaire confirms the sentence of the regulation**[*]: "Breaking off combat is most easily achieved after the successful firing of weapons."
> - During the rush to prepare for the attack it was not thought to bring large entrenching tools. Nothing could be achieved with small digging tools against the frozen, stone-hard ground. During attacks, the spade is just as important as the gun.
> - Although there was a better field of fire outward from the edge of the forest, the new position was sited 100 meters within the forest. Never again did anyone want the troops to be exposed to a shelling like the one at the edge of the Defuy Woods. This 100-meter field of fire was thoroughly sufficient for us to repel numerous French infantry attacks while causing the enemy heavy losses.
> - The casualties from enemy artillery fire during the night of Jan. 29–30 were so severe because the troops had not dug in deep enough. —*Rommel*

[*] Rommel is referring here to a standard German military maxim taught in the 1900s.

"Blood-Drenched Soil"—*In Front of Central and Bagatelle*

> ◊ Rommel writes several emotional passages referring to the losses of comrades. This is one of many passages in the book where he focused on burials and cemeteries. Later on Rommel would revisit the gravesites of friends after the war ended, and during World War II he took photos of war cemeteries in France and of the graves of his men in France and North Africa.
> ◊ In one of the most emotional passages in the book, Rommel recalls his feelings at encountering a dying young infantryman. This is one of the few times in the narrative he speaks in a very direct sense about his own feelings. He defines the experience as "unforgettable," indicating it stuck with him for life, and describes being completely overwhelmed with grief.
> ◊ Here Rommel notes attributes about fellow soldiers he admires—their sense of obedience and self-sacrifice to duty. Rommel aspired to these ideals. Later during the Third Reich, Rommel's focus on following his own duty placed him in a morally compromising position serving under Hitler.
> ◊ Rommel is indignant about being replaced by a different commander; he draws attention to the officer's age and that this man has never been in the field. He stresses he does not want to leave his men, even if he is longer in charge.

Our new positions resulted in an improvement. We were situated considerably higher and did not have to suffer from subterranean water anymore. Furthermore, the clay soil was very good to work with. We undertook to make bulletproof shelters and dugouts from four to six meters below the earth's surface. Here the continually strong and heavy French artillery fire could touch nothing. My company commander dugout, which I shared with an Uhlan officer assigned to my company, was only a sleeping space which you could only reach on all fours.

During the day we froze acutely. We were prevented from making fires due to the French artillery, which upon noticing the faintest hint of smoke would lay heavy harassing fire on the particular spot.

A 10-day relief was implemented: the front line, reserve position, and cantonment would switch places with one another. The casualties in the front line were minimal, thanks to the good positions and shelters, although the French artillery increased their harassing fire daily. From all appearances they had a cornucopia of ammunition in contrast to our guys, who could seldom shoot at all due to ammunition shortage.

I received news from the five troopers who had been severely wounded on Jan. 29 and taken prisoner that they were doing well. For the storm attack on Jan. 29, I became the first lieutenant in my regiment to be awarded the Iron Cross 1st Class, many weeks after [the action].

In February, March and April, the regiment dug itself in towards the French fortification [called] Central, in order to come to the same elevation as our neighbors on the right (Infantry Regiment 120), who had already on the evening of Jan. 29 dug in ahead of [my] Infantry Regiment 124 in front of Labordaire. The Grenadier Regiment 123 on the left worked towards Cimetiere, which formed a link eastward with Central. Continuously saps were placed forward and then connected together. In this manner the frontline was pushed forward nearer and nearer towards the enemy's main position, coming to a final stop just ahead of the French barbed wire obstacle.

French artillery and the newly introduced *Minenwerfer*[5] [a large trench mortar firing high explosive shells] caused considerable disturbances to this work. Many brave infantrymen fell in the saps. The communication trenches and paths to the rear, the command post and supply stores lay day and night under what was, to some extent, really effective French harassing fire. When the company returned to the cantonment, everyone only breathed easier after crossing the dangerous area (three to four kilometers behind the front line).

Most of all, during this relief [rotation], we had the sorrowful duty of accompanying our newly fallen comrades to their final place of rest. As time passed the relief periods grew more seldom, days in the frontlines grew more filled with losses, and the expanse of the quiet forest cemetery grew monstrously.

From the beginning of May onwards the enemy bombarded the frontmost trenches ahead of Central with light and medium *"Flügelmine"*[6]. The faint sound of their launch was very familiar to the experienced warriors of the Argonne. Indeed it was many times muffled by other battle noises, but it was enough to wake us with terror from the deepest sleep and compel us to speedily abandon our shelter.

Despite the daily sorties and the nerve-shattering combat activity in front of Central, the mood and the behavior of the infantrymen throughout was sublime beyond all praise[7].

---

5. The *Minenwerfer* denotes a short-range mortar invented by the Germans which launched impact-detonated, increased high explosive shells designed to destroy bunkers and fortifications. These were a type of trench mortar and the shells were not land mines despite using the similar term. Rommel also uses the term at times to refer to French trench mortars, which he is doing here.
6. Called *Flügelmine*, it was a type of *Minenwerfer* whose impact-detonated shell was stabilized by attached fins.
7. Rommel uses a phrase here that is usually used to refer to God in German church hymns. It is actually the title of a hymn by composer Felix Mendelssohn, *"Erhaben, O Herr, über alles Lob,"* which means generally, "O Lord, sublime beyond all praise." In this case Rommel is attributing the divine perfection to his fellow infantrymen.

Rommel (back row, on extreme right) stands for a photo with other German officers circa 1915.

Each individual performed his difficult duty with the greatest unquestioning [sense of responsibility]. More and more we grew closer together amid the blood-drenched soil of the Argonne. The bitterest thing was the perpetual departures of comrades who had to be carried, dead or severely wounded, to the rear.

Remaining unforgettable to me is one infantryman whose leg had been shot off by the splinter of a French heavy trench mortar shell. By the sinking sun, he was carried past me through a narrow trench in a bloody shelter half. The pain of the loss of such a young and so admirably tested and proven soldier overcame me completely. I gripped his hand, wanting to give him courage. But he said: "*Herr Leutnant*, this is not bad. Soon I will come back to the company, even if there's no other option but to use a wooden leg."

The courageous infantryman never saw the sun rise again. He died on the way to the hospital. His concept of duty is a definitive example of the spirit of the company.

At the beginning of May, the first wooden trench frames were delivered. Now small cut-and-cover shelters tall enough for up to one to two men were excavated and braced at the bottom of the trenches and in the trench wall facing the enemy. Relief quarters were relocated very close next to the sentry positions. The frontline was now so close to the enemy's main works that the French artillery could no longer target us without endangering their own troops. They concentrated their activity even more extensively on the rear area positions, lines of communication, reserve positions, command posts and depots.

At this time an older first lieutenant, who until this point had never yet been in the field, took command of the 9th Company in my place. The regimental commander offered me a transfer to a different company. However I refused this and stayed with the men whose company commander I had been up until this point.

In the middle of May the company had to help out alongside Infantry Regiment 67 which was positioned in the middle of the Argonne—near Bagatelle—in liaison with the 123rd Grenadiers. The adventurous "67ers" had been severely withered away in numbers by regular and severe hand grenade battles and attacks.

A different form of trench warfare prevailed among them. Little worth was placed on positions that offered protection from artillery and mine fire. The whole battle played out in shallow funnels behind measly walls of sandbags at the distance of a hand grenade toss. Nothing was to be seen here any longer of the once thick Argonne forest. The French artillery had rigorously razed the trees down here. For kilometers, only mere stumps jutted towards the heavens.

Rommel pictured in one of his many dugout shelters.

While reconnoitering the position we were to take over, accompanied by the subordinate commander, a brief hand grenade battle occurred across a broad front, which ended with the loss of numerous people. For us this was only a small foretaste!

On another day quite early in the morning, we went on the relief with mixed emotions. As had previously been our custom, we immediately deepened the position we took over and fashioned dugouts for ourselves. Heavy French artillery fire, delivered in sudden salvos, mortar explosions and hand grenade battles in every nook and cranny made this difficult for us.

In the hot weather, the position was overcome with a revolting odor of corpses. Ahead of our front and in terrain between our own works lay numerous dead Frenchmen. To bury them was impossible due to the strong enemy fire.

The nights turned out to be especially agitating. For hours at a time, hand grenade battles would be carried out across a wide front that were so wild that one never knew whether the enemy had already broken into one's own position or whether one was behind the frontline at all anymore. Throughout all of it, the barrages of various French batteries burst on the flanks. This repeated itself several times every night and gnawed severely at our nerves.

The dugout I inherited from my predecessor lay a few meters behind the frontline, on the left flank of my platoon sector. Within the span of the trench bottom—about two meters below the earth's surface—a narrow shaft was located in the wall facing the enemy which went vertically downwards. At two meters deeper than the trench, thus four meters deeper than ground level, the shaft ended in a horizontal duct that had the dimensions of an empty coffin.

Countless pieces of cork served to reinforce the walls. To protect rations and shelter, random personal belongings were stuck into the sidewalls in small niches. The whole place lacked bracing because the clay-like soil held up—but if a grenade struck in the immediate area of the entrance, one would certainly be buried alive. As soon as firing started in the area, I went forward immediately to my platoon. Also at night it was more advisable to stay at the front. The hand grenade battles kept you up on your feet for half the night anyway.

The heat in those days was unbearable. One day *Fähnrich* Möricke visited me—an especially energetic soldier. I was lying in my dugout four meters below. We talked through the vertical shaft, because the narrow abode had no room for two men. During our conversation, I expressed my anger that a person couldn't even get four meters beneath ground level to get some peace from the damn flies. Möricke replied it was no wonder, because it was totally black with flies on the edge of the trench above. He retrieved a pickaxe and hacked at it. Upon the very first swing, a half-rotted, blackened arm of a Frenchman appeared. We threw chlorinated lime and soil over it and left the dead man in peace.

Finally the hard 10 days and nights passed. Upon our speedy return afterwards to our regiment sector ahead of Central, the trench war there had taken on distinctly unfamiliar form. Due to formidably strengthened artillery and mortar fire, a mining war had started underground. On both sides sentries stood in half-roofed, strongly wired sapheads at only a few meters across from each other. During the nights, it was prevalent for very violent hand grenade battles to develop, which brought the whole trench garrison to its feet every time. Each side cut off the tunnels and position segments being driven forward. Hardly a day passed without detonations.

One day the French managed to cut off one of our saps, in which 10 men of the company were already working. Many of them were completely buried. Through many hours of work and amid continuous hand grenade battles, we managed to rescue them to the last man.

Attempts on our side to take the French sentry positions located right next to us with a surprise attack failed with severe casualties. The French had completely cocooned these sentries and the segments of trenches leading to them with barbed wire. Furthermore, at the slightest occurrence, they sprayed these obstacles from one end to the other with machine guns from blockhouses located in Central. We hoped to get out of all these circumstances, which were very distasteful in the long-term, through a storm attack on Central.

"One Last Handclasp"—Storm Attack on Central

> ◊ Young Rommel continued to use his engineering skills everywhere he goes to try to protect others around him. The excessive time and attention Rommel devoted to training his men and building shelters in attempts to shield them from harm—even in the most dire and seemingly hopeless circumstances—showed he was very concerned with preventing casualties.
>
> ◊ A sad personal loss for Rommel was the death of his friend Möricke, who shared a dugout with him. This was among many passages in which Rommel attempted to memorialize a fallen comrade. It is unique in its description of Möricke's last words to him.

Following a three-and-a-half-hour preparation with artillery and mortar [fire], the strong French positions Labordaire, Central, Cimitiere and Bagatelle were to be taken on June 30—each position having been built up like a fortress by the enemy since 1914. For weeks, the regiment worked on the meticulous preparation for this attack. Close behind the frontline, medium and heavy mortars were installed and made bulletproof. Day and night, reserve companies brought material, disassembled mortars and ammunition through the narrow communications trenches towards the front. The French harassing fire increased in ferocity. Thus many troops carrying the material fell victim to it.

As the 9th Company returned to the forest at the end of July after many days of rest in a hut camp near Binarville, we were astonished by the great numbers of medium and heavy caliber batteries which stood camouflaged from [enemy] planes beneath the fruit trees of Binarville. Also ammunition appeared to have been provided. In the happiest of moods this time, we returned to the position.

All minutiae for the five storming companies were demanded of the regiment for the attack on Central. My platoon was supposed to remain in the reserve position one kilometer north of Central during the artillery preparation, then come forward into the storm-launching position shortly before the start of the attack, follow closely during the attack of the storming troops and bring hand grenades, ammunition and entrenching tools for them.

On June 30 the artillery groups opened fire at 0515 hours. The heavy shells of 21-cm

Rommel reclines with a fox in his tent during World War I circa 1915. A photo of a woman, possibly Lucie, appears pinned to the side of the tent behind him.

and 30.5-cm howitzer batteries rolled through the sky above and past us. The effect of the shells in the clay soil of Central was monstrous. Tall fountains of dirt shot out from the ground, hole after hole appeared ahead of our positions. The strong French earthworks split apart as if battered to pieces by giant hammers. Human beings, timbers, roots, fascines, and sandbags whirled through the air. How did the defenders muster their courage [to fight]? We had never seen such a concentration of heavy fire until this point.

One hour before the launch of the storm attack, the medium and heavy *Minenwerfer* mortars began to strike blockhouses, barbed wire obstacles and walls into pieces. In vain the French artillery shot a concentration of defensive artillery fire. Our frontline was only thinly manned and was situated near the enemy's main position. Sections of French artillery plowed up the land behind [it]. Barely 100 meters ahead of me, a heavy French shell flung the mangled corpse of a Frenchman, who had been killed in January, into the branches of a high oak tree.

Constantly my glance fell to my watch. There were still 15 minutes to go before the beginning of the storm attack. Thick, blue-gray dust rising from the many shell bursts now impeded visibility. Friend and foe increased their fire.

The communications trench assigned to me had been targeted with especially heavy enemy fire throughout the entire morning. For this reason I advanced with my whole platoon—diverging from given orders—100 meters to the side of the trench over open terrain. Shells cracked all around us. We ran for our lives until we again found shelter in the dip in the ground below. Speedily we dove into the communications trench amid the French barrage and moved to the frontline. Each man lay here beside one another ready to spring. The last mortars and shells struck beyond us.

0845 hours! Across a wide front, the storming troops streamed out from cover, over craters and obstacles, towards the enemy position. French machine guns hammered away! Machine-gun fire also struck within the storm troop of the 9th Company from the right. Individuals fell, the mass of men hurried onwards, disappearing into craters and behind earthworks. My platoon followed. Everyone had complaints—whether there should have been more spades or satchels of hand grenades or ammunition. The French machine guns kept hammering on the right. Springing through their field of fire, we climbed the enemy walls, upon which the 9th Company had once stood on Jan. 29. The once proud fortifications were now nothing but heaps of rubble. Dead and wounded Frenchmen lay wedged among the wild jumble of brushwood revetments, beams, and overthrown trees. The revetment of branches set up to cover the trench walls in quiet times had now cost so many Frenchmen their lives.

To the right and ahead of us were hand grenade battles! French machine guns, shooting from rearward positions, sprayed the combat area from one end to the other and forced us into cover. The sun burned hotly. Bent over, we crept through the shot-up position towards the left, then into a communications trench towards the second French position, hot on the heels of our own 9th Company's storm troop.

In the meantime, our artillery had laid fire on the 2nd French line (Central II) located 150 meters south, which actually was supposed to have already been taken on July 1 after renewed preparations via artillery and *Minenwerfer* mortars. Yet the storm troops of the regiment, although they had not yet finished clearing the positions and bunkers in Central I, now already started storming Central II.

Violent hand grenade battles churned 30 meters ahead of us. One could see the silhouette of Central II still about 80 meters away. The French machine gun fire made an advance outside of this communications trench impossible. Our own storm troop ahead seemed to be caught in a jam.

Its young leader, *Fähnrich Möricke,* lay severely wounded in the trench. A shot through the pelvis! I wanted to have him brought back towards the rear. He refused, saying we shouldn't trouble ourselves about him, and quipping, "Weeds never die.[8]" Stretcher-bearers took him up. I gave one last handclasp to the courageous *Fähnrich*, then took command of the company at the front. The *Fähnrich* died days later in the hospital.

We stood fighting the garrison of Central II. Our own artillery was silent. Countless hand grenade salvos, followed shortly afterwards by a decisive storming [maneuver] to advance, and then we had penetrated Central II. The garrison fled, some through trenches, others across open terrain. The rest surrendered. Hurried entrenching on the

---

8. Variation of a German proverb meaning a bad person or thing is hard to get rid of. An English equivalent would be: "You can't kill a bad apple."

right and left started, while at the same time we went with the mass [of men] into a communications trench, which was three meters deep, further towards the south. Here we surprised a French battalion commander with his adjutant and staff. The taking of the prisoners was accomplished without resistance.

Hardly 100 meters ahead the communications trench ended in a large open clearing of felled trees. Ahead of us the terrain dipped at a sharp incline towards the valley near Vienne le-Chateau. A tall timber forest obscured our view towards there. Connection with our right and left was not available. To the right on the edge of the timber forest, at about 200 meters distance

Französischer Bataillons-Gefechtsstand in Central II

Rommel's illustration of a French battalion command post located in Central II.

from us, the French were visible in great numbers. We unleashed sudden fire upon them. After a short battle the enemy retreated back into the timber forest. Meanwhile over to our left, sections of the 1st Battalion had stormed ahead. At this time I established contact with them, and at the same time I reorganized the groups of men with me—enlisted men from all companies of the 3rd Battalion—and set them up in position about 300 meters south of Central II with their front facing south for defense. To advance the company further in a southern direction seemed inadvisable in light of the open right flank and the ferocity of the battle playing out in Central I and II farther back on the right. The day of our previous storm attack (Jan. 29), on which I had been so far ahead of our current frontline and was ultimately abandoned, was especially fresh in my memory.

A reconnaissance patrol established that the storm attack by our neighbors to the right had not gotten beyond Central I. During the hours that followed, heightened anxiety reigned in the flanks and the rear of those troops closed off in Central II towards the west. The French continuously attempted at these points to retake sections of their lost positions in counterthrusts. Therefore I transferred especially proven and tested warriors to these difficult posts. I reported to the battalion what had been accomplished.

On my left, companies of the 1st Battalion had pressed yet farther ahead towards the valley into the Houyette gorge. All of my security patrols located ahead of the front reported strong enemy forces in the forest 300 meters ahead of us at the edge of the slope. I discussed the situation with *Hauptmann* Ullerich, leader of the 1st Battalion advancing on the left. He agreed that the 1st Battalion would dig in on the left in cohesion with the 9th Company.

With feverish speed we went to work with shovels. I held a platoon in reserve. I used it to bring up ammunition and hand grenades and to fortify the switch position in Central II. French reconnaissance patrols that probed us were driven off. We entrenched ourselves well in the claylike earth. Soon we were more than a whole meter below ground. At this time the French artillery, which since the launch of our storm attack had no longer harassed our storming troops, began crumbling Central II behind us to ruins with heavy calibers. Apparently the French thought we were over there. Their output of ammunition was enormous. For several hours movement to the rear was prevented by their fire. The telephone connection stretching from the battalion to the company lasted only a short time. A heavy machine gun platoon was deployed to our company sector.

As evening began to fall, we were 150 centimeters below ground. Heavy French artillery fire constantly fell very close behind us. Suddenly a bugle signal sounded ahead of us, blaring loud and shrill commands. The edge of the forest then came to life. The enemy, densely massed, stormed across the distance of barely 100 meters towards our new position. Our rapid fire soon forced them to the ground. But it seemed that the edge of the valley was slightly peaked, and that we, in a crawling position, could fall upon the enemy first, if they reached about 80 meters towards us. If only we had been able to take a position nearer to Central II. Then the field of fire would have been better, but our position would then have been pounded to destruction together with Central II by the French artillery.

The French attacked with high fighting spirit this time. As night fell completely, the entire line was beset with hand grenade battles. Because our hand grenade supply was limited, we defended ourselves mostly with small arms and heavy machine guns. The night was dark. Even in the glow of ground signals there was hardly anything to be seen of the enemy through the smoke of the hand grenades. Because their hand grenades were exploding close ahead of the muzzles of our guns, they couldn't be much farther than about 50 meters ahead of us. The battle raged with varying intensity throughout the entire night. All attempts of the enemy to attack were crushed completely by the fire of the infantry troopers.

When day broke, a lengthy wall of sandbags lay 50 meters ahead of our position. Fortifications were being made behind it. After we had spent the night tensely holding our breath about the French infantry, now the French artillery seemed a relief. Fortunately the concentration of their fire roared away over and past us towards Central I and II. Only a small portion struck close behind our position, and very seldom did a shot stray amid the actual frontline. On this day we felt as if we had been rescued. By contrast, nobody envied the meal carriers and those who had to bring ammunition and material forward, as well as the fatigue details, going about their duties.

In the days that followed, we deepened our position by two meters. We began by building smaller cut-and-cover shelters braced with trench timbers for 1 to 2 men, as well as with the installation of steel-plates and embrasures using sandbags. Casualties in the frontline due to artillery fire were few, yet numerous people fell victim to enemy shells in the communications trenches which were under fire day and night.

The strong artillery groups merged together for the storm attack on June 30 were sent onward to a different

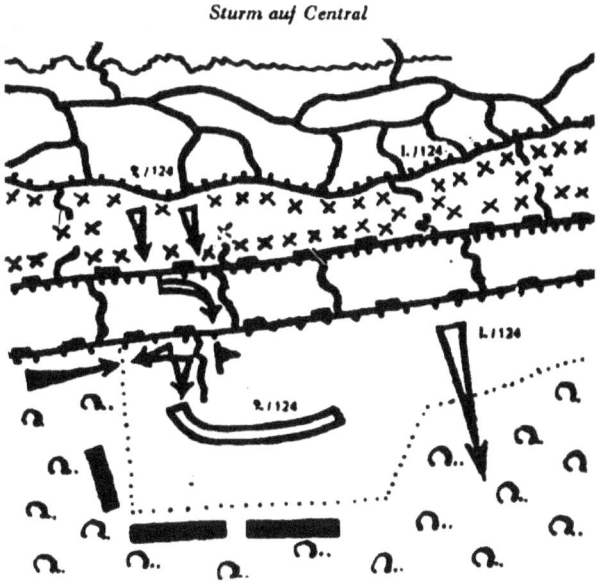

Rommel's map of the storm attack on Central.

front immediately after the storm attack was achieved. Now our weak local position artillery could oftentimes not shoot even worthwhile enemy targets due to ammunition shortages. However in the regiment's sector there was at least an artillery observer regularly in the frontline, which we infantrymen found very comforting.

At the beginning of July the enemy shot up one section of our position with daily volleys of heavy trench mortar fire. Their launching devices were so cleverly positioned that they were capable of catching extensive sections of the position with flanking fire. Due to the small width of dispersion of these launchers, which were actually of a very simple construction, direct hits regularly managed to strike in the trenches. Unfortunately we were not always able to clear out the endangered spots in time. Then the casualties were most severe. Numerous infantrymen lost their lives just due to the air pressure of the blasts of exploding, massively heavy mortars in their area.

In July I took command of 10th Company for five weeks as a substitute. In the same sector the 4th Company and 6th Company were still doing relief duty. We company commanders worked towards a unified plan to construct bulletproof shelters with multiple exits at eight meters below ground level. Work on these were done during day and night shifts by mining different positions. We officers also occasionally took shifts doing this difficult work. That bonded all of us together.

It very often happened during these days that the whole company position would be leveled within an hour by heavy French artillery. The small cut-and-cover shelters of trench timbers crumpled like cardboard boxes under the heavy shells. It was lucky that the French kept to a rigidly consistent plan during the bombardments. Most of the time they began on the left wing of the company sector. Because holding out in place amid this fire would result in heavy casualties, I ordered the endangered sections of the position to be completely cleared out every time and waited for the firing to shift sideways and rearwards.

If the French infantry chose to break into the position in coordination with the artillery fire, we would hit them back again with an immediate counterthrust; we felt ourselves vastly superior to them in a man-to-man fight.

Short saps and, as in front of Central, tunnels were directed forward at the enemy position, which was about 50 meters distance across from us. At the beginning of August, my company had to relieve the 12th Company at Martinswerk; they had suffered severe losses the day before during a storm attack following a mine detonation [of a tunnel]. The relief proceeded in the gray morning hours without incident. However the relief had barely been completed when a barrage of French artillery came upon us. Cowering together next to our dead enemies still lying all around, we experienced minutes of terror. When the fire began to subside after several minutes, we started working feverishly with spades to deepen the position. By the time we reached 180 centimeters below ground level, we had created countless tiny hollows in the forward-facing trench wall; then at least the French field guns could no longer get us. And even at this place I wanted stubbornly, if it was at all possible, to bring all of my men out safely.

Despite very heavy and frequent harassing artillery fire it was possible, thanks to diligent work, to lead the company out of the position after two days without casualties in dead or severely wounded. After handing over the company, I took my very first leave of the war for 10 days in August.

Considerations:

- **In order to deceive the enemy about the time of the attack** on June 30 against the strong Argonne position, the 3 ½ hour preparation fire of artillery and *Minenwerfer* was delivered with numerous pauses. Despite this very heavy fire, not everything in the enemy position was hit, [and] **several machine gun nests were still used in defense** during the storm attack.
- Again the **great attacking momentum of the German infantry** demonstrated itself. It did not content itself with the July 30th goal, but rather captured the closest French positions as well. It happened so fast that a French battalion commander with his adjutant located there were taken by surprise and were able to be captured. **After achieving the attack** successfully, things **rapidly shifted to defense**. In doing so we avoided utilizing French positions, because these were indeed known by the

enemy down to the smallest detail. **In anticipation** before launching the attack, **ammunition teams and entrenching teams were dispatched** among the troops. The retribution fire from the French prevented resupply to the frontline for several hours and broke the telephone connection.
- **The repulse of the enemy counterattack** in the evening and night preceding July 1, which was performed **at a very close distance**, occurred primarily **with small arms and machine guns**; hand grenades were used less.
- Before daybreak, **French infantry dug in** 50 meters ahead of our line **behind a wall of sandbags**. From all outer appearances, a portion of these sandbags had been brought up with them during the attack or, after their attack had failed, had been sent back to them by rear elements.
- **To avoid casualties, areas of the company positions** under heavy enemy fire were **cleared out for short periods of time** in the weeks following the storm attack. The currently valid [German army] regulations for defense allow for localized evacuation of sections of a rifle company during superior enemy fire upon the command of the company commander. —*Rommel*

"Fire Magic"—Storm Attack on Sept. 8, 1915

◊ Rommel describes himself as deeply saddened at being removed from command. His word choice demonstrates his strong dedication to his duties and desire to lead in battle.
◊ He uses his daring and creativity to execute a unique maneuver comparable to the Roman *testudo*, or "turtle" formation, or the Vikings' "shield wall" tactic. He orders his men, lying on their backs under cover, to pass sandbags forward and build a wall to protect themselves as they move ahead. The French witness what appears to be a "self-building" wall of sandbags.
◊ While contemporary accounts suggest Rommel was not very interested in the theater, he makes a humorous reference here to a Richard Wagner opera scene in which Wotan makes walls of flame appear around Brünnhilde.

After returning from leave, I was given command of the 4th Company, which only a few days later was supposed to take part in a storm attack on the right flank of the regiment. I assumed command of the company in the reserve position in the Charlotte Valley. After personally reconnoitering the assembly area and the attack zone, I practiced the storm attack with the company in an old position in the vicinity of the Charlotte Valley. Thus within a few days I could play the company by hand and could now set about the difficult undertaking with complete confidence. I was exceedingly saddened that my role as leader of this glorious company was only supposed to last a few days. In comparative seniority, I alone was too young to serve as a permanent company commander.

In a confident mood, my herd [of men] advanced on Sept. 5, 1915 towards the front through the connecting trenches long before daybreak. The position, which we took over from a company of Grenadier Regiment 123, was already being undermined by the French. One could distinctively hear, in various locations, the incessant work of enemy mining troops. We hoped very much that the enemy did not finish this underground work before we launched our attack. We by far preferred a straightforward man-to-man fight to being blown sky high into the air. Three long days passed in which the enemy worked diligently beneath us.

Rommel pictured at war circa 1915.

Then, on Sept. 8 at 0800 hours, a fire for effect began from our own artillery and *Minenwerfer* towards the enemy installations laying only 40 to 60 meters ahead of us. It was not far behind the preparatory fire for the attack on Central in terms of mass and momentum. The French artillery of various calibers responded with very powerful anti-infantry fire. In small, flimsily built dugouts tightly packed with three to four men huddling close together, we let the massive fire pass above us. The earth rocked continuously under the heavy impacts. A rain of dirt clods, splinters and branches came down. Strong Argonne oaks were uprooted and toppled crashing to the ground. Nothing more was to be heard of the French mining troops. Had they completely finished their work?

From time to time I rushed through the company sector in order to see the condition of my men. Repeatedly I was knocked over by the air pressure of the heavy shells and grenades making impact close ahead of our position. I took a glance over the trench wall into enemy territory: countless gigantic explosion clouds[9] jumped out of the earth; beams, sandbags, fascines and dirt clumps spiraled through the air; strong trees were uprooted; a blue-gray haze shrouded the rearward sections of the enemy installations.

---

9. *Sprengwolken*, Rommel calls them, giving quite a vivid description here. He frequently photographed smoke and explosions during World War II.

This "softening up" fire from the artillery lasted for three hours—an unspeakably long time for us within this boiling cauldron! Finally the hand of my watch ticked to 1045 hours.

Out from their dugouts, the three storm troop companies moved, ducking, into their positions. We compared the time on our watches. Upon the last impact, the storm attack was supposed to break into action on the exact second of 1100 hours. Engineer troops and carriers of ammunition and materiel advanced. Once again I showed my own troops our goal, which lay about 200 meters into enemy territory, and impressed upon them that they had to attack while making a beeline for their goal. To dispatch the enemy in the intermediate zone was the task of elements of the company following in the second line. The procedure to be taken after a successful storm attack, such as the securing of captured territory, establishing communications, and sealing areas off were once again discussed in thorough detail.

In the meantime, 21-centimeter shells and mortars of medium and heavy calibers hammered the enemy installations to pieces as the firing reached its peak. You could hardly imagine that any living creature could remain unscathed amid this tremendous fire. Only 30 seconds left! Ready to spring, the infantrymen crouched in the tunnels. Only 10 seconds left! The last shells struck close ahead of us in the enemy position.

Before their smoke had blown away, the three storm troops of the company, covering a breadth of about 250 meters across the breastworks, charged silently at the enemy like in exercises of days past, running straight through smoke and fog to their goal. A gorgeous sight!

The troops paid no attention to the flocks of Frenchmen climbing out of nearby sections of their position with their hands held high in the air and their faces distorted with fear. The prisoners were shown, with gestures as we ran past, the way toward our initial jump-off position. The storm troops ran streaming to their goal as sections following under command of the company sergeant in the second line took the prisoners into custody.

I had joined the storm troop on the right. In a few seconds we reached the set goal in a forward charge beyond the enemy trenches. Infantrymen, entrenching troops and hand grenade troops followed closely. Until this point no man was wounded. Because we had rushed forward silently, without giving the usual battle cry, the French occupants of the rearward areas of the position were totally taken by surprise in their dugouts and shelters. Convinced they had no hope of pulling through, they gave themselves up after a brief [verbal] challenge without a fight. Presently machine gun fire from a rear area of the position forced us to take cover. We rolled into the trenches on our left and established communication with the centrally located storm troop. A few minutes later, contact with the left storm troop of the company and the neighboring company on the left (2nd Company) was also established.

Feverishly we worked on orienting the captured positions for defense. Soon the sections of the trenches facing and leading towards the enemy were walled off with sandbags. Ammunition and hand grenade depots were set up. The French artillery now opened fire close behind the lines we had reached and was of such vehemence that communication with our jump-off position was cut off for hours.

Rommel (left) poses for a photo with an unidentified comrade during World War I.

French machine guns hindered every movement outside the position installations and thus totally prevented resupply.

Then the French infantry sounded a counterattack. Our field of fire was hardly 100 meters wide, but it was sufficient to bring the enemy charging at us from outside the position to a speedy standstill. Within the position itself, violent hand grenade fights developed around boxed-in areas. But also in this situation we were able, effortlessly, to assert ourselves. The terrain dipped slightly towards the enemy and therefore our hand grenades reached farther distances than those of our opponents.

During the attack itself, five men from one of our storm troops were put out of action by a carelessly thrown hand grenade. The French fire following the attack increased the casualties to a total of three dead and 15 wounded within the company. Providing for our troops became especially critical after the attack. Ammunition, material and rations had to be brought forward over terrain that was constantly being done over by French machine guns and shells. A connecting trench to our jump-off position had to be dug first. We were also missing a connection with our right.

At my suggestion to the battalion, a reserve company of 80 men under my leadership were supposed to dig a 100 meter-long trench to establish a connection over the shortest possible distance to our previous position to our right. Because this work had to be achieved across 40 to 50 meters directly in front of a continuous and manned French position, I had a plentiful supply of sandbags and steel shields prepared by my material troops. I had learned something from the French on June 30.

We began before 2200 hours. The enemy was indeed really tempestuous and riled up, shooting almost ceaselessly with machine guns and constantly illuminating their foreground [with flares]. But we had to begin, otherwise we would not complete the work in one night. First, I had a sandbag wall of about 40-centimeter high built up from both sides. The people who accomplished this work did it by laying on their backs and passing the sandbags forward until they reached the men in front, who built the wall. This work was very strenuous. Despite the enemy fire—which could not harm the teams of men lying behind the sandbags—the wall grew quickly on both sides to about 15 meters long. Then the sandbags ran out. There was then a yawning gap of about 70 meters, which I closed by arming the majority of my people with steel shields, crawling with them into the gap and forming a firing line there. As each individual took his place, he propped up his shield in front of him [for cover] and started digging in behind it. Rifles and hand grenades were lying within our reach.

Rommel's illustration of German infantrymen, drawn in white, springing up to surprise Frenchmen, silhouetted in black, on Sept. 8, 1915.

This whole proceeding did not get pulled off without any disturbance. The enemy shot numerous flares and showered us with a hail of bullets and hand grenades. The latter did not reach us, and the infantry firing could barely harm us behind the steel shields. But we did not feel altogether very well during this "fire magic"[10]. During the course of the night we nevertheless drilled into the earth behind our scant cover. As the morning

---

10. This is an ironic reference to Richard Wagner's *Feuerzauber* (Fire Magic) scene in Act III of his epic musical drama *Die Wälkure* (The Valkyrie). In the scene, heroine Brünnhilde becomes trapped on a rock as a divine punishment and is sealed behind a wall of magically conjured fire. The "Fire Magic" music occurs as the flames spread. Rommel draws a humorous parallel. He uses the term "fire magic" again later to refer to enemy fire.

of Sept. 9 broke, the connecting trench was 180-centimeter deep throughout its length. During this work, we stumbled across a dead man of the 1st Battalion, who had been lying in no-man's-land since June 30.

Just as I wanted to lay down for a rest after this difficult and nerve-wracking work, the battalion commander showed up to take a look at the new position. His visit was followed closely afterward by the regimental commander. They took the highest pleasure in the result of the 4th and 2nd Companies storm attack. The set goals had been reached. Two enemy officers and 140 men had been taken prisoner, and 16 *Minenwerfer*, two machine guns, two drilling machines and one electric motor had been captured. For the 4th Company, joy over the success was overshadowed by the death

Rommel loved horses and saved many pictures of horses he rode. In this photo he is wearing a thigh strap which probably helped him stay balanced in the saddle after his severe gunshot wound.

of *Leutnant der Reserve* Stöwe, who had been detailed to the 123rd Grenadier Regiment as a liaison officer and had his certificate for leave already in his pocket.

Shortly after the storm attack, I had to give up the 4th Company again and take over the 2nd Company for many weeks. With a heavy heart I separated myself from the 4th Company, with whom I got along exceptionally well. During a time that seemed like eternity I lived with the 2nd Company in the Feste Kronprinz[11], a bulletproof shelter and switch position located 150 meters behind the frontlines.

There I experienced my promotion to *Oberleutnant* and shortly afterwards my transfer

---

11. The *Feste Kronprinz* was one of a line of German fortresses built according to the Schlieffen Plan with the goal of repelling French attackers. Outfitted with armored batteries facing in every direction, as well as infantry quarters, a hospital, and a barbed wire perimeter, the *Feste Kronprinz* could perhaps be compared to the U.S. "fire support base" concept used in the Vietnam War.

to a new formation to be assembled in Münsingen—a skiing and mountaineering formation. Separating from an active regiment, in whose ranks I had fought through so many hard days of battle, and from the many courageous infantrymen, and from the blood-drenched, fiercely contested soil of the Argonne was difficult. The battle of the Champagne was reaching its peak when I left the woods of Binarville at the end of September.

> **Considerations:**
> - With the **newly taken-over company,** thorough **exercises** for the Sept. 8 storm attack were performed **using a practice fortification.** Timed precisely upon the second as the preparatory fire ceased, the three storm troops had to charge, overrun the nearby enemy position without battle cries and gain the set goal located about 200 meters away. Mopping up the position was the task of sections of the company following in the second and third lines.
> - Against my orders, **hand grenades were thrown** by a storm troop d**uring the charge forward** and resulted in the wounding of five men. (These were the only casualties which occurred during the storming itself). Lesson: During a forward rush, one must never throw hand grenades ahead of oneself, otherwise one's own troops will spring right into them. The **surprise of the enemy** was excellently achieved. We lunged past their frontline before they could grab their rifles. We stood in front of the rearward French shelters as if we had jumped straight out of the earth. This also resulted in proportionately large numbers of enemy prisoners.
> - After the storm attack, we rapidly turned our attention to defense, this time utilizing the available positions. The enemy counterattack which soon followed was repelled. Again after storming, the company's **connection to the rear was cut off** by artillery and machine gun fire. While **reestablishing contact** with our right close ahead of enemy small arms fire, **sandbags** and **steel shields** served well. —*Rommel*

3

*Static Warfare
in the
High Vosges 1916
&
Mobile Warfare
in Romania 1916–17*

"In Good Spirits"—The New Formation

> ◊ Here Rommel receives high-mountain warfare training that will last him a lifetime. As a *Gebirgsjäger*, he needed to maintain a very high standard of physical fitness, which he maintained after the war ended. An expert skier and mountaineer, he led training regimens that often exhausted much younger men.
> ◊ Throughout this narrative Rommel does not seem to have a high regard for the French. Here he makes a sarcastic reference to a "French mountain" implying it is less than a real mountain.

In early October 1915 in the new camp in Münsingen, Sproesser assembled the Württemberg Mountain Battalion (W.G.B.)[1] composed of six rifle companies and six mountain machine-gun platoons. Leadership of the 2nd Company was given to me. A busy life began. More than 200 young, battle-tested soldiers from different branches of service, drawn from all possible troop units along Western Front, comprised the company. In a few weeks they had to be deployed into battle amid strenuous mountain terrain. They made a very colorful sight in different uniforms, yet the men were united in good spirits from the first day. Throughout what was indeed physically strenuous and exhausting service, everyone gave their body and soul. The new uniform distributed to us later looked really smart.

At the end of November, the companies assembled for inspection by the strict commander on Gänsewag peak moved out with exquisite form. We spent that December in Arlberg in order to train with skis on the mountain. The 2nd Company was housed on the Arlberg Pass in Hospiz St. Christoph.[2] We stood from very early until well past nightfall on skis on the steep cliffs with and without rucksacks. In the evenings, the whole crowd of us sat together in the large hall of the guest lodge. The company band, led by Father Hügel, played the newest popular folk songs and mountaineer songs resoundingly. That was certainly quite different from a few months ago in the Argonne! In this manner I soon learned to get to know my men well outside of work, and the bond between leader and troops became more tightly sealed.

We relished the Austrian rations with cigarettes and wine allowances. We earned them in hard work every day. Christmas was celebrated with an especially strong [winter holiday] atmosphere. This beautiful time passed by like the wind. Four days after Christmas, a transport train led us westward—and thus not to the Italian front as

---

1. *Württembergische Gebirgsbatallion*. Often abbreviated in German as *"W.G.B."* In the German alphabet, these letters are pronounced as *veh, geh,* and *beh,* respectively, giving this acronym a catchy sound.
2. This mountain inn still exists and is now a hotel in Austria.

we had hoped. On the night of New Year's Eve, the 2nd Company took over the sector of South Hilsenfirst[3] on the Vosges Front from Bavarian reservists. It rained and stormed.

The new company sector was 1,800 meters wide and had a height difference of 150 meters between the right and left flanks. Ahead of our front were strong obstacles, among them one barrier which was charged with electricity at night. A continuous occupation of this position was impossible in such a broad area. One especially dominating high point in the position [we held] was fortified as a base. Each individual created a small fortress for himself with defensive advantages on all sides, and with stockpiles of ammunition, rations and water. In constructing tunnels, the vast experience from the Argonne was utilized: at least two exits, and very strong cover! Unlike in the Argonne, the enemy position was not situated within throwing distance of hand grenades, but drew near only to our right flank and the middle of our lines (near the so-called peak of the French mountain) across a vast distance of many hundred meters. The remaining elements of enemy positions were spread wide apart along the edge of connecting forested terrain.

Top: Rommel created a picture postcard of himself skiing to send to Lucie. Bottom: German mountain ranger troops practice skiing down a steep precipice while carrying rucksacks.

Apart from a few shells and harassing machine gun fire from time to time, we detected hardly anything of the enemy. However we suffered greatly from misfortunes caused by weather forces. During the course of spring and summer, we got to know the mountain areas—the *Sudelkopf,* the *Sengernkopf, Ilienkopf* and the *Mättle* position[4]. During this time, the numerous officer candidates zealously pursued their training.

In September the company arrived in the open position located on the northern slope of the Hilsenfirst mountain. Here the French were lying across from us nearby. Artillery and heavy shellfire besieged us fiercely every day.

3. A mountain in France.
4. Colloquial regional names for various mountain peaks, with *Kopf,* meaning "head," denoting a peak.

"Extraordinary Difficulties"—*Assault Troop Operation Pinetree Peak*

◊ By now Rommel is more confident in his ability as a leader. Foreshadowing his future career, he creates detailed sketches of maps to discuss with subordinates. He mingles with men along the frontline and points across the terrain to demonstrate what he wanted done. He now has both means and ability to realize larger plans, and freedom to operate more independently. However the young man is still learning. Although his mission turns out successful, he experiences more than a few awkward moments. In this anecdote, an almost darkly comical scene unfolds as Rommel and two subordinates attempt to remove unforeseen enemy obstacles next to a sentry post without making noise.

◊ Such adventures were characteristic of Rommel's military life. During World War II, Rommel often became lost while exploring and was at times threatened by friendly fire. In many instances, he narrowly evaded capture. None of these close calls however seem to have put him off or changed his plucky and almost impudent style of bravery.

At the beginning of October 1916, various companies of the battalion, including the 2nd Company, were ordered to prepare for missions with the goal of capturing prisoners. Until this point I had been extremely hesitant to use my company in this capacity, since I knew well from battles in the Argonne that such undertakings can become extraordinarily difficult and in most cases incur lots of casualties. Yet now after it was ordered of me, I went about the task briskly.

The next thing I did was reconnoiter the possibilities to approach the enemy position one evening in the right area of the company sector, together with *Vizefeldwebel* of the reserves Büttler and Kollmar. We stalked and crawled through the fir forest, which was tall and very dense in some spots, towards a French outpost position, which had been established on the upper end of a forest path inclining in the direction of the enemy. Cautiously we crossed the path overgrown with tall grass and weeds about 50 meters ahead of the enemy outpost and then crawled slowly like slugs in the roadside ditch towards the enemy wire obstacle. With exceptional care, we split the tangle of barbed wire using wire cutters.

It was already growing dark. Now and again we heard the stirring of the French sentries, now standing only a few meters beyond us. We could neither see them nor their outpost due to a thick hedge lying between us and them. Slowly we penetrated the expansive obstacle. Now we just had to cut through the lowermost wires.

Now all three of us were in the middle of the obstacle. Barbed wire wrapped around

A German infantryman wearing white to blend in with the snow is shown in a mountain trench.

us like spiderwebs. Suddenly the French sentry just ahead on our left became restless. He cleared his throat and coughed many times. Was he afraid? Had he heard something of our activity? If he were to throw a hand grenade into the ditch, it would be all over for us three. We could not even twitch inside the barbed wire obstacle, much less defend ourselves. We held our breath.

Minutes of terror stretched on. As the sentry quieted down again, I slowly withdrew the reconnaissance patrol. Meanwhile it had become totally dark.

As we crept back amid the dense underbrush, a pair of twigs cracked. The enemy immediately alarmed the whole garrison and sprayed the area between their position and ours with machine gun and rifle fire for minutes. Pressed flat against the ground, we let the hailstorm of bullets pass over us. Finally we reached our own position again without injury. This much had become clear—extraordinary difficulties faced any undertaking of ours in that forested area.

The following day, I scouted possibilities of approach to the enemy position on the so-called "Pinetree Peak." Here the available circumstances were much more opportune. In darkness, one could approach right up to the enemy obstacles across the forest clearing, which was overgrown with grass. The obstacles here were however particularly strong and consisted of three rows. Cutting through these obstacles would demand severely strenuous and hours-long work. This enemy position itself lay about 150 meters beyond these rows of obstacles. Following lengthy periods of observing from different spots by day and night, we established that there were two enemy dispositions on Pinetree

Peak. One was situated somewhere in the middle of the forest clearing in a camouflaged sentry post. The other was 60 meters to the left and above on a rocky outcrop, which had a commanding view of the surrounding terrain and provided good opportunities for surveillance and shooting. Rarely did the enemy deliver harassing fire with machine guns from this area of the outpost.

Our planned mission could only be carried out on a very dark night due to the lack of cover in the meadow and the light color of the grass.

In the following days and nights, we scouted opportunities for approaching the enemy position on Pinetree Peak and observed the routines of both enemy posts. We meticulously avoided doing anything that could alert the enemy to our planned operation.

Based on interpretation of our reconnaissance findings, I developed the plan for the operation. This time, I did not want to creep up to the enemy posts, but instead overcome the obstacles located

Rommel cuddles a dog on a snowy hillside during World War I.

in the space between the enemy positions, seize the enemy trenches and then attack the posts from the side—effectively heaving the enemy out from the rear. In order to do this, my own storm troop had to be strengthened to 20 men, because they would have to be divided after reaching the enemy position. To enable our retreat after our storm troop achieved breakthrough, in case a fight developed with stronger enemy elements occupying the trenches, I would deploy a wire-cutting troop to the enemy sentry posts. From there, they would slink stealthily up to the enemy obstacle and remain there on standby until either the storm troop started clearing the enemy trenches with pistols and hand grenades, or—if the desired silent takeover of enemy sentry posts was successful—wait until a signal was given from the captured sentry stations. Then these troops would begin to cut narrow passages in the wire obstacle and thus enable the storm troops to retreat in the quickest possible distance.

I discussed the operation in depth with subordinate commanders, showing them sketches and sometimes demonstrating to them by hand while looking out across the terrain from our trenches. Individual troops prepared themselves by practicing their tasks close behind our position.

The Oct. 4, 1916 was a cold, inhospitable day. Fierce northwest winds drove shredded clouds around our position, which was at about 1,000 meters elevation. Before evening the wind turned into a storm. Cloud-shattering rain clattered down upon us. This was exactly the kind of weather I had wished for the mission. Now, the French sentries would surely tuck their heads in, turn their coat collars up high, and embed themselves in the deepest, most protected corners of their posts and would thus become hard of hearing. Moreover, the howling wind would drown out many noises of us sneaking up and cutting through the wire. I reported my view to *Major* Sproesser that we should carry out the attack in the approaching night and received his permission.

Three hours before midnight, I left our position with my three troops. A totally pitch-black night—it stormed and rained! Crawling, we moved very slowly ahead towards the enemy position. Soon the wire-cutting troop under *Vizefeldwebel* Kollmar and *Gefreiter* Stetter branched out left and right. Accompanying the storm troop, *Leutnant*

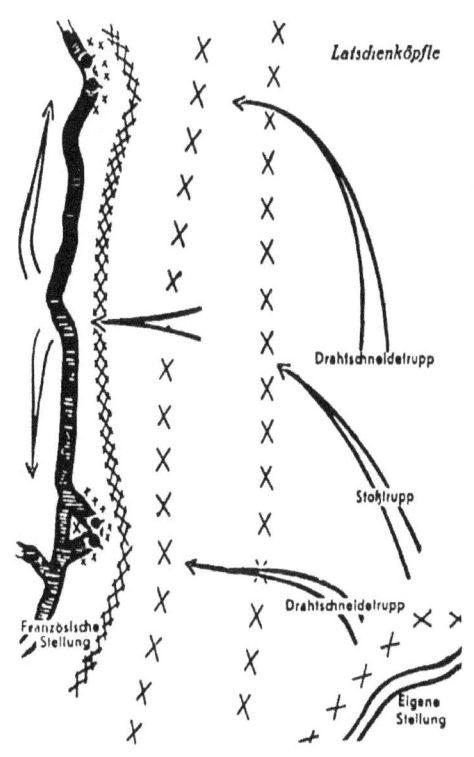

Rommel illustrates how his three wire-cutting troops approached and penetrated the enemy's position after setting out from their own position on the lower right.

Schafferdt, *Vizefeldwebel* Pfeiffer and myself crept ahead in front with wire cutters. The additional 20 men followed in rows with three steps distance between each man. Soundlessly we groped on all fours towards the enemy. The wind howled and whipped rain into our faces. Soon we were totally soaking wet. Tensely we listened in the night. Sporadic shots rang out overhead to our left. Now and then a flare flickered through the darkness. The enemy ahead of us was behaving quietly. The night was so dark that we could only recognize the edges of boulders lying around the area when we came within five meters of them.

Presently we were at the first obstacle. Difficult work began. One of us three would wrap each wire with a piece of cloth and then would fasten the wire cutters onto the wire. The others relieved pressure from the wire before he finally clipped through. The ends of the severed wire were held and carefully bent backwards. Under absolutely no circumstances could we rush through this, because it would make noise. All of this had been thoroughly practiced in detail beforehand. Every now and again we would take a break and strain to listen in the night. Then we would take up the tiring work again. Centimeter after centimeter, we cut forward through the tall, broad and densely woven French obstacle. We accustomed ourselves to cutting a way through the lowermost wires in this manner.

Hours of grueling work! Occasionally a wire squeaked despite our efforts. Then we ceased the work for several minutes and listened closely to the darkness ahead of us. It was already approaching midnight by the time we broke through two obstacles. Unfortunately the rain and storm had subsided somewhat and things were getting brighter. Ahead of us lay high "Spanish rider"[1] obstacles in a continuous row. The individual support frames were long and heavy, and their numerous wires so thick, that our wire cutters proved to be too weak. We crawled a long distance to the right and attempted to pull two Spanish riders apart from each other. This resulted in loud noise. It pierced us through marrow and bone with fear. If the enemy sentries, located only 30 to 50 meters away from us, were not fast asleep, they would certainly have been alarmed.

The minutes of waiting were terrifying! Everything remained quiet over at the sentry post. I refused to make any further attempt to separate the Spanish riders, which were seemingly anchored firmly together. After a brief search, we found a way to pass beneath the row of Spanish riders through a shell hole over to our left. Carefully we squeezed though and reached the edge of the enemy position only a few meters beyond.

Another rainstorm began. We three found ourselves between the enemy position and the rows of obstacles. The water splashed on the trenches and rippled across the stepping stones towards the valley. Carefully the first man of our storm troop squeezed underneath the Spanish riders and came across. The rest of them were still in the back, some still under the first and second wire obstacles.

Suddenly we heard the sound of footsteps in the enemy trench ahead to our left. Multiple Frenchmen were coming towards us downhill through the trench. Their footsteps beat slowly and regularly through the night. Apparently they had no clue that we were here. It didn't sound like there were many of them. I estimated three or four men. Was it a trench patrol?

What should we do? Let the enemy pass or ambush them? A fight with them would probably not go over without making noise. It would come to man-to-man grappling. One shout, one shot would alarm the whole enemy garrison. I could not deploy my own storm troops; they were mostly still lying stuck under the obstacles. We could certainly overpower the enemy patrol—but then the other enemies occupying the trenches would pitch into the fight and start spraying gunfire on the obstacles. A retreat would

---

1. A type of *cheval-de-frise* defensive obstacle, sometimes known as a "knife rest." The obstacle would typically include a wooden frame with projecting X-shaped spikes (sometimes sharpened logs or stakes) strung with a web of barbed wire.

A photo of Rommel standing at the head of his assembled 2nd Company, which he has written as "my 2." Rommel took a similar photo in World War II with his troops in 1940, shown at the beginning of this chapter.

only be possible with heavy casualties and bringing back any prisoners was highly doubtful. Quickly I weighed the factors for and against one another, and then finally decided to allow the enemy to pass by.

The two men accompanying me, Schafferdt and Pfeiffer, were in agreement. We took full cover along the edge of the enemy trench. Above all, our hands and faces needed to be hidden. The Spanish riders prevented us from crawling back where we came from. If the Frenchmen coming down toward us took a good look around, they would definitely spot us. Ready to lunge in case, contrary to our intentions, it came down to a fight, we waited for the enemies approaching nearer and nearer. Their footsteps beat in unison. They spoke softly. Tense seconds crawled by! Without any hesitation the French patrol stepped towards us—and past us. Beyond and below their steps echoed in the trench. We took deep breaths.

We waited a few more minutes to make sure the patrol did not come back. Then we slipped, one man at a time, into the enemy position. The rain had stopped. Only the wind continued to whistle over the bare hillside. As many men climbed down into the trench, earth and stones broke away from the trench wall and rolled noisily away across the stepping stones. Again some tense minutes passed by.

Finally the whole assault group was in the trench. The force divided. *Leutnant* Schafferdt went towards the valley, and *Vizefeldwebel* Schropp went towards the mountainside, each taking 10 men. I went with Schropp. Quietly we felt our way forward and upward through the steeply sloping trench. Only a few dozen steps separated us from our goal, the sentry post on the rocky outcrop. Had the enemy already noticed something? We stopped and listened!

There—to the left something clattered into the obstacle, and at the same time something clattered

Rommel's sketch of German infantrymen dashing through trenches with barbed wire obstacles and explosive bursts visible on either side of them.

on top of the trench wall to our right. Hand grenades burst with loud explosions! The leading men of the assault group recoiled. The storm troop got backed up behind them. The next hand grenade salvo would strike between them. If we didn't get out of here at once, we would be lost. So, we ran for it! We charged at the enemy and ran amongst them despite their hand grenade firing.

My horse groom, Stierle, who had only come along for this mission, was grabbed by the throat and choked by a Frenchman. *Unteroffizier* Nothacker dispatched this enemy with a pistol. Shortly afterwards, we overpowered two additional Frenchmen in the sentry position. One Frenchman managed to escape to the rear.

As fast as we could, we searched the trench walls with flashlights, looking for dugouts. One we found empty. A second we came across was full of Frenchmen. With *Unteroffizier* Quandte, I crawled through the opening which was only 60 centimeters high, clutching a pistol in my right hand and a flashlight in my left. Seven Frenchmen, armed and ready for battle, sat against the wall. After short negotiations all of them laid down their weapons.

It would have made things less dangerous for us if we just dispatched all of them in the dugout using a hand grenade. However this contradicted our orders to take prisoners.

Leutnant Schafferdt took two prisoners without taking any casualties. The wire-cutting troops had worked feverishly during the fight and exit routes were ready for the troops.

Because the mission had fully achieved its goal, I gave the order to withdraw. We had to get away before French reserves entered the fight.

Unmolested by the enemy, we reached our own position with 11 prisoners. What was even more worth rejoicing over was that not a single casualty had been incurred on our side. Only *Gefreiter* Stierle had sustained a very small scratch from the splinter of a hand grenade. Recognition from our superiors was of course forthcoming.

Unfortunately the following day claimed a victim. In a quiet area of our company sector, the tried and true *Vizefeldwebel* Kollmar was shot and killed by a French sharpshooter. This significantly dampened our joy over the success at Pinetree Peak.

Afterwards our days in the open area position were numbered. The *Oberste Heeresleitung* [Supreme Army Command] foresaw other tasks for the Württemberg Mountain Battalion. During the second half of October, we traveled east.

"Horrifying Night"—At Skurduk Pass

> ◊ By this time, Rommel has changed a lot from the bumbling young soldier who left Swabia on a train. He has begun to perceive himself as part of an elite unit and has consequently gained a sense of bravado. After encountering Bavarian troopers who fearfully told him about the Romanians' horrifying ferocity in battle, Rommel sums up his response with a cocky statement. "*Na*," he says, using a little German expression that is so sloppy and versatile it can mean all things to all people. Here it means skepticism. "We wanted to see for ourselves," he adds. His swaggering reply was omitted from the U.S. Army's translation. Ironically the Romanians proved to be a thorn in Rommel's side as time went on.

In August 1916 the front of the Central Powers withstood a mighty onslaught by the Entente. At the Somme, Englanders and Frenchmen strove for the ultimate outcome amid the use of monstrous force. The fire flickered anew upon the blood-saturated meadows around Verdun. In the east, the front continued to waver following the Brusilov Offensive, which had cost our Austrian allies alone half a million soldiers. In Macedonia, 500,000 combatants under command of General Sarrail stood ready to attack. And on the Italian front, the 6th battle of the Isonzo had ended with the loss of the Görz[2] bridgehead as

---
2. Present-day Gorizia.

well as the city. The enemy here was already mustering for a new attack.

At this point in time, Romania entered the grand scheme of things as yet another enemy. Romania believed the time had come to enter the war and thus assure a quick victory for the Entente, and hoped for a rich reward from their allies. On Aug. 27, 1916, Romania declared war on the Central Powers. Half a million Romanian soldiers crossed the border passes in order to strike Siebenbürgen.[3]

By the time the Württemberg Mountain Battalion rolled into Siebenbürgen at the end of October, magnificent victories had already been fought out in Dobruja as well as Hermannstadt[4] and Kronstadt[5] which had thrown the Romanians back across their borders. However, the final decisive blow had not yet been made. Russian aid strengthened the Romanian army, which only a few weeks prior had crossed the borders with the boldest hopes.

The Württemberg Mountain Battalion disembarked at Puy on the detonated railway line towards Petrosani. We made a strenuous march on crammed, bad roads, which were backed up with columns of all kinds, to Petrosani. In order to get ahead, the following methods proved expedient: the foremost groups of our company marched with bayonets fixed. They created space among the confused masses that constantly blocked the way. The company's vehicles were accompanied on the left and right by riflemen. Whenever horsepower threatened to collapse in the muddy earth, the arms of strong men intervened. Thus we pressed slowly but inexorably forwards. Romanian POWs with high, peaked fur caps on their heads passed by us.

It was shortly before midnight when the 2nd Company reached Petrosani and rested for only a few hours on bare floorboards inside a schoolhouse. Our feet burned after the long march. Yet even before daybreak, the 2nd Company and 5th Company were loaded into a large truck. We drove through Lupeni towards the southwest, onward to the imperiled mountain frontline.

Only a few days ago, the forward thrust of the 11th Bavarian Division across the Vulcan and Skurduk passes had miscarried. During bitter fighting after emerging from the mountains, segments of the infantry and artillery were repelled and dispersed by enemies in the passes. At the moment it seemed that Schmettov Cavalry Corps held the border crest. If the Romanians attacked further, our weak powers could hardly hold fast.

After many hours of truck driving, we were unloaded in Hobicaurikani[6]. There, the cavalry brigade we were now supporting sent us marching toward the border crest, in the direction of Hill 1794. We climbed up narrow footpaths. Our packs, containing four-days

---

3. Now called Transylvania.
4. Today's Sibiu.
5. Brasov.
6. Present-day Uricani.

Rommel, right, stands at the head of his mountain troops among other officers.

worth of raw food materials, weighed heavily upon us. We had no pack animals or winter mountaineering equipment with us. Even the officers carried their own rucksacks.

For hours, we climbed into the mountains over steep cliffs. Some people, such as an officer who had been with the Bavarian troop units fighting in the great beyond over the mountains, came towards us from the opposite direction. They looked as though their nerves were torn to shreds. According to their explanations, they had barely come out alive from a battle in the fog. Most of their comrades had been killed by the Romanians in close-quarter fighting. For days they, the few survivors, had wandered around lost and starving in the seemingly endless mountain woods and had then finally broken through across the border crest. They described the Romanians as extremely wild and dangerous opponents. Well…we weren't convinced. We wanted to see for ourselves.

In the late afternoon we arrived at a sector command post (under *Oberst* R.) situated at 1200 meters elevation. As the companies unloaded and cooked some food, *Hauptmann* Gößler (leader of the 5th Company) and I were briefed on the situation and received orders to begin the march as immediately as possible in order to reach Hill 1794 by evening, occupy the position located above it and reconnoiter south via Muncelul and Prislop. Nothing had been heard from a reconnaissance squadron sent south through Muncelul for two days. A telephone station and a troop of horses tied to a hitching post were supposed to still be located on Hill 1794. Nothing was known of the situation with neighboring troops on the right and left.

It rained as soon as we set out. Without being led by anyone who knew the terrain, we ascended towards Hill 1794. It rained harder. Night began to fall. Soon it was pitch dark. The cold rain became a downpour and soaked us down to the very last threads of our clothing. It was impossible to march further across the jagged, stony cliffs. We camped on both sides of the mule track at about 1,500 meters elevation. It was impossible to sleep in our wet clothing due to the extreme cold. Every attempt to make a fire using surrounding scrub brush failed in the pouring rain. Shivering from the cold, we huddled close together in blankets and shelter halves. As soon as the rain lessened, we again attempted to make fire. However the wet scrub brush fizzled out without producing any warmth.

Every single minute of this horrible night seemed to stretch on forever. After midnight, the rain stopped completely—but instead a mighty, ice-cold wind made it impossible to even sit still in our wet clothes. Freezing, we stamped our feet around the smoking fire. At last it became light enough for us to continue our ascent to Hill 1794. Soon we came to an area containing snow.

As we reached the hill, our clothing and our packs were frozen to our backs. The temperature was below zero. An icy wind swept over Hill 1794, which was covered in deep snow. The positions that had been described to us were not there. A miniscule foxhole, hardly big enough for 10 men to sit in, contained the telephone section. Over to the right stood about 50 tethered horses, trembling from the cold. Shortly after our arrival, a snowstorm blew over the plateau. We could only see a few meters ahead of us.

Hauptmann Gößler described the situation on the hill to the section commander on the hill and tried his best to have our two companies withdrawn. However all representations made by this experienced alpinist, and even the objections of our doctor, explaining that for troops to remain in a snowstorm in wet clothes, without any available shelter, without fire, and without warm food would inevitably result in frostbite and most severe illness within mere hours, were in vain. We were threatened with court-martial if we yielded even one step of ground.

In order to establish the whereabouts of the missing reconnaissance squadron, *Vizefeldwebel* Büttler was sent through Muncelul in the direction of Stersura. The mountain infantrymen pitched tents in the snow. It was impossible to make fire. Despite numerous cases of high fever and vomiting at the onset of nightfall, new attempts to convince the section commander to relieve us availed nothing. A horrifying night began. The cold became more and more biting. Soon the men could no longer bear sitting in tents and again tried—as in the previous night—to keep themselves warm through motion. It was a long, long winter night! As day broke, the doctor had to send 40 men to the hospital.

I volunteered myself to *Hauptmann* Gößler to go to the section commander and personally describe the condition of the troops. I achieved, at the very least, that our request for immediate relief was forwarded on to higher authorities. As I returned to Hill 1794, *Hauptmann* Gößler arrived at the firm decision to withdraw the companies at once, no matter the consequences. At this point 90% of the men were receiving medical care for frostbite and cold symptoms.

At about noon we were relieved by a fresh troop equipped with pack animals and wood, and the weather cleared up again. Meanwhile Büttler's reconnaissance patrol located the reconnaissance squadron on the southern foothills of the mountain. The temperature over there was bearable at 1,100 meters elevation. There was no sign of the Romanians.

After three days the company was again ready for action. Amid a noticeably more comfortable climate, we climbed the Muncelul mountain. After camping at 1,800 meters elevation, we headed towards Stersura, an outlying summit of the Valcan Mountains projecting vertically north and northwest. The company occupied outposts some 1,000 meters north of Stersura. As they nested themselves in all-around defense formation on a forested knoll, secured by three outguards, things livened up on Stersura. Romanians numbering in about battalion strength entrenched themselves there in multiple overlapping positions.

Clashes with weak enemy forces occurred over the next few days without casualties on our side. We lived in tents right next to our positions. Pack animals brought daily food supplies from the valley beyond the mountain crest. A telephone link existed between Group Sproesser and the outguards. Over to the right lay the Arkanului mountain. On its steep southeast slopes, one could see the artillery of the 11th Division that had been left standing there. Other units of the Württemberg Mountain Battalion were situated on the nearest mountain ridges some two kilometers east of us.

Fog covered the lowlands far beneath us and surged like an ocean against the sunlit mountain ranges of the Transylvanian Alps. A gorgeous sight!

> **Considerations:**
> - The **insertion into Hill 1794** demonstrates the great extent to which **climate in high altitudes** can impair the **capabilities and stamina of troops**, particularly when their equipment is incomplete and not appropriate for their task and supplies are lacking. On the other hand, we see what the soldier is capable of bearing in the face of the enemy. Conditions may demand that dry wood or charcoal be brought to troops located at 1,800 meters elevation. On the southern slops of the Valcan Mountains, we heated our tents a few days later with small charcoal fires hung up in tin cans. —*Rommel*

"Like Rushing Water"—The Storming of Lesului

At the beginning of November the Romanians stood armed against a forward thrust in the direction of Bucharest from the German military powers around Kronstadt. They held the majority of their reserves together north of Ploiesti and did not notice that a new German attack echelon was assembling on the Valcan and Skurduk passes under *General* Kühne whose purpose was to break into Wallachia and advance towards Bucharest from the west.

During the first days of November—even before the beginning of the main attack on Nov. 11, 1916—sections of the Württemberg Mountain Battalion placed on the right flank of Group Kühne captured and occupied the mountain ridges Prislop, Cepilul, and Gruba Mare after bitter fighting and thus secured passage out of the mountains for the main forces. They held the conquered territory against enemy counterattacks. The Romanians put up a really good fight in all of these battles. On Stersura, the Romanians fortified their positions with wire. On Nov. 10, the 2nd Company was sent to Gruba Mare without one platoon, which stayed behind for security against the enemy on Stersura. On Nov. 11, the attack of Group Kühne began.

The Württemberg Mountain Battalion was to take Lesului, a mountain with an especially commanding view situated at 1,191 meters elevation which bordered

Wallachia on its southern slopes. The Romanians had strongly fortified this mountain. Multiple and continuous enemy positions, sited one behind the other, lay within the long saddle between Gruba Mare and Lesului, which was only sparsely covered throughout with bushes. The attack of the Württemberg Mountain Battalion was supported by a mountain artillery battery and would consist of four and a half companies, including 2nd Company. Detachment Gößler was assigned to mount the frontal attack while Detachment Lieb, consisting of two and a half companies, would mount an encircling attack from the east against the enemy positions. The frontal attack was only supposed to begin after the encircling attack of Detachment Lieb had drawn the enemy's attention.

Rommel's illustration of the action at Lesului shows Lieb's detachment on the upper left, with the 5th and 2nd Companies visible at the base of the forested summit.

At daybreak on Nov. 11, the 2nd Company, strengthened by a machine gun platoon, inserted itself in the front line on the right and positioned itself for attack 200 meters in front of the first Romanian position on the hillside sloping towards Lesului. While moving into the assembly area, a clash resulted with a Romanian patrol on our right flank. The Romanians were repelled after a brief firefight and numerous prisoners were left in our hands. We had no casualties on our side.

Now the Romanians recognized our preparations for the attack. Throughout the whole morning, they sprayed the area in which we were located with rifle and artillery fire. Because there was sufficient cover to be found everywhere, we had no casualties. For our part, we seldom shot, instead eagerly taking the opportunity to reconnoiter the enemy positions and diligently prepare our fire support for our own attack. A mountain artillery battery was brought into position between some rock walls on our left close behind the frontline. Numerous observation posts scanned the enemy terrain sharply.

Hours passed. Just around noon, Detachment Lieb attacked the flank of the enemy situated across from us, while Gößler's detachment stepped in for the frontal attack.

Within the 2nd Company, *Leutnant* Grau swept the enemy positions ahead of us with his heavy machine gun from a somewhat elevated position. Then the company moved to launch the storm attack. Like rushing water bursting from streams, the individual groups broke out from the bushes and sped downhill. Contrary to our expectations, no close quarter fighting occurred at all. The momentum of the attacking infantrymen flushed the enemy out of all their positions in the saddle between Gruba Mare and Lesului within just a few minutes, and our men reached Lesului, located at about 700 meters distance.

We did not take many prisoners because the Romanians, with most extraordinary agility, disappeared into the ravines on both sides of the saddle.

Soon we also had taken the peak of Lesului. We encamped there in tents in the evening. Our joy over our success was great, especially since the 2nd Company had only had one man lightly injured during the frontal attack.

As night fell, reconnaissance patrols moved into the lowlands towards the south to determine the disposition of the enemy and to fetch meat and bread. Food in the mountains had been truly meager and monotonous in recent times.

Early in the morning on Nov. 12, the reconnaissance patrols returned. Nowhere had they run into the enemy. They brought live cattle as well as some already slaughtered. Soon, the sight of meat roasting on spits over open fires could be seen everywhere. The streaming rays of the November sun made it easy to forget the cold night in the tent.

Considerations:

- The assembly position for the attack on Nov. 11, 1917 was about 200 meters in front of the enemy position—a reverse slope position, consisting of multiple lines. The enemy committed an oversight by failing to hinder our approach from such a near distance to the frontline of his main defensive area.
- After multiple hours of preparations (recognized and drawing fire from the enemy), the frontal attack followed, introduced by heavy machine gun fire that opened fire from 200 meters away from the frontline. The terrain offered no other possibilities for fire support.
- The individual heavy machine gun then, through continuous fire, forced the enemy into cover in locations where our rifle platoons wanted to make breakthroughs. After about 30 seconds, as the rifle platoons stormed onwards, the machine gun shifted its fire and pinned down the remaining elements of the enemy position. After our breakthrough goal was achieved, the machine gun followed us speedily and was then used from elevated positions to support our attack through the lengthy saddle pass right up until we reached the peak of Lesului. The enemy was totally surprised by

> this manner of conducting battle, although he had already been aware for hours of our intention to attack.
> • Our success could have been even greater if the frontal attack had taken place half an hour later. Then Detachment Lieb, deployed to encircle the enemy, could have been standing at the enemy's rear. —*Rommel*

"Terrifying Minutes": **Battle at Kurpenul-Valarii**

◊ Rommel is now a very seasoned combat infantryman. He plots ambushes and refers to his enemies as "rewarding targets." He is now also emotionally hardened. Although painful for him, he demonstrates he can decide not to help comrades in trouble for the sake of a strategic purpose. As time goes on, this is not always the case. He will later act according to emotion at times, particularly when feeling moved by compassion or loyalty. He will rescue a POW and stake the fate of his whole unit in a desperate bid to save surrounded comrades.
◊ His rather cold passage suggesting the Germans could wait silently to surprise their approaching enemies was omitted from the 1944 English translation.

In the late afternoon of Nov. 12, 1916 the company received orders to descend from the eastern slope of Lesului and, strengthened by a machine gun platoon, take the small town of Valarii. At the same time the remainder of the battalion went off in two columns down the western slope of the mountain to achieve the same goal. Magnificent sunlight reigned on the peak of Lesului, but while descending my column very soon found itself surrounded by a dense fog. I marched with my compass and stood on a path leading towards the valley. Soon I heard the obscured hum of voices emanating distinctly from the valley. Was it commandos?

Then, below to our left, a Romanian battery began firing shot after shot towards the Vulcan Pass. The origin of the noise was not far from us at all. Thus at any moment we could run right into the enemy in the thick fog. Therefore I ordered my force to stalk onwards toward the valley with extreme caution. Our advance guard point, our flanking patrols and our rear point were secure as we climbed down through the turf as soundlessly as possible and without speaking.

Soon it started to get dark and the fog thinned. We saw a long, stretching village some 1,000 meters ahead of us in the valley consisting of noisy, detached farm buildings. Was it Valarii or Kurpenul? Small bands of men were recognizable through binoculars in different locations—most likely soldiers. It appeared that sentries were standing at the entry and exit points of the town. If we marched onward, we would reach the town in 10 minutes.

However, I thought it was inadvisable to continue marching or to attack without having contact on the right or left and without forces behind us. I preferred instead to prepare myself to make an attack against the village and await the arrival of neighboring

A view of Kurpenul from the north.

units on the right. I abstained from sending reconnaissance patrols to scout the location because I did not want to give away to the enemy the fact that I was in such close proximity. Also, it was possible to learn a great deal through sharp observation from a distance.

Until the onset of total darkness, we kept ourselves well-camouflaged in small hollows and clumps of bushes, ready to attack the town, and waited—in vain—for our neighboring column to arrive on the right. Eventually I let the company form a position of all-around defense in a suitable spot and ordered them to rest until further notice. Our sentries were directed to wake us up just as soon as anything could be remotely heard of the neighboring units, or as soon as anything that seemed suspicious became noticeable. And then the infantrymen rested for countless hours with their rifles in their arms.

It was already almost midnight when the noise became audible of our neighboring units of the Württemberg Mountain Battalion descending over on the slope to our right. Quickly I ordered my men to their feet. In the bright moonlight we crept through low-lying bushes towards the town of Kurpenul—Valarii with our machine gun platoon inserted on the side to our left for covering fire. Unchallenged, the foremost sections of the company reached the edge of the town. The enemy was nowhere to be seen. However, single shots rang out near our neighboring column on the right. Cautiously I moved into the town with the company. The machine gun platoon was brought up after us.

The various detached farm buildings were fully occupied. Near huge ovens and on stove benches, countless family members of any gender and age imaginable were sleeping under blankets and fur pelts. The air in these rooms was so thick you could cut it with a knife. To try to reach any mutual understanding through communication with these people was exceptionally difficult. Nowhere did we come across armed enemies. I took the company to spend the rest of the night in the schoolhouse and in two farm buildings nearby that were ideally suited for defense. Security detachments were sent out. Then I presented myself with numerous action reports in the western section of this widely sprawling town to report my doings to *Major* Sproesser. The remaining elements of the battalion had moved into the western side of the town. Weak enemy forces had run for the hills from this spot after the first shots.

Major Sproesser assigned individual companies to secure sectors. The 2nd Company was given the eastern half of the town with a front facing south. To the right was 3rd Company and on the left we were supposed to establish a connection with Infantry Regiment 156 at daybreak. Nothing was known of the enemy.

At about 0300 hours I arrived back at the company. It was a pitch-dark night! The infantrymen slept in the schoolhouse. I reconnoitered the sector assigned to the company with subordinate commanders. To the very near east of our lodgings, a wooden bridge led across the shallow Kurpenul stream which was about 30 to 60 meters wide. Poplars and low-lying willow bushes stood on the riverbanks. On both sides of the stream, paths led southward. The path on the east was wider and, according to maps, also more significant. A few isolated farm buildings stood in the area close to the bridge. West of the stream, the village extended yet another 100 meters south. Our security would consist of a subordinate commander's post west of the Kurpenul stream on the path through the village leading south, and an outguard in the area of the bridge east of the stream. However before I deployed this security, a dense fog enveloped us completely, exactly as had happened to us the previous day. I sent off reconnaissance patrols to the south on both sides of Kurpenul and attempted to make contact with the 3rd Company on our right and Infantry Regiment 156 located to our left rear. Slowly it grew light, but you could only see about 50 meters ahead through the thick fog.

Before contact was established with our neighbors, the reconnaissance patrol of *Gefreiter* Brückner, sent south on the eastern side of the Kurpenul stream, reported that they had run into a company of Romanians massed in close formation in the fog about 800 meters south of the outguard station. The Romanians had their bayonets fixed, yet did not recognize Brückner's patrol. I had barely reported this to the battalion, with which we now had a telephone connection, when a report came from the outguard near the bridge: "A Romanian patrol, 6 to 8 men strong, is located in the fog 50 meters to the rear of the outpost. Should we open fire?"

As the company readied itself for combat, I hurried to the outpost myself. After I established without any doubt that there were actually Romanians—recognizable by their tall fur caps—roaming around in the rear of the outpost, I opened fire, accompanied by a pair of good marksmen from the company. As soon as the first shots rang out, a group of the enemy patrolmen fell. The rest of them vanished rapidly into the fog without shooting. Minutes later, a fierce firefight developed by our neighbors left to our rear.

Additional patrols dispatched south now brought reports that, east of the stream, a strong Romanian detachment was marching towards our outpost and was by now, at the time we received the report, only a few hundred meters away. Hastily I brought one of the heavy machine guns assigned to me for support to the outguard position. I ordered the machine gun to be fired into the fog on both sides of the path. From the enemy side came a few scattered shots. Then everything was quiet.

Until this point we had not yet achieved contact with 3rd Company on our right. All outward appearances suggested that there was a yawning gap of hundreds of meters between the companies. Fierce shooting was also now developing over on our right. It seemed the enemy was advancing in a broad front straight at Valarii—Kurpenul.

To close the gap between ourselves and 3rd Company, I marched our company, consisting of two platoons with one heavy machine gun, along the western bank of the Kurpenul stream south through the long and winding village. Our outpost, armed with a heavy machine gun, remained east of the bridge to protect our flank and rear. I wanted to reach the southern edge of the village. There I hoped to find an advantageous field of fire and quickly establish contact with our neighbors on the right across open country.

I went in the point of the column, which was formed by a selected group. The company followed at a distance of about 150 meters. The fog began to surge intermittently thick and thin. Sometimes the visibility ahead was about 100 meters, then restricted to about 30 meters again.

Shortly before the point of our column reached the southern end of the village, we collided straight into a Romanian column in closed formation coming in the opposite direction. Within seconds, a most violent firefight developed at scarcely 50 meters distance. The first shots were fired in a fixed, standing position. Then every infantrymen sought cover from the rapidly incoming enemy fire. The Romanians outnumbered us by at least 10 to one. Rapid fire kept them at bay.

New enemies popped up from hedges and bushes right and left of the path, creeping and shooting closer. The disposition of our column point became divided. We held a farm building on the right of the path. The rest of our company appeared to have taken full cover in the farm buildings about 150 meters to our rear. Due to the fog, they couldn't support us with fire. Should I bring the company up forward or take my point men back? Due to the fact that we were facing overwhelming enemy forces and in view of the fog, the latter option seemed to me the more suitable. I ordered the other point men to hold the farm building for an additional five minutes, then retreat on the right side of the street and through the building on the right of the street to join the company which was disposed to provide covering fire. Then I sprang across the street to join the rest of the company in the back. The thick fog quickly foiled the bullets aimed at me by the Romanians. Immediately I ordered a platoon of the company and a heavy machine gun to open fire in the space to the left of the street. Soon, under this covering fire, the point men returned. Our men had no choice but to leave the severely wounded infantryman Kentner lying where he fell.

Now, ahead and halfway to our left, human figures rose up from the stream. Soon it was teeming with Romanians. At the same time, our outpost over to our left was embroiled in heavy fighting. On the left it was completely open and could easily be enveloped. Over on the right at a rather far distance away yet another intense firefight was in progress. There was still no contact with 3rd Company. If the enemy also enveloped us on the right, then our company would be totally encircled. The stories I heard from the Bavarian soldiers during our climb to Hill 1794 came into my memory. Things must have happened the same way with them!

Yet I continued to command: "Platoon 1, hold the position under all circumstances! Second Platoon, remain at my disposal behind the right flank of Platoon 2!" Then I hurried off to the right with numerous couriers in order to personally establish contact with 3rd Company. For about 200 meters, we ran alongside hedgerows and across open fields, and had crossed nearly an acre when shots were fired at us from a knoll about 50 to 80 meters ahead on our right. Carbine shots! You could clearly tell by hearing the sharp crack. Thus these were our own troops! The furrows in the plowed field provided makeshift cover, but we were unable neither by shouting nor waving to clarify the misperception. Thankfully those aiming at us were bad shots.

After terrifying minutes, a thick fog released us from this awkward predicament. We hurried immediately back to the company. I gave up on all further attempts to establish a connection with 3rd Company. It was now clear to me where some elements of it were located. I hoped to easily close the gap of about 250 meters wide with a reserve platoon. However, things would turn out very differently.

When I arrived at the village road, I came to ascertain that Platoon 1, together with the heavy machine gun, had launched an attack—contrary to my orders. According to the noises of battle, the platoon was now located somewhere on the southern edge of the village. If I was extremely aware of the audacity of the platoon leader and his men, I was also just as extremely aware that it was completely impossible to hold the southern edge of Kurpenul against the overwhelming enemy force in the fog without any defense on either my right or left. The only good thing was that the reserve platoon had stayed in the place I had commanded it.

The din of battle grew louder all around us. With a sense of foreboding, I hurried ahead toward Platoon 1. Soon the platoon leader, already on his way, ran into me breathlessly when I was halfway there and reported: "Platoon 1 has driven the Romanians back to 300 meters south of the village and has destroyed two Romanian guns. Suddenly the platoon was pressed very hard by a strong enemy force located a few meters distance away. The platoon is nearly surrounded. Our heavy machine gun is shot up and destroyed and its crew dead or wounded. Help must come to us right away. Otherwise the platoon is finished."

I was not in the least pleased about this course of events. Why had the platoon not remained in its position as I had commanded? Should I now, as per the request of the platoon leader, send my very last reserve forward? Would not all of us then meet the fate of being encircled by superior forces and squashed? Then, would not the left flank of the entire Württemberg Mountain Battalion also be caved in?

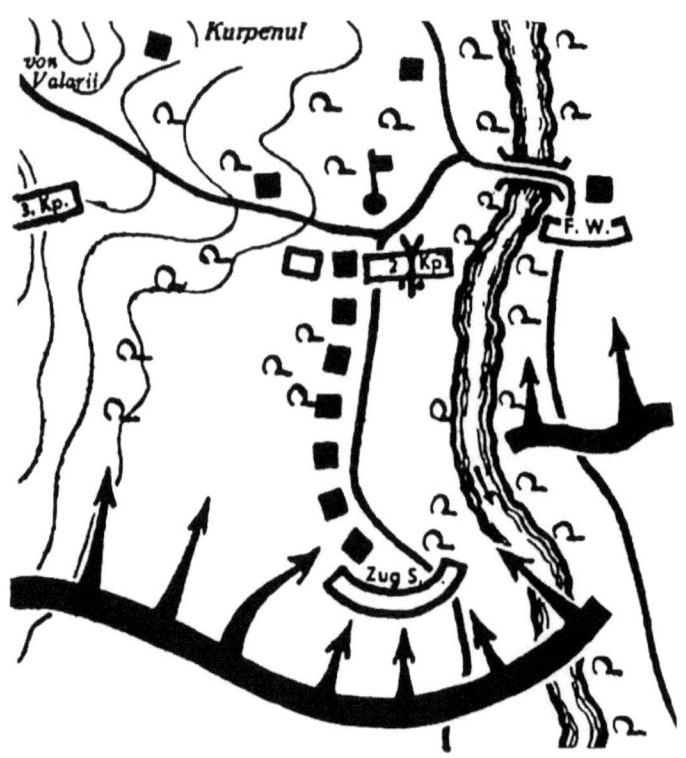

An additional map of Kurpenul shows the locations of the 2nd and 3rd Companies and the platoon (*Zug*) that ventured too far.

The answer was No. As difficult as it was for me, I could not help Platoon 1 in these circumstances. I gave orders that Platoon 1 should immediately disengage the enemy and return via a route parallel to the village road. I wanted to deploy the rest of the company as a covering force for Platoon 1. It was hardly convenient for the disengagement that now the sun was tearing increasingly through the fog and had cleared visibility for several hundred yards.

What harrowing moments! Platoon 2 went running into position through the middle of the village, shooting at dense masses of Romanians who were storming onto the riverbanks from the left. At the same time, the remnants of Platoon 1 shot their way back from their position ahead, pursued by a black multitude of Romanians. Rapid fire at their whole line brought sections of the hard-pressing enemy force to a standstill. Yet on the right and left, the enemy masses swelled nearer. On our side we were now missing the heavy machine gun, now lying ahead of us shot to pieces. Whatever scarce men returned from Platoon 1 were immediately brought up into our firing line. At top speed I rushed to the outpost across the bridge, found everything in order, took the heavy machine gun needlessly sitting there with me, and put it to use in the endangered position in the village.

But still the Romanians didn't let up. Despite heavy casualties, they kept storming at us again and again. Now our company headquarters staff had joined in the firing line. Its leader, *Feldwebel* Dallinger, fell mortally wounded with a shot to the head. The fog weakened more and more, and now we could see for the first time just how strong the enemy across from us was. Would our ammunition last? Our left flank remained constantly open.

I reported the situation to *Major* Sproesser via telephone and asked for additional backup forces with the utmost urgency. A few minutes later *Leutnant* Kohl arrived, running at full speed, with about 50 men. I had the platoon move behind our left flank with the task of overseeing

This image of Rommel and Lucie appears to have been taken in Danzig just after their November 1916 wedding. Rommel makes no mention of his wartime marriage in his memoir, which occurred around this time in the narrative.

protection of the left flank while I kept the newly arrived mass of men under my control. Soon afterwards the 6th Company arrived also. In the same manner, I echeloned them rearwards on the left and kept them at my disposal. Now there was nothing more to be afraid of.

Meanwhile, the 2nd Company, lying in the field of fire, had dug in. Slowly the enemy withdrew from our front of well-aimed carbine shots and the heavy machine gun fire. Cautiously I went to probe after the enemy with a reconnaissance patrol. Now visibility was completely clear. Again we came to the southern edge of the village. There we found the severely wounded men of Platoon 1. They had been robbed by the enemy of their few personal possessions, such as pocketwatches and knives. However, nothing else had happened to them.

In clear visibility, the southern edge of the village proved to be a supremely good position which commanded the terrain far and wide. For this reason I now brought the company back over here, reorganized it anew and ordered the men to dig in here. An additional heavy machine gun arrived. The enemy had vanished. Only from across great distances on our left did we get any rifle fire. Over on the right lay the enemy battery that had been shot up and destroyed by Platoon 1. It was later proved that other units of the battalion had also shot at it. Since no enemies were in the foreground, I roamed across the forward area accompanied by many of my infantrymen and surveyed the battery. Krupp guns! German workmanship!

Soon, however, Romanian firing lines popped up in the south and came at us. They were still more than 2,000 meters away. Wave after wave of them rose up out of the furrows in the earth. All sections of the company now took full cover. We could just wait quietly for the enemy attack. Just as the first wave of enemies came within 500 meters of us, I ordered my men to fire at will. At once the entire enemy assault broke down. In the firefight that presently developed, our side sustained no casualties. The crowds of enemies presented very rewarding targets for our heavy machine guns. The enemy withdrew as it began to grow dark. Elements of the company continued to take in numerous dozens of prisoners from the forward area as we settled in for the night. Patrols sent out ahead encountered no enemies. The company dug in. Some infantrymen went looking around for meat to roast.

Our company's casualties were painful and deeply affected us—17 men were wounded and three men were dead.

Like the 2nd Company, all the other elements of the Württemberg Mountain Battalion had their men stand fast, and did their part at Valarii-Kurpenul on the right flank of Group Kühne in order to make this forward thrust over the mountains a complete success. On the Romanian side, hundreds of dead completely covered the battlefield. Even the Romanian division commander had come to his death. After the battle, the

Lucie, wearing the same hat as in the previous photo, shows a rare grin as she and Rommel peek out of a train, possibly going on honeymoon. Rommel doesn't mention taking leave for the wedding, instead skipping his narrative straight to mid-December 1916.

way into Wallachia was opened to us. The beaten enemy was pursued. Two days later, the Württemberg Mountain Battalion moved into Targiu Jiu.

> **Considerations:**
> - The strengthened 2nd Company descended on the afternoon of Nov. 12 **in fog with security on all sides** (point, flankguards, and rearguard). The situation was totally unclear. An unexpected clash with the enemy could have occurred any second. For the protection of the troops, they **rested in tactical grouping ready for action** (all-around defense, rifles in their arms, security patrols ahead of the front).
> - The **importance of battle reconnaissance and the establishment of liaison with neighboring troops** is quite compellingly proven by the events of Nov. 13. Without the early recognition of the advance of strong Romanian forces, the reinforced 2nd Company would easily have been crushed by the enemy masses in the fog.
> - At the first outpost, **heavy machine gun fire opened up early into the fog** in the direction of the reportedly approaching enemy. This quickly **clarified** the situation and **gave time** to the 2nd Company to close the large gap to the right.
> - During the **clash in the dense fog** between our point men and the enemy column on the southern edge of Kurpenul, **it did not come to a bayonet fight**, but a firefight instead. Why? Due to the inferiority of our numbers, a bayonet fight was not advisable. We would have been stabbed to death and shot to death by the overwhelming forces. However, **rapid fire by a few riflemen delayed the storming attack of enemies outnumbering us 10 to one** by several minutes.
> - Both the **point men** and also later **Platoon 1 shot their way back through fog** to join units in position toward the rear. While doing so they received very strong support from the **firing** of these units **into the fog** in the area between the village street and the Kurpenul stream, and which was **swept very closely along the path of their retreat**.
> - It is very **easy** during a **fog battle to be shot at by your own troops**. Here again, as previously at La Briere farm, neither shouting out nor signals could achieve an adjustment of fire.
> - The **extraordinarily difficult situation** during the fight in the village with a strongly superior enemy force **was mastered by us** because, at the **focal point** of the attack, defense was mounted **down to the last man** and **forces from other, less endangered positions** were **brought over** during the battle. The **leader** must, in such situations, be extremely **mobile**. —*Rommel*

"Storm Into the Unknown": **Hill 1001, Magura Odobesti**

◊ Rommel suffers another personal loss with the death of his friend Eppler. His description of Eppler's burial was omitted from the U.S. Army translation.

◊ At this battle, Rommel fights near someone who will also become a Field Marshal during the Third Reich. Ferdinand Schörner served in the Royal Bavarian Infantry Lifeguards Regiment and also received the *Pour le Mérite*. Schörner was an early and enthusiastic admirer of Hitler who joined the Nazi Party in 1943. In February 1944, Schörner became a Nazi political instructor. He developed a reputation for inflicting unusually harsh punishments and being eager to mete out death sentences. He acquired the epithet *"Schrecken der Eismeerstraße"* (Terror of the Arctic Passage) and was reputedly called *"Blütiger Ferdinand"* (Bloody Ferdinand). Schörner was so fanatical about Nazi ideals that he had his soldiers hanged for alleged desertion in 1945, and had 22 German soldiers shot for "standing around without orders." He executed German civilians for imagined wrongdoings and allegedly shot dogs for barking too loud. He was responsible for atrocities on the Eastern Front. In his last testament, Hitler named Schörner as Commander in Chief of the German Army. After murdering many for alleged "cowardice," Schörner deserted his men, fleeing in disguise before being captured.

◊ Although both men shared similar backgrounds, Rommel was very different. Rommel's son Manfred later wrote that Rommel's political views, when expressed, tended to be liberal and moderate. Rommel chose not to join the Nazi Party although he could have gotten away with openly supporting Nazism when it was encouraged everywhere in German society. Rommel treated his soldiers with firmness but respect. He also demonstrated mercy to captured enemies. Although Rommel was promoted by Hitler due to his military talents, his career quickly plummeted as soon as he began to challenge members of Hitler's High Command about the conduct of the war. Although he enjoyed brief military success under the Nazis, Rommel prioritized his inner convictions; his conscience led him to oppose Hitler and his regime, and this cost Rommel his life.

◊ Schörner and Rommel met occasionally before and during war, as they were both involved in training troops. For now, neither young man could anticipate their very different futures. In this battle, they were allies working together on a joint mission, without any idea that one day a toxic ideology would turn their nation against itself.

◊ Readers will have noticed by this time Rommel developed a liking for using fog to

> confuse enemies. This was something he would do again in North Africa using sandstorms and dust clouds.

In the middle of December we marched over Mirzil—Merei—Gura Niscopului—Sapoca in the Slanicul valley and there joined the Alpine Corps. Resistance in the lowlands from Romanian forces, strengthened by Russian divisions, became increasingly stiff. Only slowly and in battles sustaining considerable casualties did the 9th Army succeed in taking ground at Ramnicu Sarat and the stronghold of Focsani. The Alpine Corps received the order to relieve forces fighting in the lowlands by clearing out the enemy from the pathless massifs [mountain masses] between the Slanicul and Putna valleys, while at the same time preventing an enemy thrust from the mountains towards our troops advancing towards Focsani.

We spent Christmas Eve deep in the mountains in as uncomfortable circumstances as you can imagine. Afterwards the 2nd Company marched to Mera as a reserve of the Alpine Corps from Bisoca through Dumitresti—De Lung—Petreanu. On Jan. 4, 1917, the company retreated to join the battalion, whose staff were located in Sindilari. That same day, the company, strengthened by a heavy machine gun platoon commanded by Kreuzer, occupied Hill 627 about 2.5 kilometers northwest of Sindilari. To guard the way to Foscani, strong groups of Romanians held the vast, wildly craggy, and mostly forested mountain massif of Magura Odobesti, which was 1,001 meters high.

We were supposed to capture this mountain massif on Jan. 5, 1917. The Royal Bavarian Infantry Lifeguards Regiment was launched from the south and southwest, while the Württemberg Mountain Battalion was launched from the southwest and west.

My strengthened company was given the task to launch an attack on Hill 1001 from across Hill 523 (2.5 kilometers northeast of Sindilari) without direct support on our right or left. The Bavarian Lifeguards were on our right with their left flank about six kilometers southeast in the area of Hill 479. On the left was Detachment Lieb on a mountain ridge located about four kilometers from Hill 627 which inclined from the west towards Hill 1001. Both were ready to attack.

As per orders I advanced in the early dawn hours with my detachment. After crossing several very steep and mostly forested valleys, we reached Hill 523 by sunrise. A binocular "scissors telescope"[7] left standing there put itself to good use. As the company rested under cover, I scanned with utmost meticulousness every hill and valley using the telescope. In doing so I quickly gathered a clear picture of the disposition and strength of the enemy forces positioned across from me.

7. This device was called a *Scherenfernrohr*. In German it is a "scissors telescope" while it is sometimes known in English as a "rabbit-ear telescope." It consists of two conjoined lenses forming long vertical binoculars that project above a person's head like a pair of antennae. It increased depth of perception and allowed users to peer into contested areas without risking getting shot in the head. The Germans used it in both World War I and II.

Off to the right, in the direction where the Bavarian Lifeguards were supposed to be, my line of view was unfortunately blocked. I could reconnoiter nothing of these neighbors at all. At about 1,000 meters distance ahead of me in a northeastward direction, Romanian reconnaissance patrols were wandering in the valley. Behind them, a range of hills positioned in the foreground of our target, Hill 1001, and running parallel to it in a north-south direction, was totally occupied by Romanians from one end to the other. Portions of entrenched positions were clearly visible through the light forest.

It was impossible to make a concealed advance through the broad, treeless valley ahead of this position during the day. Over to our left, Romanian outguards of about platoon strength were positioned on the range of hills north of Hill 523, which was crowned only by isolated farm buildings and small patches of forest. These outguards had fortified positions with an overall front facing west at their disposal.

From what I could see, the most promising opportunity for approaching Magura Odobesti was undoubtedly the range of hills inclining towards the peak from the west, where Detachment Lieb was positioned. I decided that for my company to approach the strong enemy forces in a northeast direction without cohesion on our right and left appeared futile. Thus I would approach close to Detachment Lieb and operate together with them. However, we were separated by about four kilometers as the crow flies from these neighbors, whom I also could not see and whose location I could only estimate.

I dispatched numerous patrols towards the enemy position in the northeast, giving them orders to distract this enemy's attention from my intended line of attack north and then find their way back to the company after about two hours. Shortly afterwards it transpired that two enemy outposts were attacked and the enemies there were repelled back towards their main position without any casualties on our side.

We reached a continuous forested terrain and were now only two kilometers away from the ridge where we estimated Detachment Lieb to be. I turned northeast intending to take the range of hills running north-south in the foreground of Magura Odobesti [which were fully occupied by Romanians] at the spot where it connected with the mountain ridge inclining towards Hill 1001 from the west.

I personally went forward with the point men through the light beech forest. The strengthened company followed at an interval of about 150 meters distance. Following a road, we descended into a gorge. As our security patrols reached the deepest part of the gorge, we observed movement on the steep cliff beyond us. A Romanian column with countless beasts of burden was climbing down the zigzagging path straight towards us. The front of their column was barely 100 meters away from us. Their total strength could not be estimated. What should we do?

It seemed the enemy had still not noticed us. Rapidly I withdrew the point men sideways into the bushes, then pulled about 50 meters back and lay there on the lookout. At the same time I brought up a courier to send orders to our leading platoon to develop itself. Before this had been carried out, bullets fired by Romanian infantry struck among us. The point men returned fire. Soon, Platoon 1 joined in the firefight right alongside the point men. Our position was on unfavorable terrain, because the enemy, whose strength was difficult to estimate, was firing down from the high cliff. During a long protracted firefight it would be impossible to prevent severe casualties on our side. Therefore I decided that I would rather just storm into the unknown. The success was greater than we expected—the enemy fled as we charged at them with battle cries, seven Romanians and numerous pack animals remained in our hands, and we sustained no casualties.

Rommel, center, is pictured with Ferdinand Schörner, left, at a skiing event in the 1930s. Although both fought in the same location during World War I and won the *Pour le Mérite*, both took different attitudes to Nazism during the Third Reich.

We stormed uphill after the fleeing enemy and arrived, gasping for breath, at the cliff crest. But we were met by heavy fire there. My brave and good courier Eppler fell to the left of me with a fatal shot to the head.

After bringing up the heavy machine gun platoon and two rifle platoons, I attacked in a northern direction through the forest on both sides of the road. We managed to move forward only slowly and arduously. Not the slightest trace of the enemy was to be seen, yet his strong fire crashed all around our ears. By all outward appearances, the firing got stronger as we took more ground. Ultimately we found ourselves in a lightly forested area across from a fortified enemy position 250 meters across from us. The resistance coming from it was so strong that further attacks seemed futile. Between us and the enemy position was a shallow dip in the terrain. We lay in a hardly convenient position on its forward slope.

To avoid unnecessary casualties, I ordered the rifle platoons to make a fighting

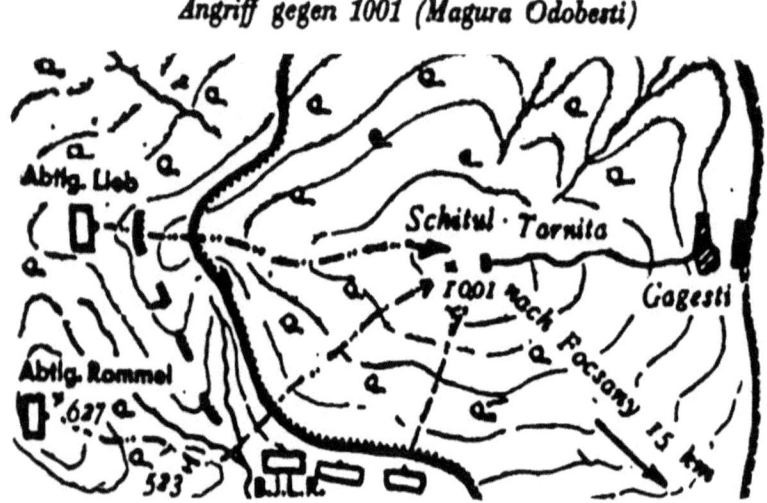

Rommel's map of the attack on Hill 1001 shows his detachment and Detachment Lieb to the left and the Bavarian Life Guards (B.I.L.R.) at the bottom center-left.

retreat to the nearest hill under protective fire of the heavy machine gun platoon. Now we were positioned on a small knoll 400 meters across from the enemy. The firefight died down slowly. Only sporadic shots rang out. Because there was no connection with our neighbors right and left, we occupied the forested knoll in an all-around defense position and lightly dug in. The reserve platoon and the heavy machine gun platoon were placed in the middle.

The day waned as we buried poor Eppler—the only death that the fight had cost us.

Before complete darkness fell, we unexpectedly discovered elements of Detachment Lieb in a forest meadow over on our left, only about 700 meters away. Soon I was connected to them via telephone. I discussed the situation with *Oberleutnant* Lieb and later also with *Major* Sproesser. A frontal attack by Detachments Lieb and Rommel against the Romanians' strongly fortified forest position would hardly guarantee success. Whether or not the position could be enveloped from the southeast was something we needed to reconnoiter at once.

At my command, *Vizefeldwebel* Schropp carried out this reconnaissance upon the south flank of the enemy position during the night. It was an extraordinarily difficult task due to the terrain, which was laced with canyons. Many long hours before daybreak, he brought back the exemplary report: "Without encountering the enemy, the reconnaissance patrol, advancing in a northeasterly direction, crossed a deep ravine and reached the crest in the rear of the enemy position across from us. There we located and crossed a path which appeared to be used regularly by the Romanians."

I reported the results of the reconnaissance immediately to *Major* Sproesser and received orders to carry out the enveloping attack against the enemy's forest position with 2 and a half companies at daybreak. Detachment Lieb was tasked with the frontal assault—however, he was only supposed to attack when my detachment was fully engaged in fighting.

It began to snow quite heavily. At daybreak there were 10 centimeters of fallen snow; the weather was cloudy and wet. Clouds of snow drew together thickly above the mountain crests. The 6th Company arrived to reinforce me. I left Hügel's rifle platoon behind in our previous position and gave them the task of pinning down the enemy with frontal fire during our encircling march and to draw their attention away from us. Next, accompanied by one company and 2/3 of another plus the machine gun platoon, I climbed off in an eastern direction through a very deep gorge. Schropp led the way, retracing his path from the night before.

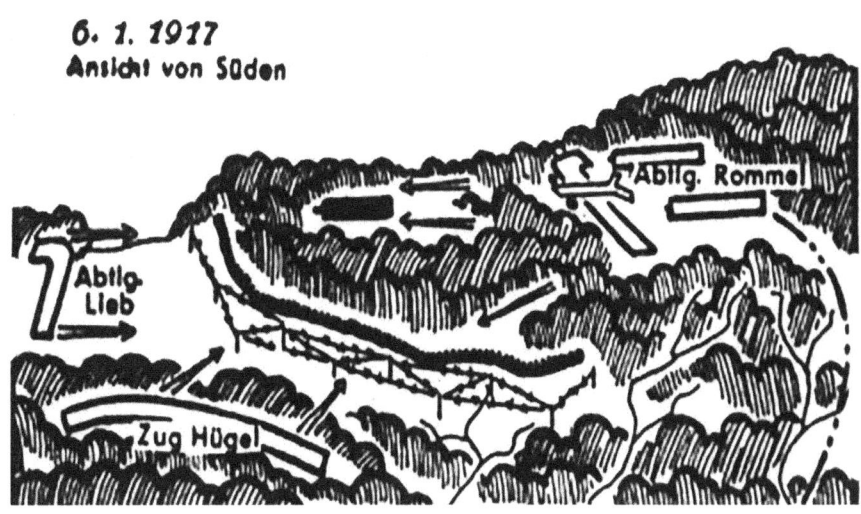

A view from the south demonstrates the movements of Rommel's and Lieb's detachments as well as Hügel's platoon.

As Hügel's platoon opened fire from our former position, and the Romanians—fearing an assault—reacted fiercely against it, we soundlessly crossed the ravine and then climbed in a northeasterly direction. After strenuous climbing, we reached the high crest and there discovered freshly laid tracks from a Romanian detachment in the snow.

Fog restricted our vision to barely 40 meters distance now. We could stumble into the enemy within the blink of an eye at any second. I ordered the 2nd Company to set aside their heavy packs and echeloned the detachment quickly to prepare for attack: 2nd Company and the heavy machine gun platoon in the frontline, 6th Company in the second line at my disposal. Over to the left, the firefight started by Hügel's platoon had gone silent. Only single shots rang out now and again.

Cautiously we stalked along both sides of the narrow path through the wintry forest, going west toward the enemy's rear. Suddenly we heard voices ahead of us in the fog. I ordered everyone to halt and commanded them to prepare the heavy machine gun to fire. Then cautiously we slithered nearer. Soon we came across an abandoned campsite. The campfire was still smoking. Not a single Romanian could be seen anywhere. We kept going.

A forest clearing suddenly appeared ahead of us. Several Romanians moved across it, unaware of our presence. How strong could this enemy force ahead of us be? Maybe there were only a few of them, but there could also very well be a whole battalion ahead of us. Either way we were capable of dealing with them, so I had the heavy machine platoon launch a surprise fire at the figures moving in the fog. A few seconds later, our whole detachment rushed forward towards the enemy with a thunderous battle cry.

It now became apparent that only a few Romanians were standing across from us. They no longer attempted to shoot. Instead they tried to save themselves by fleeing rapidly downhill. We didn't worry about them. Instead we stormed westward down the path, drew enemy fire for a few minutes—without being able to see the enemy—and then distinctly heard the battle cry of Detachment Lieb approaching.

Now we had to make sure that the two detachments of Rommel and Lieb, coming together from opposite directions, did not shoot each other in the fog and forest. This extremely precarious task soon resolved itself without a problem. The enemy force between both detachments was pulverized. Enemies escaped being taken prisoner by rapid flight downhill through the forest. Only 26 Romanians fell into the 2nd Company's hands. But these survivors who fled also did not escape their fate. They reappeared from the forest days later when we were already standing on the Putna. After brief negotiations, a Romanian battalion of 500 men in closed formation surrendered itself to the leader of one of our pack animal trains.

Following our wonderfully successful attack—casualties on our side were negligible—Detachment Lieb ascended Hill 1001. I ordered the 2nd Company to again take up their packs and joined the advance. Snow flurries set in. The fog thickened.

In the area of the peak of Hill 1001, Detachment Lieb came upon Romanian reserves who had positioned themselves in areas protected from the wind. A short fight developed. The determined speed and tenacity of our mountain rangers quickly put an end to it. The Romanians took casualties and cleared off the summit. They would never again occupy their positions, which were now covered in snow.

Cold wind swept over and across Hill 1001. Ice crystals stuck into our faces like needles. Amid these conditions, the detachments Lieb and Rommel hurried to reach the monastery of Schitul Tarnita, which was situated on the eastern slope of Hill 1001 a few meters below the peak. No enemy guarded the way. The monastery severely disappointed our expectations—especially in terms of spaciousness and provisions—but it at least offered a little bit of protection against the troublesome weather conditions.

After an hour, elements of the Bavarian Infantry Lifeguards Regiment arrived at Schitul Tarnita and claimed the monastery as their shelter. Because the Bavarians claimed higher rank and authority, we had to give in. Lieb's detachment, however, asserted themselves and remained within the monastery. But my detachment was kicked outside and quartered in windy, abysmal earthen huts, which had no capacity for retaining heat, in the surrounding area outside of the monastery. There we survived a most bitterly cold night.

During this time I came to a decision to seek out the more habitable areas in the valley.

> **Considerations:**
> - The **thoroughly detailed observation of the battlefield through the telescope brought clarity about the disposition and location of the enemy** during the advance of the 2nd Company, which was strengthened but only loosely flanked at it moved against the enemy on the forested mountain massif of Hill 1001. This was just as important as the combat reconnaissance of patrols.
> - During the clash with the enemy in the forested gorge, the **fighting strength** of the mountain rangers allowed them **to master** this **extremely difficult situation**.
> - The **attack came to a standstill** during the evening 250 meters in front of the enemy's main fortified position. To prevent casualties, I ordered the rifle platoons lying in the lightly timbered high forest to make **a fighting retreat** under the **covering fire** of the heavy machine gun to a more favorable position. Thus no casualties occurred. Today, in a similar situation, this could also be done with the **use of artificial fog.** Thus the enemy would immediately shoot very fiercely into the fog, but would soon grow weary of the fog and cease firing. Then would be exact time to disengage.
> - The **exemplary results of the combat reconnaissance during the winter night** (by *Vizefeldwebel* Schropp) enabled us to advance in the enemy's rear on Jan. 6, 1917. Lesson: Scouts must be active while troops are resting during the night.
> - **To deceive, divert and to pin down the enemy,** Hügel's platoon started **a long firefight** during the enveloping march of Detachment Rommel.
> - During the **surprise fire** and **enveloping storm attack in the thick fog against the enemy,** whose strength could not be established, the heavy machine gun platoon was positioned *in the frontline.* Their fire sprayed the ridges that we were supposed to attack and quickly freed this area from the enemy.
> - **Romanian reserves** stood in driving snow on a slope of Hill 1001 shielded from the wind—**without connection** to their frontline and **without security.** Thus Detachment Lieb was able to surprise this strong enemy and throw them back. —*Rommel*

"Bitter Cold"—*Gagesti*

> ◊ Rommel makes a rapid and stealthy withdrawal after dark. He would perform similar maneuvers in North Africa, such as at the Battle of El Agheila in late 1942.
> ◊ He navigates by starlight. This skill would serve him well in the deserts of North Africa.
> ◊ A different kind of enemy emerges during Rommel's nighttime adventure—an inhuman one. Wolves begin stalking Rommel and his men in the winter darkness.

In the very early hours of Jan. 7, 1917 I sent reconnaissance patrols to the Putna valley on both sides of Gagesti. Dense fog was prevalent amid extremely biting cold and 30 centimeters of snow. At about 1000 hours, *Unteroffizier* Pfäffle, in charge of food provisions for our troop animals, brought the report: He had ridden about four kilometers towards the valley without encountering the enemy. There he heard the sounds coming from the valley of the vehicles of numerous columns and loud shouts. The enemy seemed to be withdrawing from there. He could see nothing due to the foggy weather.

While passing on the report via telephone to *Major* Sproesser I requested permission for the strengthened 2nd Company to probe the way forward towards Gagesti. One hour later we went off. Walking single file we went through a lightly wooded high forest. In the fog we could only see up to 80 to 100 meters far. A group led by the brave *Vizefeldwebel* Hügel provided security for the march 100 meters ahead. In addition, flanking patrols secured the advance on the right and left. The heavy machine gun company was echeloned in the center of the company and the guns were loaded on pack animals.

After a half hour march we came out of the high forest. From there, a narrow footpath led through a nursery of young trees which were barely meters high but very thickly planted. I marched now close behind the point men. The fog had lessened.

Suddenly shots rang out ahead. Shortly afterwards Hügel could be heard giving commands. Then he reported that he had run into a Romanian patrol on the narrow path. The first shots had taken out the Romanians walking in front. Afterwards the remaining seven surrendered. Meanwhile the company had moved into closed formation. Caution was now the order of things. Perhaps the prisoners were providing security for an enemy column. Hügel went back with the point men. A few minutes later, he reported reaching the eastern edge of the tree nursery and the approach of a line of enemy riflemen of at least company strength, now at a distance of only about 100 meters away. Rapidly I had the foremost platoon occupy both sides of the path at the edge of the tree nursery and open fire. We were answered by a clattering of heavy enemy gunfire through the low vegetation and we immediately dived to the ground.

The deployment of the heavy machine gun platoon introduced difficulties. Its leader

German mountain ranger troops are shown on the move across snow-covered terrain.

reported that his rifles were frozen solid. He had to thaw them first before they could be used. The firefight a few meters east on the edge of the tree nursery became very fierce—it seemed that the superior enemy force was fully attacking.

In a tiny basin in the earth, the heavy machine gun platoon was feverishly trying to use liquor to defrost the guns. The enemy cone of fire burst through the sparse saplings with a pattering noise like rain. It was utterly exasperating that the heavy machine guns could not join in the fight at this dire moment. If the enemy outflanked us on the right or left, we would have to retreat. The 2nd and 3rd Platoons therefore provided security in these directions.

Finally at last the first machine gun was in working order and went into position. But there was no more shooting. In the now thickening fog, the enemy broke off the fight, and soon we no longer had viable targets. To shoot into the fog would be a waste of ammunition, which our troops could not afford due to the difficult resupply conditions in the mountains. Covered by the heavy machine gun, my platoon and I seized a small hill with a little house on it, which stood in a fenced vineyard. By now there was not a single shot being fired.

Opposite, on a bare hill to the south, a large number of apparently leaderless Romanians were milling around by themselves. We waved to them with handkerchiefs and in this way soon had taken 20 men prisoner without fire being exchanged. The Romanians had by all appearance had enough of the war, which up to this point had been going truly badly for them. A few of the prisoners then helped to gather and bring in some of their additional comrades. The remainder of my company was brought up. Because enemies could approach us from all sides, I arranged the company in a circular formation and sent out security patrols and scouts 500 meters in all directions. Soon this resulted in additional Romanian prisoners being brought in. By himself, *Gefreiter* Brückner surprised five Romanians in a vineyard house, rendering them flabbergasted.

Accompanied by *Leutnant* Haußer I personally roamed the foreground to reconnoiter a more suitable place, a farm building if possible, to assemble the company. It was about 10 degrees below zero. We were extremely freezing. My stomach was growling.

Although there was no farm building in the surrounding area to be found, there was an even more advantageously located place for the company very close to the north of a deep ravine. It was a little house in yet another fenced vineyard. Here in the single room, which had no capacity to retain heat, lay a severely wounded man—a Romanian left behind by his compatriots. Dr. Lenz took him under his care. Yet there was very little hope to bring him through alive. Meanwhile the company quartered itself here.

The deep ravine led through the valley towards Gagesti. The terrain around our new position was unoccupied and had open visibility for numerous hundreds of meters to the east and north. On the remaining sides it was covered with light scrubland. Waves of fog rolled in and out constantly. At times visibility was restricted to 200 meters.

On a hill to our left we heard the sound of voices. Dr. Lenz and I stalked over there. At a distance of about 800 to 1,000 meters from the company, we discovered a group of resting Romanian troops—probably the size of a battalion—behind a fruit orchard on an open field. Hundreds of men, horses and vehicles stood jammed together here within a tight space. Campfires burned in their midst.

Although the fog was ideal for a concealed approach, the terrain was not suitable for a surprise fire; thus I abandoned the idea. It was already 1400 hours. In an hour and a half it would be dark. In this piercing cold it would be impossible for the company to spend the night in the open. Where was Gagesti? We had to commandeer a civilian farmhouse in the area to sleep in, as we did not want to make a penitent return to Schitul Tarnita monastery! In addition, we had to capture something edible. Our excruciating hunger made us enterprising.

Then I went from our company position with Dr. Lenz and his orderly towards the east, along the left side of a 3-meter deep gully. On the right of the same precipice, *Vizefeldwebel* Pfeiffer went ahead with two to three men at 50 meters distance from us.

Hardly 300 meters from our company's quarters, we discovered a large number of Romanians at a little house on the north side of the gully. Perhaps an outguard? Although we had only one carbine at our disposal on the north side and four on the south side, we approached the enemies shouting and waving handkerchiefs and demanded they surrender. The group did not move...but they didn't start shooting, either. Now it was totally impossible for us to turn an about-face and leave. Soon we were only 30 meters apart from them. Inwardly I trembled with fear of the outcome. The group across stood with their rifles at the ready, talking and gesticulating amongst themselves, yet they did us the courtesy of not shooting. At last we were among them and talked them down. I gave them a long talk, saying something about the end of the war, and then turned the 30 prisoners over to Pfeiffer's patrol.

Three of us went further east towards the valley. After about 150 meters the silhouette of an approaching company materialized in the fog. Should we risk it? A distance of only 50 meters separated us from the group. Now they had recognized us, so—we went for it! Waving handkerchiefs and shouting, we continued to walk closer. The men of the company recoiled. Their officers furiously bellowed: *"Foc, foc!"* ("Fire, fire!") and starting beating up their men who seemed like they would rather have laid down their weapons.

Our situation grew precarious. The company aimed and fired. A hail of lead burst out and passed by us. In this moment we dived to the ground and then Dr. Lenz and I sprang backwards. Dr. Lenz's boys fired off a few rounds from their carbines before he had even commanded them to. The fog concealed us from rapidly aimed fire. However, enemy units pursued us as others shot blindly at us.

Hard pressed by the enemy, we reached Pfeiffer's patrol. There were still 30 captured Romanians standing here with their weapons very close by. We quickly herded the prisoners into a ravine that offered cover from our pursuers' fire and then drove them ahead of us, running, towards the company position. Had our pursuers directed sweeping fire along the length of the ravine, we would have been forced to abandon it. However the enemies chasing us were bad shots. We passed into the company positions with our prisoners without any casualties. Shortly afterwards our company's fire brought our enemies, pursuing us in a broad skirmishing line, to a standstill. A fierce firefight developed at 100 meters distance. Our heavy machine gun gave us considerable superiority in firepower.

Should I attack? No. The loss of each fallen mountain ranger would be felt painfully. Soon night fell. Both sides kept firing shots at each other from time to time just to show they were still there. There was little hope of shelter and a warm bite to eat in the bitter cold! *Leutnant* Kohl of the 3rd Company came over on a horse to check on us. He took control of the 80 POWs we had captured up to this point and took them away with him. Additionally, he took a report to Schitul Tarnita of my decision to press on towards Gagesti during the night.

Over the last few hours the weather became clearer, but also colder. Soon stars were blazing in the heavens. Bushes and trees appeared black against the white snow as they jutted up from the plain. I sent last regards to our enemies across from us with rifles and machine guns, then rapidly withdrew my fighting force from the enemy. Soundlessly we went uphill on a narrow path in a northwesterly direction. Point men and rear guards secured the march. The heavy machine gun platoon was in the center. The heavy machine gun, still warm from shooting, was covered with blankets and shelter halves to protect it from freezing.

After 500 meters on the path, I turned

a: Feindl. Posten
b: Feind am Lagerfeuer

Arrows show Rommel's approach. A) the enemy sentry position on the hilltop; B) the group of enemy soldiers crowded around the campfire.

north. The polar stars replaced the compass. We slinked forwards alongside a black, thorny hedge. In this way we would not stand out in the snowy environment. Not a word was spoken.

Then the rearguard reported that a strong detachment of Romanians was following us. At this point I ordered a halt next to a dark row of bushes and had the heavy machine gun prepared for action. These measures proved superfluous, because the leader of the rearguard soon personally reported that he prepared an ambush for the enemy in a suitable spot and had captured them without any shooting—25 Romanians! I could not use them and quickly shoved them off under guard to Schitul Tarnita.

We went further north. After about 800 meters I turned back towards the east again. I had studied the map in detail before the march. We had to arrive precisely straight at our goal, which was the northern entrance to Gagesti. The company deployed itself soundlessly and advanced with all three platoons in columns next to each other. The heavy machine gun and myself were with the middle platoon. In this manner we felt our way forwards from one cluster of bushes to another. The terrain dipped gradually towards the Putna valley. We halted many times and surveyed our surroundings sharply using binoculars.

As the moon rose high on our right, the glow of a fire became visible on the ground ahead of us to the left. Soon we saw there, at about 600 meters distance away, several dozens of Romanians standing around a huge campfire. Behind them, an enemy detachment marched from the left to the right, probably towards Gagesti. The area was obscured by the long stretch of a bare hill, upon which only a few clumps of trees were recognizable through binoculars. Half to the right, vast fruit plantations hindered our range of vision.

Hungry wolves were at the same time stalking closer to the infantrymen in the cold winter night. Should I start heading towards the enemy on the left, or should I leave him there and keep going straight towards Gagesti?

The latter option seemed more correct to me. Slowly and with exceptionally great caution, our three columns crept within 200 meters of the bare hilltop, keeping pressed closely against the black hedgerows. The enemies on the hill were now about 30 meters above us. On our left flank, only 300 meters away, about 50 Romanians were sitting around the fire. At this point, some of my guys insisted that they had recognized definite movements in the clusters of trees on the hill ahead of us. I could not clearly establish any such thing using my binoculars.

Moving even slower than before, we sidled alongside the hedges as we went forward, and at last gained the lower half of the hill, which could not be observed from the top. As I quickly prepared the company for attack, scouts moving farther ahead observed the edge of the hill. They recognized Romanian sentries about 100 meters ahead of us. Should I deploy the heavy machine gun? Since there were just a couple of enemy sentries it seemed unnecessary to me. I wanted to take the hill under my possession with a surprising and soundless storm attack—without any shooting, if at all possible. The attack on the northern section of Gagesti, which I had to assume was heavily occupied, needed in any case to come as a surprise to the enemy.

My subordinates were quick to receive their orders. Then the strengthened company stormed silently ahead. No whistle, no shout of command, no battle cry! The infantrymen materialized in front of the Romanian sentries as if they had sprung out of the earth. It happened so fast that the enemy no longer had the wherewithal to fire any alarm shots. The enemy sentries vanished hurriedly downhill.

The hill was ours. The rooftops of Gagesti gleamed in the moonlight ahead of us and to our right; the village was about a kilometer long. The northernmost buildings were barely 200 meters away and some 30 meters steeper from us. There were great distances in between the individual groups of houses.

Alarm bells then began to sound in the northern part of Gagesti. Soldiers rushed into the streets and converged in tightly bunched groups. I waited for them to come storming out in a dense mass in the moments that followed in order to reclaim the hill they had just lost. They should go ahead and come, as far as I was concerned. Our heavy machine gun was loaded up for full-automatic fire and my riflemen, armed with carbines, were in position across a 200-meter front. A platoon stood ready at my disposal behind my left flank.

Minutes stretched by. Things became quiet again in the village below. Apparently because we had concealed ourselves and were not to be seen on the hill, and also were not shooting at them, the alarmed troops went right back into their warm quarters—which they had probably only emerged from very begrudgingly. We were astonished! We were also surprised that not once did the Romanian sentries make any attempt to regain their old position. It looked now as though they were standing in the village below between the buildings. In the meantime it was now 2200 hours. We froze and lay there hungry in front of the warm huts of Gagesti. Something needed to change here. The decision: we would tear the northernmost buildings of this huge village away from the enemy, as we intended to use them to settle ourselves in, warm up, get some nourishment and rest at the very least until daybreak.

I sent *Vizefeldwebel* Hügel with an assault troop, consisting of two groups from our right flank, to attack one of the buildings. He was ordered to approach along one side of a dark hedge, to return fire if shot at, and then take the building across from him in conjunction with the left platoon under the protective fire of the remainder of our strengthened company. The individual groups were briefed about their tasks. Then Hügel went forward. The assault troop approached within 50 meters of the building. There they were fired upon. Immediately all our machine guns as well as Janner's platoon started firing, while at the same time, our left platoon rushed at the place with a raucous battle cry. The infantrymen had already reached the building before the Romanians inside had a chance to come out. Now Hügel launched a storm attack from the other side. The remainder of our strengthened company bellowed out a battle cry from the very depths of their bodies, which roared across the night like the sound of a whole battalion. At this point our heavy machine gun platoon could no longer continue shooting without endangering our own troops among the buildings on the northern edge of Gagesti. Thus the platoon directed its fire to the right and sprayed the roofs of the expansive village for several minutes.

Suddenly all went quiet below on the northern edge of the village. Only a few shots were exchanged here and there. Seemingly the Romanians had quickly cast aside their weapons. I immediately hurried over there with an additional platoon and a heavy machine gun. Just as I arrived amid the buildings, the prisoners we captured were being assembled. There were over 100 men. What was even more gladdening was that no one among us had been wounded during the firefight. No more shots were fired from the buildings around us. Only our machine gun platoon occasionally sprayed fire over the roofs to our right. Since everything had gone so well so far, I took the company to the right, going from building to building. We took the Romanians quartered there as prisoners. All of them surrendered to their fate without offering resistance.

With our position secured on all sides, and the prisoners and the heavy machine gun platoon in the center, I soon marched south with the whole company along the main village road. There were 200 prisoners! There was no end to them. Everywhere infantrymen were pounding on the doors and hauling new prisoners out. Our troops providing us with fire support were drawn in closer. We neared the church. Now the number of prisoners was thrice as much as our own—360 men!

a : Stoßtrupp Hügel

The 2nd Company is positioned on the overhang, with Hügel's assault troop (marked A) slipping among the houses from the right.

The church was situated in a small knoll which sloped sharply into a cliff towards the lower village, which was only 200 meters away. Occupied buildings formed a half-circle around the church. This looked to me like the ideal place to have a secure shelter for the rest of the night. The prisoners were put into the church, while our company occupied the buildings all around it. Then, with units of the company, I undertook a further exploration of the lower village, which contained the road linking Odobesti and Vidra. We came across no Romanian soldiers there. They appeared to have hastily exchanged their quarters for the east bank of the Putna river after hearing the noise of battle from the upper village.

Instead I met the community leader, who communicated to me through a German-speaking Jew that he wanted to hand over the keys of the village assembly hall to me. Furthermore, members of the community had baked 300 loaves of bread for the arriving German troops, slaughtered multiple cows and had numerous barrels of wine ready for the troops at our disposal. I selected what we could use and had it brought to the church area in the upper village. Billets had been set up there in the meantime. Midnight had passed by the time the last units of our strengthened company had settled in. Sentries provided security for the resting troops.

Given that we were about six kilometers ahead of our own lines, and without support on our right or left, I felt safe and snug in Gagesti only as long as it was nighttime. To be on the safe side, I wanted to move to a nearby hill commanding the terrain east of Gagesti at daybreak. Then we would see where the remaining enemies were.

The troops ate and rested. I prepared a short combat report which was taken via courier rider to Schitul Tarnita at about 0230 hours. The courier also took along a Romanian *"Logele"*—a 3-liter wooden cask—containing delicious red wine for *Oberleutnant* Lieb.

The rest of the night passed without disturbance. Shortly before daybreak on January 8th, I moved with my whole group to the hill just east of the Gagesti church. As the day dawned, we firmly established that the snow-covered terrain all around us was free of the enemy. Only in the east did we see enemy troops engaged in spade work beyond the Putna river. Thus I moved back into our former quarters around the church and sent out reconnaissance patrols in different directions.

I took a morning ride through the lower village in the direction of Odobesti accompanied by Pfäffle, who was in charge of feeding our troop animals. We had sent the pack animals back to Schitul Tarnita the previous evening. By whinnying, the animals would have given away our troops advancing to Gagesti. At daybreak, Pfäffle had brought the staff up. By riding towards Odobesti I wanted to establish contact with our troops west of the Putna.

As we trotted through the lower village of Gagesti, there were no shots to be heard for far and wide. Riding in the cool freshness of the morning was invigorating. I let my horse, named Sultan, gallop strongly and more freely, and indeed I was more preoccupied with the horse than with my surroundings. Pfäffle was riding 10 meters behind me. We were about 1,000 meters away from the Gagesti church.

Suddenly something moved on the street ahead of my horse. I looked up and was more than a little shocked to see a Romanian patrol of about 15 men with fixed bayonets standing right in front of me. It was too late to turn around and gallop away. By the time I could manage to turn the horse around on the narrow street and get him into a gallop, I would surely catch a few bullets in my body. I made the quick decision to keep trotting ahead without changing my tempo, go straight up to the patrol, greet them with friendliness and give them to understand that they were prisoners and had to disarm, and should get marching to the Gagesti church where 400 of their comrades were assembled. I was extremely uncertain whether any of the Romanians could understand my language. My hand gestures and my quiet, friendly tone and manners, however, worked with convincing effect. The 15 men left their weapons behind on the street and went wandering off the road in the direction I had given them. I then rode ahead for only 100 meters before galloping back to my company on the quickest way I could. If this happened to me a second time, I would surely not be lucky enough to come across such simple-minded enemies.

Throughout midmorning, 1st Company and the 3rd Machine Gun Co. were assigned to me as reinforcements and put under my command. Until further notice, Detachment Rommel now consisted of two rifle companies and a machine gun company. *Leutnant* Haußer became my adjutant.

A reconnaissance patrol brought additional prisoners back from their patrol rounds. At about 0900 hours, "the war began again." Romanians and perhaps additional Russian artillery laid very heavy harassing fire on Gagesti from positions on the hills east of Putna. We cleared out areas that were in particular danger. In the wide expanse of the town we certainly had adequate space. Also, fortunately no casualties occurred at this time.

In the afternoon the enemy fire accelerated to a level of great ferocity. It brought back memories to us of the western theater of the war: shells falling all around us. Bullets burst through the roof of the house in which our detachment command post had been set up. In this place also—as often before—the heavy firing likely came as a result of the brisk traffic of our couriers. The situation became extremely uncomfortable. The detachment occupied the outskirts of Gagesti and dug in. Did the enemy intend to attack?

During the heavy firing, *Major* Sproesser arrived in Gagesti on horseback and set up his command post in the front line along the road between Odobesti and Vidra. The enemy artillery fire persisted with undiminished forcefulness until the onset of darkness. We expected a night attack, a tactic loved by the Russians, and took special care in securing our unsupported flank.

Considerations:

- During the **collision of our point** with the Romanian patrol on a narrow path in a plantation, a few shots of the leading point man quickly decided the fight. It is important in such circumstances that a soldier creeps forward towards the enemy with **their weapon ready to fire** (rifle safety off, light machine guns being carried in firing position for shooting). Because whoever is able to shoot first and unleash the most fire will win.
- A few minutes later **during the firefight** with the strong enemy force th**e heavy machine gun failed at a critical moment because it was frozen**. It had to be warmed over a few meters behind the frontline with flames lit from alcohol. Afterwards the heavy machine gun was kept warm with blankets during the further course of the battle.
- It was possible **to break off the fight at nightfall** after a short, forceful surprise fire on the nearby enemy.
- The night attack on the northern sector of Gagesti by moonlight and a dense blanket

> of snow was brought about from two different directions **under heavy protective fire by the heavy machine gun platoon.** After the successful storm attack, this platoon supported the advance into the town's lengthy expanse by firing over the houses. In reality this was not meant to hit anything, but **the moral effect** of this upon the enemies quartered in the warm houses was so extreme that they allowed themselves to be taken prisoners without putting up any further struggle.
> - No **casualties** occurred on our side during the fight in Gagesti.—*Rommel*

"The Air of Winzenheim"—At Vidra

At about midnight we were relieved by elements of the Alpine Corps and moved northwards amid bright moonlight along the valley road. We marched for 10 kilometers parallel to newly dug Romanian and Russian positions, sometimes hardly 1,000 meters in front of them, without being harassed. Our own troops did not settle down to fight the enemy here. At daybreak, staff of the Württemberg Mountain Battalion and Detachment Rommel met in Vidra and found comfortable quarters for the first time in long time.

Just as soon as I was making myself comfortable, the following order from the battalion reached me: "The enemy has broken through on the mountains north of Vidra. Detachment Rommel is to ready itself with utmost speed and must advance north to Hill 625 north of Vidra, where it will place itself under the authority of the reserves of Infantry Regiment 256."

This demand just about exceeded the limits of human ability. For four days, my detachment had fought continuously amid the most difficult circumstances and had marched for the entirety of the previous night. At this moment the infantrymen were

> ◊ Here we get a sense of Rommel's love for being a teacher. When describing training troops, he writes, "I especially loved this duty." The word here, *"lieben,"* literally means "to love." It is not a word Rommel uses often in writing; when he does, he means it.
> ◊ Rommel takes personal ownership of what he calls "his" own school—*"meine Schule."* He is obviously thrilled with the idea of education and being in charge of his own teaching program. Rommel's word choices in writing can sometimes be impersonal. His very personal and warm word choices here stand out.
> ◊ Rommel references one of the mountain rangers' favorite marching songs, *"Die Kaiserjäger."* It's an extremely hearty song combining love of wide-open spaces with fierce military pride. Since he mentions this march specifically it must have been special to him. The lyrics of the last stanza would probably have very much appealed to Rommel; here is an English translation:

dead tired and had already settled into their quarters. Now they were being hurled into battle on the snowy mountains north of Vidra.

At the assembly place, I briefed the companies about our new task using only a few words. Then the detachment moved north into the mountains. I galloped ahead with *Leutnant* Haußer, *Unteroffizier* Pfäffle and a dispatch rider. The untiring legs of our horses carried us quickly across vast, snowy mountain meadows into the danger zone. Since there were enough of our own troops in the second line there upon our arrival,

> "When we stand before the foe,
> Our chests swelling with mettle,
> All before us to shattered pieces will surely go,
> Amid our lust for battle.
> There is no shrinking back among us,
> We stand man-to-man,
> Strong as German oak trees,
> Which to break, nobody can.
> If one among us falls in battle,
> With a bullet burning through his breast,
> He dies a *Kaiserjäger*,
> Of the first Regiment!"
> Another one of Rommel's favorite marches was "*Ein Jäger Aus Kurpfalz.*"

the deployment of my detachment was no longer necessary. After a cold night around campfires in deep snow, we received an order from the battalion to return to Vidra. In the happiest of moods, the troops returned quickly to the good quarters there, in which mail from home awaited us.

The Württemberg Mountain Battalion stood at the disposal of the Supreme Army Command[8] and moved the next night—again marching past the enemy front near Gagesti—back to Odobesti. Over the next few days we marched through the fortress of Focsani, which in the meantime had fallen, and onward through Ramnicu Sarat into the Buzau area. Snowstorms delayed our transport for departure. Then, in biting cold in mostly unheated railway cars, we rolled westward again in a 10-day train journey. In the Vosges we served for seemingly endless weeks as an army reserve, then went forward into the position sector of Stoßweiher—Mönchberg—Reichsackerkopf.

A third of the battalion (two rifle companies and one machine gun company) became corps reserves in Winzenheim and were under my command. I was directed by *Major* Sproesser to bring up our standards of training in exercises and combat maneuvers to their previously high level.

I especially loved this duty. In the course of a few weeks, all companies of the battalion passed through my school. Nightly drills with subsequent practice, assault troop raids on dummy fortifications in the style of storm troop battalions, and fights of all types ensured that the troops were fresh and ready for battle.

In May 1917 I took over a subsector at Hilsenfirst. At the beginning of June, the French

---

8. The *Oberste Heeresleitung,* also known as the OHL.

Rommel, on horseback with two officers riding on either side of him, parades through a town to the sound of marching music played on drums and brass instruments by soldiers in the lead.

pummeled us for two days across a wide front. Within a few hours, positions that had taken a year to build up were leveled flat. But the enemy did not launch an infantry attack. Our repetitive barrages that were requested seemed to sap the enemy of their desire to attack.

The work to repair the demolished positions was not completely over when the battalion was called away on a new deployment. Energetically primed for action, our troops—who by that time had reached the highest peak of their training— bid farewell to the High Vosges. Yet again, the air of Winzenheim rang with the favorite song of the Württemberg Mountain Battalion: *"Die Kaiserjäger."*[9]

A close-up image of Rommel riding behind a drummer.

---

9. The *Kaiserjäger* march, also known as *"Mir sein die Kaiserjäger,"* is of Austrian origin. The lyrics were written by Max Depolo. It remains a beloved marching song among German-speaking mountain soldiers.

4

Battles in the Southeast Carpathians, August 1917

"Surprise Attack"—*March to the Carpathian Front*

A grimy-looking Rommel (third from left) is shown among other officers during the war.

Although the enemy's eastern frontline had been sent reeling by the outbreak of the Russian Revolution, it continued to tie down significant forces of the German Army in summer 1917. In order to free up these forces for the ultimate decisive action in the west, the enemy frontline in the east had to be brought to a collapse. To accomplish this, the southern flank of the Russo-Rumanian front would be attacked from a generally northern direction by the [German] 9th Army, positioned between the lower section of the Siret river and the edge of the mountains 30 kilometers northwest of Focsani, and also by the Gerok Group[1] positioned in coordination in the mountains on the left, and attacking in an easterly direction.

After an eight-day train journey in glowing summer heat from Colmar through Heilbronn, Nuremberg, Chemnitz, Breslau, Budapest, Arad, and Kronstadt[2], the troop transport I was in charge of (containing 1st, 2nd and 3rd Companies) became the second last to arrive from the battalion on Aug. 7, 1917 in Bretcu around noon. At the train station, I found out that Gerok Group was already supposed to begin attacking the hills on both sides of the Oituz valley in the morning on Aug. 8.

---

1. Gen. Friedrich von Gerok of Württemberg.
2. Present-day Brasov.

Angriff am Ojtoz-Paß

Rommel's map of the attack on the Oituz pass.

The three companies grabbed canned rations and were taken in trucks to join them in a three-hour drive without a supply train. We went through the Oituz pass towards Sosmezö (Poiana Sarata), which lay very close to what was then the Hungarian-Romanian border. Our food and fighting supplies were supposed to be sent along immediately and in close order to Sosmezö after they were unloaded.

In Sosmezö we met the personnel in the valley who had already marched sections of the battalion into the mountain north of the Oituz valley before noon. The telephone connection with the staff was broken down at the moment. A mess sergeant passed on the battalion order to us verbally: Detachment Rommel was to move up with the rest of the battalion through Harja, Hill 1020 to Hill 764 (Bolohan).

The whole valley was teeming with Austrians, Hungarians and Bavarians. Numerous batteries, including some of heavy caliber, stood positioned on either side of the valley road. Because I could only begin the march into the mountains after the arrival of our supply train, I had my detachment encamp in their tents in a very tight space.

Austrian sentries with fixed bayonets stood watch so that none of my infantrymen helped themselves to the potatoes in the local commanders' potato field. This was a legitimate measure to take in this time of extraordinary scarcity of food.

Dusk fell. The battalion music played on amid the campfires for another hour or so. We faced the coming days with a completely confident attitude. Scenes from our past winter campaign in Romania were still fresh in our minds. At about 2200 hours the fires were put out. The troops slept. It was necessary, because the coming days would surely make extreme demands on their capabilities.

Our night's rest only lasted a few hours. Our supply train arrived already around midnight. Shortly thereafter I ordered the men to wake up, break camp, distribute four days' worth of rations and get the companies in marching order. Because all vehicles would remain behind in Sosmezö, the companies and my detachment's staff took a few pack animals from the supply train to carry ammunition, rations and baggage. Then the detachment began the march through Harja. Soundlessly the columns swept forward in the bright, warm moonlit night. I wanted to leave behind the positions in the valley and the route ascending towards Hill 1020 before daybreak, since these had almost certainly been reconnoitered by the enemy. From Harja outwards, the path led for the most part through forests, and was steep and slippery. By daybreak the companies still had opportunity to test their strength by hauling along an Austrian gun battery, which would also be put to use in the fight.

During the course of the morning prior to noon, artillery on both sides did a lot of shooting. We already feared that the breakthrough of the 15th Bavarian Reserve Infantry Brigade—to which the Württemberg Mountain Battalion was attached—would come too late. Despite our very fleet-footed marching steps, we only managed to reach the wooded Hill 764 just before noon.

As my detachment rested, I reported our arrival via telephone to *Major* Sproesser and received orders to move ahead as a brigade reserve to Hill 672, where Sproesser's staff was also located. Upon my arrival, six companies and later also the three machine gun companies were placed under my command. During the course of the attack, it became known that the Bavarian Reserve Infantry Regiment 18 had taken the initial Romanian defense systems on Ungureana after a very tough brawl. In this instance the Romanians, despite all expectations to the contrary, were said to have hit back very bravely and defended every trench segment and every shelter with extraordinary, tooth-and-nail tenacity. No breakthrough in the enemy front had occurred.

My forces were already settled in for the night, had set up their tents, and were cooking when new orders came to move ahead with three rifle companies and one machine gun company until reaching an area close to the west of Ungureana (779). *Major* Sproesser would go in the lead while I would lead the four companies behind him.

The forest was pitch black with darkness. We tramped along man behind man upon a soggy, narrow footpath. On the nearest mountain crests ahead of us, flares shot high in the air, machine guns clattered time and again, and shells burst. Soon we arrived at our goal. I reported our arrival and received orders to settle down for the night with my companies in a hollow close to the north of the mule path.

I had only just assigned each individual company commander the areas and duties for their units, as the detachment remained standing in a long row on the narrow path, when shells struck to our right and left on the hillside. A Romanian surprise attack!

The darkness all around us was shaken by the bright flashes of bursting shells and the zipping and whirring of shell splinters as it rained dirt and stones. Pack animals tore themselves away and galloped off into the dark night, taking their loads with them. My riflemen lay pressed flat against the slope and let the fire pass over them until the enemy guns went silent after 10 minutes. Thankfully there were no casualties.

The companies moved quickly into their previously assigned places. After the stresses and strains of the day, we slept exceedingly well on the grassy ground, wrapped up in our coats and shelter halves, despite the onset of heavy rain.

"An Especially Valorous Deed": Attack on Trassenknie on Aug. 9, 1917

> ◊ At this point in his memoir, Rommel is very focused on operations and includes fewer personal anecdotes.
>
> ◊ As in other armies during World War I, the Germans sometimes made up their own names for geographical features. The *"Trassenknie"* to which Rommel referred to is a bend in a mountain trail.
>
> ◊ The young soldier displays the same spirit as he would as a general during World War II, but it would take many years for him to perfect his methods. Here, he attempts what his opponents in World War II would have recognized as a "signature Rommel" outflanking maneuver. It goes off somewhat clumsily. The canny Romanians keep withdrawing, thwarting the young commander's outflanking attempt before he has a chance to get it into full swing. Young Rommel veers forwards and sideways in an awkward dance as he tries to rope the enemy in. He chases the enemy to a point he hadn't planned on—which causes him some worry.
>
> ◊ Rommel uses his engineering skills to analyze enemy construction work. His powers of deduction in this arena are exceptional—he makes inferences about the enemy's defensive abilities based on their construction methods.
>
> ◊ He writes with particular emotion about two comrades who take extraordinary measures to save their sergeant. He was clearly deeply moved by this event and wished to record it for posterity, citing it as an "example" for his intended readership of future soldiers.

We were awakened very rudely before daybreak by a renewed surprise fire of the Romanian artillery. *Leutnant* Haußer, my adjutant, and I had cleared an area right above a small hollow and bivouacked there. Some shells burst in the hollow right next to the pack animals standing there. The animals tore themselves free and galloped right over us away into the darkness. Now shell after shell struck all around us in a circle. The rounds whisked past us by a hairsbreadth. As soon as the furious firing subsided a little, we decided to spring a few meters into the hollow, which offered more cover.

The enemy fire soon ceased again. This time a few of our people had been wounded by shell splinters. Senior physician Dr. Lenz took care of them. As day broke, I sought out the battalion command post and recovered from the terrors of the night with some warm coffee. At about 0500 hours, we received orders to advance across the southern slope of Ungureana at the same level as the Bavarian Reserve Infantry Regiment 18 and to proceed to carry out the attack.

Amid very heavy harassing fire, we crossed the western slope of Ungureana in approach trenches, springing from crater to crater, and were happy to reach the southwest slope of the mountain, which was heavily forested and hardly coming under any fire.

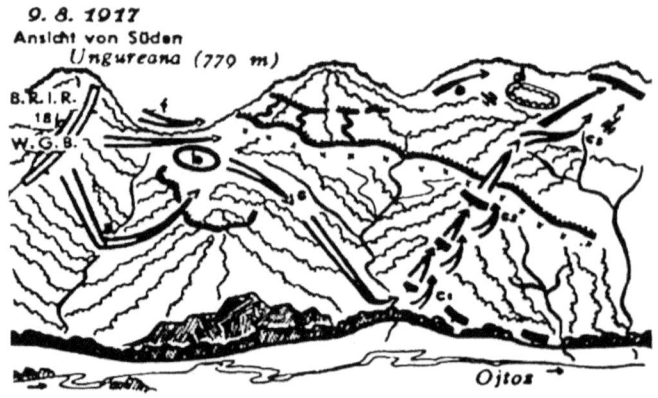

A view of Ungureana from the south showing Rommel's starting point, B) rest before noon; C) afternoon attack; D) position reached by evening; E) enemy counterattack and F) the joint attack by the Württembergers and the Bavarians.

There I was given the task of driving the enemy from the forested plateau 800 meters south of the Ungureana peak. Right afterwards I undertook to establish communication with the Bavarian Reserve Infantry Regiment 18, which since yesterday had been dug in barely 100 meters on the slope above us. Unfortunately I could not at this point establish where the Romanian position across from us was situated in this wooded terrain. No reconnaissance had yet been conducted in the direction of the plateau. Now for the first time I saw for myself the terrain that I was supposed to advance upon as I looked outward from the hill and scrupulously analyzed every detail of the map. We were separated from the plateau by a deep ravine. Just like the plateau itself, this was covered with a timber forest with thick underbrush.

In order to make a rapid determination as to whether and where the plateau was occupied by the enemy, I sent out an NCO with 10 men as a reconnaissance patrol and gave him a telephone troop to take with him. After only about 15 minutes, a report came back: "Strong fortified installations on the plateau have been evacuated by the enemy."

Then, at this point, I moved with both companies in columns following the line of telephone communication through the bushes into the abandoned enemy positions on the plateau and installed my men there in all-around defense position. I had to reckon with the fact that enemy forces could come from any direction intending to occupy this well-fortified position yet again. When I verbally delivered my report to *Major* Sproesser, hardly 30 minutes had passed from the time I had first been given orders to complete my task.

During the day before noon, our main action consisted of reconnoitering in the practically pathless forest towards the south (in the direction of the Oituz valley) and the east. As a result of this activity, two prisoners were brought in. At around midday Royal Hungarian Honved infantry[3] approaching from the west, relieved us on the plateau. Upon the battalion's command, Detachment Rommel—strengthened in the meantime by the 3rd Company—deployed north using similar security measures as before noon (reconnaissance patrol with telephone communication) through the timber forest to the mountain crest 400 meters southeast of Ungureana. My detachment nested there, as before, in all-around defense position, because no direct connection [to our comrades] was readily available to us on either right or left in the forest and I wanted to prevent unpleasant surprises. It was now known that the enemy was firmly established in very strong positions on the ridgeline about 800 meters east and northeast of Ungureana.

These enemy positions were attacked at about 1500 hours following a short artillery fire for effect and the enemy was driven back just beyond the bend of the "trail" 1.4 kilometers east of Ungureana. The Bavarian Reserve Infantry Regiment 18 was supposed to attack on the crestline and the Württemberg Mountain Battalion was supposed to attack south very close by. My detachment was also intended to participate in the attack in the frontline.

As the companies rested and cooked meals in the deep trenches nearby in the west, I sent out more reconnaissance patrols, each with a telephone, to approach the positions we were supposed to attack in the afternoon. *Vizefeldwebel* Pfeiffer and 10 men went forward as the southernmost reconnaissance patrol. He had to establish whether, where and how strong the enemy force was that occupied the crest running south from the bend in the road.

Based on analyzing the manner of construction of enemy fortifications located on the plateau 800 meters south of Ungureana's peak, I concluded that the enemy on the eastern slopes ahead could hardly have the wherewithal to construct a built-up position there. It seemed most likely to me that only the crest and valley positions were especially strong, while the slope positions were, by contrast, few and shoddily built. Here, by all outward appearance, was the weak point of the enemy defense—here, for a bold

---

3. This was a Hungarian unit in the army of Germany's principal Central Powers' ally, the Austro-Hungarian Empire.

troop, was the path to rapid and far-reaching success.

The reconnaissance troops committed north ultimately came across positions fortified with wire obstacles, while Pfeiffer in contrast reported capturing 75 Romanians with five machine guns about half an hour after setting out. But how was this possible? Until this point no shots had even been heard in that direction. Pfeiffer gave a concise verbal report via telephone: "The enemy was quickly taken by surprise while resting in a ravine without security 500 meters southeast of the division's campsite. We discovered them as we climbed down, approached silently with 10 carbines and then called out to the Romanians demanding that they give themselves up. Because the Romanians had set their weapons down in a place beside their resting spot, they were defenseless and, come rain or shine, had to allow themselves to be led away by 10 men."

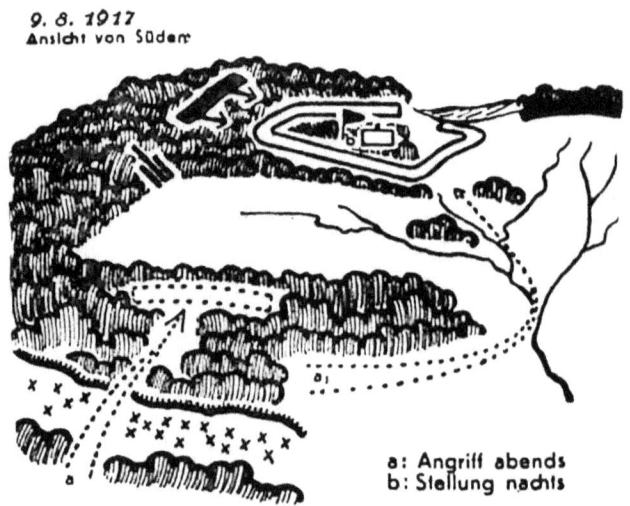

A view from the south shows A) Rommel's attack in the evening and B) position by nightfall.

I reported this success of Pfeiffer's reconnaissance patrol via telephone to *Major* Sproesser and suggested that during the impending attack that I myself break through the enemy position on the southern slope, which was probably poorly constructed, with an assault troop at the same time that the frontal attack against the crest was taking place. The combined efforts would result in a surprise thrust from the south against the ridgeline by the bend in the trail—and thereby in the rear of the strong enemy positions 800 meters east of Ungureana. Thus it would compel the enemy to give up his position systems between Ungureana and *"Trassenknie"* (the trail bend). *Major* Sproesser passed the suggestion on to the brigade. Shortly afterwards I received orders to carry out the suggested attack against the slope position with 2nd and 3rd Companies. Unfortunately I was not given any heavy machine guns to bring along. Soon the detachment was marching soundlessly in columns parallel to the telephone lines set up by Pfeiffer's scouting patrol, which now took over the point position of our force. Pfeiffer had no longer come across the enemy during further advances. We climbed along a steep cliff towards the valley through a forest of tall, leafy trees with many branches and thick undergrowth.[4] Against my will, I had to resign myself to following the path selected by Pfeiffer's patrol. He led us down and downward into the Oituz valley—we lost 350 meters of elevation.

I was only able to reel Pfeiffer in once we reached a forest 100 meters north of the Oituz valley road. He now received the order to *climb*[5] in a northeasterly direction towards *Trassenknie*. Accompanied by *Leutnant* Haußer and numerous couriers, I walked close behind the point of our column.

Soon a signal came from ahead that something was not in order. I went forward. Pfeiffer pointed out to me Romanian sentries in an illuminated section of the forest some 150 to 200 meters away from us. Behind them were other Romanians in position. The enemy seemed to be focusing his full attention on the open terrain on both sides of the valley road. We left him in peace and climbed up a narrow footpath on the rugged and densely forested western cliff that led to a crest in the direction of *Trassenknie*.

It was clear to me that we would inevitably come across further Romanian positions during this ascent. Thus I gave orders to the point men to immediately take cover and conceal themselves in the event of any run-in with the enemy, to secure the advance of the detachment, and only to shoot if the enemy attacked. The Romanians were supposed to believe that they had only come across a small reconnaissance patrol. Thus I wanted to gain time to concentrate my detachment and prepare them for the attack, which would hopefully come as a surprise.

At about 150 meters above the valley floor, the point men were fired upon from a position on the hill above and— in accordance with their orders—took full cover without shooting back. Quickly I arranged the detachment in ready position for the attack. On the right was 3rd Company while 2nd Company was on the left. Assembly and preparation were executed in total silence in a few minutes within the thick bushes on the wooded hill without disturbance from the enemy. For the attack, I ordered: "2nd Company, from astride the narrow footpath, attack the enemy on the hill above only for show—bind him down with fire, deceive him by throwing hand grenades and yelling battle cries towards the western side of the slope beyond the main direction of our attack and provoke the enemy to deploy his forces here. Do this from complete cover to avoid casualties. At the same time 3rd Company will launch an enveloping attack from the right. I am with 3rd Company."

However our preparations were not over before the fight began, because Romanian patrols probed down from the hill into the area where we were preparing. They were driven off. Now I ordered 2nd Company to immediately attack. The company encountered

---

4. Rommel strikes an ironic and slightly humorous tone in this anecdote about having to follow an overzealous patrolman leading the men along a wild "bushwhacking" route whilst forgetting to ascend to their goal.
5. Rommel's emphasis on the word "climb."

an occupied position 50 meters above on the hill. During the firefight which now began, in addition to the hand grenade fight developing on the west hill, I climbed a few hundred meters east through dense bushes with 3rd Company. We successfully reached the flank of the enemy, who was at about platoon strength and whose full attention was held by the frontal firefight.

As we closed in, the enemy speedily cleared out from his positions and withdrew uphill. I could not follow them up there with 3rd Company, because in doing so I would come into the line of fire of 2nd Company's frontal attack. So I held back again with 3rd Company on the right. The 2nd Company began pursuing the withdrawing enemy and, now facing strong resistance for the first time, followed through with the same conduct as during their first attack. So also did 3rd Company remain true to their given task. The retreating enemy could hardly sit down someplace before the bullets of 2nd Company were already flying around his ears, hand grenades were crackling and 3rd Company was rushing with utmost speed to a new flanking path on the right.

For mountain rangers loaded with heavy backpacks, the continuous running on this hillside, as steep as a rooftop, was a colossal feat. On top of that there was a burning August heat. Various people collapsed in faints due to overexertion.

Five times in this manner we chased the enemy, whose numbers continuously grew stronger, from his positions. At last only *Leutnant* Haußer and I, with 10 or 12 men, were close on the enemy's heels. With continuous firing, battle cries and by throwing hand grenades off to one side—so as not to endanger ourselves by throwing them ahead of us—we kept running after the Romanians, who were retreating close together in packs through the underbrush. In this manner we managed to drive the enemy back across a constructed and seemingly continuous position fortified with obstacles and to prevent them from taking a stand here.

After we broke through the position, the forest ahead of us diminished. Our path led ever onwards uphill, although it was not as steep. Presently we reached a forest clearing bordered on the right by slopes covered with long grass. About two enemy companies were going across these in a broad front in a northeasterly direction back to the mountain crest. Over on the right was a Romanian mountain battery, which was in the middle of changing its position towards the rear. They hurried as fast as they could to get to safety with their pack animals. We rapidly fired from within a thicket at the fleeing enemy. It was good they couldn't see how few we were. As the enemy vanished into the nearest part of the forest and into dips in the terrain, I gave *Leutnant* Haußer orders to press on after them with our few people.

As we left the edge of the forest, a Romanian mountain battery began showering my mountain rangers with shells from the left side of the northwest corner of the clearing from about 400 meters away. The shells rattled, hailed and splintered through the trees!

We concealed ourselves behind thick bushes. Shortly afterwards, men of 2nd and 3rd Companies began arriving, having wheezed their way breathlessly up the hill. I moved them to the right into a dip in the terrain that offered cover.

We were now about only about 800 meters away from the goal of our attack—the crest in the area of *"Trassenknie."* The enemy was collapsing before us. Now it was necessary to rapidly seize the moment without regard to the exhaustion of my troops.

Very fierce sounds of battle had been going on for a long time from over on our left near Ungureana. It seemed like the attack of the Bavarians and remaining elements of the Württemberg Mountain Battalion were underway there. As we advanced further towards the crest, fire from rifles and machine guns obstructed our climb. It seemed that already during the few minutes we took to catch our breath, the enemy commanders had taken the opportunity to regain control of their troops and create a front.

Because I didn't have my own machine guns within my two companies, my riflemen had to succeed with their strength alone. Through clever use of the tiniest furrows in the terrain, we succeeded in prowling nearer and nearer to the enemy on the crest, who was seemingly well-aware of the importance of his position. Wherever any of us was spotted by the enemy, fierce fire struck immediately. In this manner *Vizefeldwebel* Büttler got shot in the stomach close beside me while observing the enemy.

Twilight now benefited our advance. Shortly before darkness fell, Detachment Rommel took the crest just west of the Romanian crest position, which until this point had been difficult for us to accomplish. In a small saddle dip in the terrain which was 60 meters from the muzzles of Romanian machine guns but concealed from their fire, units of my riflemen nested themselves with fronts east and north. Meanwhile other units seized the oak forest just to the west and faced enemies to the north and west.

Yet the Romanians attempted to drive us down from the summit again through counterattacks. Only energetic carbine fire forced the attackers back into their jump-off positions. Since we had pushed in wedge formation across the crest road, our line of communication was restricted between Romanians in positions east and west of us. Our telephone connection to the battalion, which through tiresome efforts had been stretched along with us during the whole advance and the fighting, was now cut off. Therefore I had to report that we had reached our goal using countless light signals.

I organized my units soundlessly in the darkness. The detachment entrenched itself in all-around defense formation because we had to reckon with the possibility of enemy attacks from all sides. I kept a platoon at my disposal close beside my command post in the oak forest. Security detachments were sent forward to areas where the enemy was not in close proximity across from us. We had no connection to the battalion. By all outward appearances the frontal attack there in the afternoon had not had the expected results. Between *Trassenknie,* which we were about 500 meters east of, and Ungureana there was only persistent fighting. This meant we were one kilometer behind the enemy frontline.

Beneath the canvas of a tent I dictated the combat report to *Leutnant* Haußer by the glow of a flashlight. Light could not be allowed to be seen anywhere, otherwise our position would immediately draw Romanian fire.

During this time, two mountain rangers performed an especially valorous deed. The *Gefreite* Schuhmacher of 2nd Company and another of his comrades transported the severely wounded *Vizefeldwebel* Büttler in a shelter half down to the Oituz valley, which was far off and some 350 meters below our elevated position. They carried their sergeant all night long until they reached a doctor they found in Sosmezo, who operated on him immediately and thus saved him. Their action, carried out on a very dark night, across difficult terrain and a very long route (about 13 kilometers as the crow flies) was a most extraordinary accomplishment, and a magnificent example of the loyalty between comrades.

Before my combat report was finished, I was relieved of my acute worry about what would happen at daybreak on Aug. 10. One of our patrols sent west had established contact with units of the Bavarian Reserve Infantry Regiment 18. Together with the remaining elements of the Württemberg Mountain Battalion, the Bavarians had launched a frontal attack with artillery support in the afternoon. However, they had not been able to make much headway against the enemy, who were most tenaciously defensive of their positions. Then, amid the sounds of battle and later through the light signals relaying the results of Detachment Rommel's attack, it became clear to them who were enemies and friends. The Romanians, to prevent themselves from being cut off, had evacuated their positions between Ungureana and Trassenknie under cover of full darkness and withdrawn in a northeasterly direction towards the hills declining in the direction of the Slanic valley.

My combat report was sent before midnight per courier along the shortest path to the battalion at Ungureana. At the same time I ordered that a new telephone connection be laid from our position to the battalion. The night was cool, and I froze so intensely in my clothes, which were so permeated with sweat that they were completely wet, that at 0200 hours I preferred to get up and moving.

I went off with *Leutnant* Haußer to the front line and reconnoitered the enemy position lying between 60 to 90 meters across from us on a small forested knoll in the east (in what you might call a small oak forest).

Because I had forbidden unnecessary shooting in view of our difficult supply situation, the enemy certainly acted recklessly. Their sentries marched all around on their positions as if in supreme peace and moved against the brightening eastern horizon with distinct silhouettes. To gun them down would be easy as pie but I wanted to postpone this to a later time. When it became completely light, we recognized that the Romanians across from us in the east were in a broad front in nearly linked positions which ran from Petrei Hill across the small oak forest towards the north.

> **Considerations:**
> - The surprise fire of Romanian artillery in the night of Aug. 8–9 in the place where Detachment Rommel was lying in reserve caused some casualties. Had the troops dug in, these could easily have been avoided.
> - In the heavily wooded low mountain range on Aug. 9, **combat reconnaissance via scouting patrols, who had taken a telephone connection along with them**, especially proved its worth. I could call the patrols at any time during the advance, obtained reports in only a few minutes, could give new orders or withdraw part of the patrol—or I could, advancing parallel to the telephone line laid by the successful patrol, quickly occupy the terrain the patrol had reached with my troops. Courier paths, which in the mountains were so time-consuming to travel, were used only sparingly. However a perquisite was having a sufficient supply of telephone equipment.
> - **During the difficult attack in the forest up the steep hill**, the enemy in the position above was **deceived about the focal point of the attack** through fierce fire, battle cries and the throwing of hand grenades, and was **provoked into wrongly deploying his reserves.** The clash with 3rd Company in the flank and rear thus led speedily to success. In the same manner, five similar positions occupied by a total of two companies were taken one after the other. **The attacks unfolded so quickly one after another** that no time was left to the enemy to regroup his strength.
> - **Despite the enemy's superiority in numbers and weapons**—the Romanians had a more than plentiful amount of machine guns and mountain guns at their disposal—Detachment Rommel succeeded, by the use of the tiniest furrows in the terrain, **in taking the crest one kilometer behind the enemy frontline and to hold fast against enemy counterattacks.** This **forced the enemy to abandon his positions** across from the Bavarian Reserve Infantry Regiment 18 and the Württemberg Mountain Battalion.
> - **After its successful attack Detachment Rommel quickly entrenched itself in all-around defense position.** Had it not dug in, the detachment would have sustained heavy casualties from enemy surprise fire and counterattacks.
> - Our casualties: 2 dead, 5 severely wounded, 10 lightly wounded. —*Rommel*

"Exhausted From Blood Loss": **The Attack on Aug. 10, 1917**

> ◊ Young Rommel perfects techniques of concealment he would later use in North Africa. He also launches a masterful assault combining complexity and hard-hitting precision. Not content to watch from a distance, Rommel and his staff—incredibly—follow the storm troops into the thick of the action.
> ◊ Rommel is shot again, leading him to consider giving up command. Although he is severely hampered, he writes that the idea of giving up command "burdens" him.
> ◊ A deadly encounter with a French officer occurs. While the U.S. Army translation alludes to the French officer urging Romanians under his command to kill Germans, it does not render his sentence in full: "Knock the German dogs dead!" That translation also omits Rommel's ironic passage describing this French commander being hit by a bullet from "point-blank range" afterward. The passage implies it was a fatal shot. While Rommel often attributes "coup de grace" actions to individual troopers, he makes no mention of who shot this French officer.
> ◊ Someone who plays an important role in Rommel's life makes an appearance in this chapter—Hermann Aldinger. A fellow Swabian, Aldinger was a landscape architect by profession. He remained a close friend of Rommel and accompanied him on campaigns in France, North Africa, and was appointed to Rommel's staff in August 1944. He was present at Rommel's death—by his side even up to the last moment when Rommel was taken from his home and forced to commit suicide. Rommel's death hit Aldinger hard. He later recovered Rommel's service cap, which had been taken to Hitler's headquarters as a trophy, and returned it to Rommel's family.

On Aug. 10 at about 1800 hours a telephone connection was established with the battalion. I learned from the *Ordonnanzoffizier* that *Major* Sproesser had received my combat report and as of one hour ago was already on the march towards *"Trassenknie"* with all units of the battalion.

At about 0700 hours *Major* Sproesser arrived with the remaining companies of the Württemberg Mountain Battalion; he addressed Detachment Rommel to convey his highest appreciation for our resounding success which began on Aug. 9 and continued to have results. Then I oriented myself towards the situation facing my detachment from the east. There the Romanian sentries were behaving with total lack of caution even in the brightness of daytime. Indeed even Romanian units occupying the position could be seen sunning themselves close beside the positions they had dug in the nighttime between Petrei Hill and the small oak forest. No such thing with us. All sentries and occupying forces of Detachment Rommel were well camouflaged and had strict orders to never allow themselves to be seen anywhere and only to shoot in the event of an enemy attack.

The enemy positions stretched lengthwise along the bare western slopes of Petrei Hill (693) across the ridge rising towards the small oak forest; this ridge was only sparsely

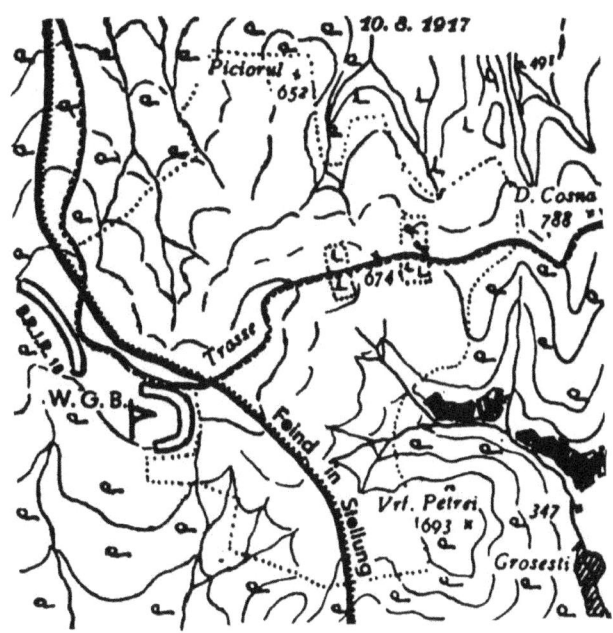

Rommel's map of the situation on Aug. 10, 1917.

covered by clumps of bushes. The small oak forest itself seemed to be heavily fortified. It commanded its surroundings to the south, west and north. North of this small oak forest, the enemy positions stretched through low-lying thickets towards the valley to the deep Slanic gorge. The positions consisted of single nests and larger points of strength, which could provide each other with flanking support on both sides and dominated the barren slopes in their foreground for far and wide.

According to brigade orders issued shortly after 0700 hours, the mountain battalion was supposed to reach the fork in the road 350 meters west of Hill 674 in an attack that day. The enemy was also to be pushed yet again from his positions. However we could count on no support from our own artillery in this endeavor, because these units were currently changing positions to come towards the front. *Major* Sproesser tasked me with preparing and carrying out the attack and placed under my command the 1st, 3rd, and 6th Mountain Companies, as well as the 2nd and 3rd Machine Gun Companies—a mighty fine fighting force.

My plan of attack was: to suddenly strike our unsuspecting enemy around noon with a powerful blow of machine-gun fire, and thus drive the enemy garrison from 400 meters south to 300 meters north of the oak forest and there keep them pinned down. At the same time we would break into the area around the oak forest, coil around segments of the enemy positions near the oak forest on the right and left and cut them off, and attack en masse with a thrust towards the crest that would break through to the east and reach Hill 674.

The preparations were exhausting and time-consuming. Throughout the course of the morning I personally brought forward 10 heavy machine guns over long detour paths in order for them not to be seen by the enemy. Some of these were arranged on the wooded crest right behind our frontline and others were put in the nooks and crannies of the southern slope in completely camouflaged positions. I personally showed every single gun its target and instructed the crews about how to fire before, during and after the attack. I set the time for opening fire at 1200 hours sharp. I designated the platoon positioned close to the trail as the base platoon.

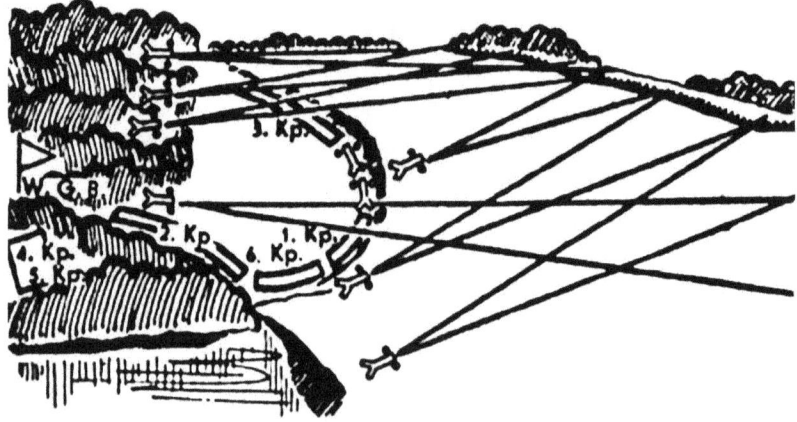

Rommel's map of his firing plan for the attack on Aug. 10, 1917.

At about 1100 hours the remaining elements of Detachment Rommel were ready in position. I had chosen the southern edge of the small oak forest as our break-in point. Now the hollows in the ground 80 meters southwest of the oak forest began filling silently with storm troops: 3rd, 1st and 6th Companies and one heavy machine gun platoon. I gave specific instructions to the troop carrying out the storm attack (3rd Company), for a feint attack over and above to the left (to be done by units of 3rd Company) and to the troops carrying out the main attack.

Mail arrived 10 minutes before the attack began. This was hurriedly distributed.

At exactly 1200 hours I gave the machine gun base platoon the agreed signal to open fire. Just a few seconds later, all 10 machine guns opened fire. The oak forest was covered with especially heavy fire. To divert our opponents and induce them into taking overly hasty measures, the machine gun in the platoon on the left of the 3rd Company bellowed battle yells with all their might at the same time they opened fire, and flung countless hand grenades ahead of the northwest corner of the oak forest. All of this was done from complete cover to prevent casualties from occurring on our side once the strong counterfire from the Romanians was aimed at this area.

Amid this ear-shattering din, the storm troop of 3rd Company launched their attack towards the southwest corner of the oak forest 100 meters south of the trail. They were somewhat camouflaged by the smoke and fumes of the hand grenades which blew sideways through the air. The heavy machine gun firing from behind had worked particularly strongly on this section of the enemy position. Now, in accord with my orders, this gun redirected its fire more to the right and left, and thus created a narrow path free of fire for the storm troop to pass through. The storm troop rushed forwards silently, firmly determined to do a complete job here. I followed the storm troop with my staff from close behind, and behind us came the remaining elements of 3rd Company with the heavy machine gun. Shots rang out and cracked from all sides.

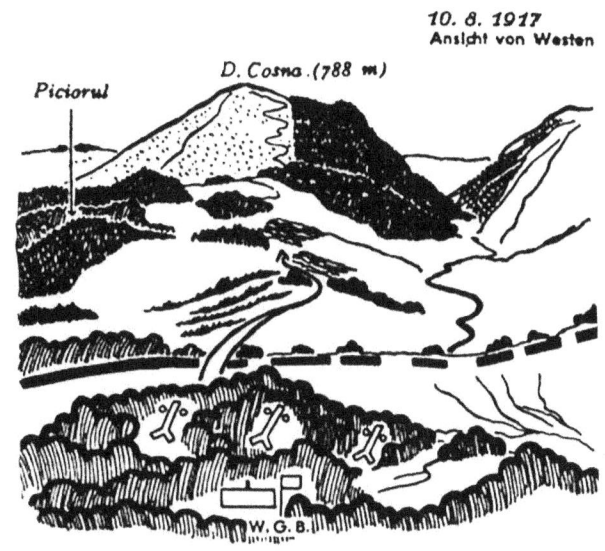

A sketch depicting Mount Cosna from the west on Aug. 10, 1917.

Only about two minutes had passed since we had first opened fire. Our 10 machine guns hammered incessantly, while wild battle sounds whirled near the trail on our left. Presently the storm troop broke into the oak forest. A fight at first ensued in the enemy trenches. But the mountain ranger troops made quick work of it. Since there was no path forward in the trenches they came out of cover to continue their storm attack forward. The machine gun platoons firing at the oak forest supported this attack exceptionally well because their firing had driven the enemy into cover; the enemy was only a few meters away and now shot at the storm troops from the left.

One of my orderlies took out a Romanian aiming at me from 15 meters away on the left with a shot to the head. No sooner had we come into possession of the enemy position in the oak forest than did a strong enemy counterattack sweep in towards us from the northeast. Our heavy machine gun was not yet in position and the guns located to our rear could not fix their sights on this new enemy due to the curve in the northeast hill. Soon the enemy was throwing hand grenades at us. In a hard, stressful battle with carbines and hand grenades—even my detachment staff had to use their weapons—we fiercely maintained our hold on the land we had taken against a strongly superior force. A few minutes later, one of our heavy machine guns entered the fight to our advantage. I could again turn my attention to the task of leading.

The break-in point in the small oak forest was secured by elements of the 3rd Company and the heavy machine gun platoon to the south and north. I charged all remaining troops (1st and 6th Companies, as well as elements of both machine gun companies which had been freed up by our breakthrough onto the crest) with breaking through onto the crest in the direction of Hill 674. As heavy machine gun units kept the enemy pinned down in his positions on both sides of the oak forest, and while other units filled gaps that had been breached in the enemy positions, the main force of Detachment Rommel stormed en masse forwards to the crest with complete disregard of the firing all around them. For these men, there was only one goal—the terrain near Hill 674. Echeloned in deep formation, with 1st Company at the point, all went forward as fast as possible.

Without encountering enemy resistance, the foremost units of 1st Company reached the small knoll 400 meters west of 674. Following closely behind them, I had just crossed a small hollow. A sheaf of enemy machine gun fire ripped into the hollow and forced me flat to the ground. The spray of fire chopped holes in the grassy turf all around me. The fire was coming from the slope 800 meters southeast of Hill 674, from about 1,200 meters distance. I had located reasonably good cover behind a small bulge in the ground and I wanted to spring forward just as soon as the machine gun fire stopped.

Then suddenly I took a shot from behind in my left forearm. Blood squirted. Looking all around me I discovered a detachment of Romanians firing from a cluster of bushes 80 meters behind me, shooting at me and some other people of the 1st Company. To remove myself from the line of fire of this dangerous enemy, I sprang in zigzags across the knoll ahead of me. The units of 1st Company I found there must have defended themselves from all sides for 10 minutes solid. Then the Romanians persisting from the west were dispatched in a close-quarter fight with the rest of our men following up from behind.

The French officer commanding the Romanian detachment screamed, "Knock the German dogs dead!"—until a bullet got him from a pointblank range.

Meanwhile a very turbulent fight had developed further to the rear. The Romanians had recovered from their initial fright and tried, through counterthrusts with local reserves, to tear the land we had gained out of our hands. Thanks to the unparalleled courage of all the mountain rangers and the forceful energy of their leader, things were ultimately decided in our favor in all locations. Soon, 1st and 6th Companies had taken the terrain in the area of 674 in further advances without encountering further serious resistance.

In the meantime I had Dr. Lenz bandage me up. Then I gave orders to the Detachment to occupy the conquered area in the following groupings and to organize themselves thus: "6th Company to strengthen Aldinger's heavy machine gun platoon on Hill 674. All remaining units to remain at my disposal in the wide valley just north of the trail, 350 meters west of 674."

Despite terrible pain and exhaustion from loss of blood, I did not give up command of the detachment. *Major* Sproesser was informed about the result of the attack via telephone. At this point a large column on the mountain path could be seen coming towards us from the direction of Mount Cosna[6]. We oriented ourselves towards defense, and the spade was used to its fullest capacity. I urgently requested artillery fire against the approaching enemy forces. However, the entirety of our own artillery was still not in position or ready to fire due to its changing of position. The enemy, unhindered, came closer still.

Hauptmann Gößler came forward with the remaining companies of the Württemberg Mountain Battalion. We divided ourselves into detached units. Detachment Rommel now consisted of the 5th and 6th Companies with Machine Gun Platoon Aldinger, now deployed in the frontline, along with 2nd and 3rd Companies and the 3rd Machine Gun Company. In the meantime 1st and 4th Company, as well as with the 1st Machine Gun Company, went over to Detachment Gößler. These units nested themselves 300 meters west of Hill 674, just south of the mountain path.

Contrary to our expectation, the Romanians marching in from the direction of Mount Cosna did not launch a counterattack against our new line at Hill 674. They only probed ahead with strong patrol troops. As we effortlessly repelled these patrols, the Romanians took possession of the heights 800 meters across from 5th and 6th Companies which sprawled on both sides of the mountain path in an expanse of two kilometers from north to south. Under these circumstances there was no justifiable reason for us to send more forces to the frontline. The front formed by the 5th and 6th Companies together had a breadth of about 600 meters. Their unexposed flanks were folded back. In conjunction with 6th Company, Detachment Gößler secured the southern slopes. The remaining companies of Detachment Rommel, in conjunction with 5th Company, secured the northern slopes. Each force, in the same manner, established outpost positions (outguards and outguard sentries), and the occupation of the conquered areas was organized in depth. At about 1500 hours the Romanians withdrew reserve forces from their line formed from the western slopes of Petrei Hill, through the small oak forest and to the western edge of Slanic. However it was impossible [for us] to establish contact with our neighbors on the right and left.

Romanian artillery fire now began with great ferocity. It ripped apart wire connections, hamstrung every courier movement and hacked up the terrain between the oak forest and 674 on both sides of the mountain road. The telephone connections to 5th and 6th Companies were continuously being restored to working order, which was a difficult and most dangerous task for the telephone troops. All throughout the whole afternoon, the firing persisted with undiminished violence. Most fortunately, the companies in the front line and the area of land in which the reserves had nested themselves were barely

---

6. Known today as Varful Cosna or VF Cosna.

affected. Not until late afternoon did the Austrian artillery enter the battle. Among other things, a 30.5-centimeter shell hit the peak of Mount Cosna right in the middle of a clustered bunch of people (into a group of Romanian and French officers, as it later turned out). The casualties of my detachment during the attack and the subsequent artillery bombardment were fortunately very few. At my command post on a bluff 350 meters west of Hill 674, I dictated, during the artillery barrage, the combat report for the attack from the small oak forest to Hill 674. Only just before evening did the enemy artillery fire cease. Pack animals brought food and ammunition.

Because I was completely exhausted from blood loss and was severely hindered in my movements due to my tightly bandaged arm and the overcoat hanging over me, I was burdened by the thought of giving up my command. However the difficult situation of my detachment demanded that I did not give up my post for the time being.

Additional troops were now placed under the command of *Major* Sproesser. His command post was in the oak forest two kilometers southwest of Hill 674. The reserve troops of Group Sproesser (elements of the Bavarian Reserve Infantry Regiment 18) were located there. Additionally the communications officers of various artillery groups had set up their observation points there.

Night fell.

Considerations:

- The **attack** of Detachment Rommel on Aug. 10, 1917 against the **dominant, fortified Romanian position** had to go forward **without the support of artillery or** *Minenwerfer* [mortars]. Only heavy machine guns were available to support the attack. By **forming a point of concentration* of machine gun fire** there, the storm troop of 3rd Company could break in and **pin down the additional enemy positions. After the storm assault**, the main attack had been so well-prepared that it led to success with noticeably few casualties.

- The Romanians on Aug. 10 did not make their previous day's **mistake** of **neglecting the hill position**. A **break-in** into the enemy position on the hill promised little success on Aug. 10, because the attack area was an open space and such a break-in could easily be blocked by machine gun fire from the hills all around. We needed to grapple with the enemy head-on right on the mountain ridge.

- Combat reconnaissance: Keen observation of the enemy territory brought excellent results during the night of Aug. 10 and the earliest morning hours. The foremost enemy outworks and the behavior of the occupants were zeroed in on with precision. Patrols were not sent out from our side in order not to disturb the enemy and thus alert them to our attack preparations. The enemy actually made a huge mistake in not keeping the area ahead of their position under surveillance—indeed they even behaved themselves in a totally unwarlike way (standing sentries, garrison outside of covered shelters). So our surprise invading attack struck them like a hammer blow.

- **The storm troop** of 3rd Company had **a way** paved for them through the oak forest by multiple heavy machine guns. These fired from positions in the timber forest some 150 to 200 meters west of the break-in point and gave combined fire at the enemy in the oak forest. Because their fire was spaced out on the right and left in such a manner, the storming troop of 3rd Company was not endangered by it. In the further course of the attack, these same heavy machine guns did an exceptionally good job of supporting the rolling up** of the enemy positions at the oak forest by laying their fire at the area close ahead of our storm troop.
- **The feint attack 100 meters left of the break-in point** was played out from complete cover using hand grenades and battle yelling. This was done to draw the counterfire of the enemy in the oak forest in a wrong direction and possibly prompt them to send in their reserves. This attack fully succeeded in achieving its purpose of helping the storm troop get forward, and it was done without any casualties. Probably the enemy had **reacted quickly** to our break-in to the oak forest and l**ed a counterattack from the northeast**, but the superior fighting abilities of our mountain rangers again prevailed in defense in this situation.
- **The Romanians had occupied the crest to the rear of the continuous positions with reserves,** but these were for the most part not ready for defense by the time of our successful surprise breakthrough, and they were overrun while in their shelters. Wherever they mounted defense or took steps towards counterblows, they were quickly overpowered by the superiority of the mountain rangers. This happened because five of our companies pushed one after the other through our breakthrough point and were quickly followed by Detachment Gößler with an additional four companies. Thus our **surprise attack** was carried off with the **necessary momentum**.
- **After reaching our attack goal,** we rapidly **turned to defense.** The companies were well-camouflaged as they dug into a frontline. The unsupported flanks in the south and north were secured by outguards from the reserve companies. **To send out reconnaissance patrols to further distances** was inadvisable. They would have easily been gunned down by [enemy] troops occupying the rear Romanian positions or

captured. Instead the enemy territory was scanned with the utmost attention to detail from various observation posts (within the Detachment and companies in the front line). Shortly after reaching our attack goal, the **crest between the oak forest and Hill 674 was again left vacant by our own troops.** These troops had nested themselves into furrows in the terrain off to the side. The very strong enemy artillery fire in the afternoon could hardly touch them here.

- Due to the attack on the crest by Detachment Rommel, **the enemy was compelled to vacate his breached position** in the afternoon and to withdraw his forces into a new rearward position.
- **The enemy leadership was barely mobile in mindset.** They limited themselves only to defense and did not decide on any counterattack unless many reserve forces and strong artillery were available for them and the terrain in the north and south were totally convenient for a counterattack. —*Rommel*

* Literally, *Schwerpunkt*.
** *Aufrollen*.

"The Situation Looked Desperate"—The Storming of Mount Cosna on Aug. 11, 1917

All remained quiet at the front. Not once did Romanian patrols bother the companies in the frontline. At about 2200 hours *Major* Sproesser disclosed to me that on the following day the brigade was commanded to attack Mount Cosna at 1100 hours with artillery support, and he demanded a suggestion from me for this.

Judging by the terrain, an attack from the west and northwest appeared potentially rewarding to me, because the highest parts of the mountain crest were not forested. Thus support from artillery and heavy machine guns for the infantry attack would be easy to set up. Furthermore, the numerous dips in the terrain north of the high road offered good possible approaches for the attacking

◊ Readers will notice that Rommel's inability to sleep has become a regular theme. Although soldiers can find it difficult to sleep in combat situations, infantrymen can sleep out of sheer tiredness. Nothing, however, seems to help Rommel get any sleep. His restlessness was so acute he attempts to stay awake and active all the time, with intermittent "crash" periods. This is something I address in greater detail at the end of the book.

◊ Rommel stresses not wanting to withdraw from an area due to unwillingness to leave the bodies of dead companions.

◊ He does not allow subordinates much room to improvise; he wants to be everywhere at once, and his plans troops. At this, *Major* Sproesser pleaded with me to remain for another day despite my wound and assume leadership of the attack groups from the west and northwest. I was to be given command of the 2nd, 3rd, 5th and 6th mountain companies, the 3rd Machine Gun Company, and the 1st Machine Gun Company of Reserve Infantry Regiment 11. At the same time, the southern attack group under the command of *Hauptmann* Gößler, consisting of the 1st and 4th mountain companies, 1st Machine Gun Company, and 2nd and 3rd Battalions of the Bavarian Reserve Infantry Regiment 18, would attack Cosna from the South, approaching from across Hill 374 to Hill 692. This new and gravely important task stirred my spirit. I stayed.

In the further course of the night, I could hardly sleep. My wound stung. My nerves were keyed up due to the stresses and strains of the last few days. Also my brain was busy with thoughts of my new task. Before daybreak I woke *Leutnant* Haußer.

> largely depend on his person. If Rommel had been killed or incapacitated, would his troops been able to carry on without him? Probably not very well.
>
> ◊ Troops are given little opportunity to rest; they are constantly exhausted. Rommel drives his men hard. His lack of care about his own wellness arguably has a negative impact on others; his merciless attitude about rest will have drastic consequences later.
>
> ◊ Young Rommel develops more of his "signature moves" here. Creeping past enemy forces, he deals destruction in deft outflanking movements. Like a grappling hook launched into enemy territory, Rommel penetrates enemy areas with a thin line of advance, embeds his troops, seizes territory ahead and shreds enemy forces with encircling backstrokes. He would use these techniques again in World War II.

We went together to 5th and 6th Companies in the frontline, reconnoitered in the early morning light the terrain we would attack, and firmed up our attack plan. The enemy was 800 meters east of our foremost position situated on the next mountain range on both sides of the mountain road. Enemy sentries stood hidden behind trees or in bushes. North of the road there seemed to be a dense line of infantry in a newly dug position. Elements of the occupying force stood together in groups. No shots from either side broke the stillness of the breaking day. Our position was well-camouflaged and was hardly recognizable from the enemy side.

Opportunities to approach the new enemy force were hardly as favorable as I had imagined. Ahead of our front and to the south, bare grass slopes offered zero cover from enemy fire. The terrain 600 to 800 meters north of the mountain road appeared more favorable. Numerous clusters of tall, thick bushes were strewn across the grassy hills of the ridge leading towards the Piciorul. The Piciorul (Hill 652), lying 1.5 kilometers north of the mountain road in the flank of 5th Company, had a tall broad-leaf forest.

The peak of Mount Cosna thrust itself into the rays of the rising sun on the horizon with stark majesty—the goal of our Aug. 11 attack. Would we accomplish it? We had to! I forgot my wounded arm. I had six companies to lead against the enemy. With self-assurance and new strength I approached this sobering task which demanded much responsibility.

Using the companies which until this point were stationed in the frontline, I wanted to pin the enemy down in his positions starting at 0800 hours, distract him and impede his view of the ravines northwest of his positions. During the course of the morning I wanted to creep forwards with the bulk of my detachment under cover through the bushes south of the Piciorul towards the enemy position north of the mountain road and get within storming distance. Then, with the promised artillery support, we would break into the enemy position at about 1100 hours and if possible push through in a column streaming towards Mount Cosna. At the same time the enemy elements at Hill 674 would have to be grappled with in a frontal attack.

I placed the 5th and 6th Companies with Aldinger's machine gun platoon under the command of *Leutnant* Jung, whom I relayed instructions to through *Leutnant* Haußer about his own scope of action and the tasks of his group during the attack against Mount Cosna. I left *Leutnant* Haußer with Detachment Jung in order to maintain contact with Group Sproesser and to ensure close cooperation with the artillery.

At 0600 hours I moved off with the remaining four companies in rows towards the north through complex networks of thick bushes. A telephone connection to Jung's battle group was laid down at the same time. After about 600 meters, I turned the point of my group east and then drew nearer to the enemy, approaching through a shallow dip in the terrain rising towards the ridge between 674 and Picorul, which was sparsely covered with trees and bushes. From time to time we halted and observed. Soon, using binoculars, I realized with great astonishment that this ridge was occupied in its entire length by enemy sentries. The Romanians had also established combat outposts ahead of their new positions. Neither my 5th Company, located to the left of these sentries, nor the reconnaissance patrols of my reserve companies had ascertained the existence of these enemies.

A surprise attack against the main Romanian position from the northwest looked under these circumstances to be near impossible. If I attacked the enemy outposts, the enemy in the main position east of Hill 674 would be alerted. My attack therefore would no longer be a surprise. And thus the prospects of success would be greatly diminished.

We remained halted there under cover from enemy sight. Observing the terrain all around led me to decide to play a trick[1] on the enemy outposts. Just as we had gone unseen by the enemy in our approach, so we now went unnoticed as we retreated some distance back along the same path. We turned north and reached the dense forest area on the northwestern slope of the Piciorul without running into the enemy. Then we turned east again and afterwards began approaching the Romanian outposts through the thick underbrush of the broad-leafed forest. From this point onward I echeloned my security in even greater depth. An especially skillful *Vizefeldwebel* of the 3rd Company crept ahead of us at a very far distance. I

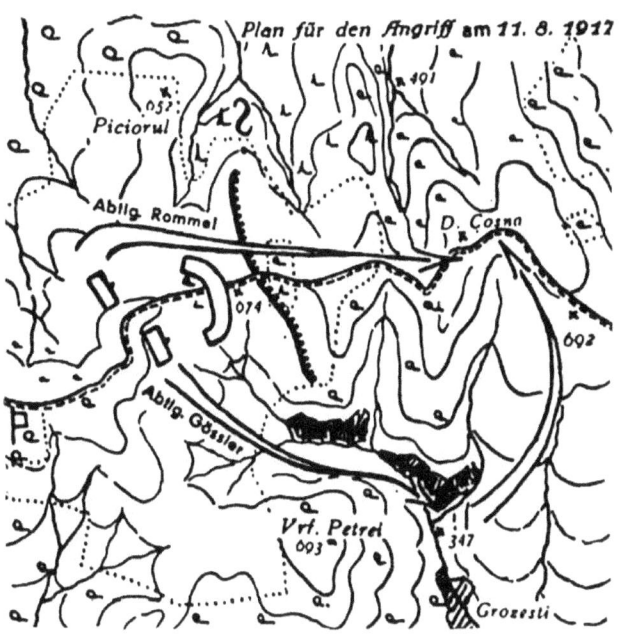

Plan for the attack on Aug. 11, 1917.

personally showed him the way through hand signals and by calling out in a faint voice. His platoon leader, *Leutnant* Hummel, had taken this man's heavy rucksack onto his back at my request. I marched a few meters behind the *Vizefeldwebel.* Next behind me came the remaining 10 people of our column point with 10 steps distance between each man. About 150 meters further behind them followed the four companies in rows. This spacing was arranged so that I, halting the point men with a hand signal, could not hear the slightest noise of the marching rows still approaching at that moment. Obviously my entire detachment—now spread in length to about 800 meters—was being totally silent. Each individual infantryman was strictly ordered to be careful not to make even the tiniest noises. The troops knew that everything now depended on approaching the enemy outposts without being noticed.

At my hand signals, we halted and then advanced again. By listening intently for several long minutes we managed to determine the precise locations of two Romanian outposts. The enemy sentries chatted, cleared their throats, coughed and whistled. Meter by meter, we came nearer. The enemy outposts stood apart by a distance of between 100 and 150 meters. Not the slightest thing could be seen of them through the thick underbrush. I moved with the point men into the middle of the space between the two enemy sentries. Now we were on a level plane with them. We held our breath. The enemies on our right and left let nothing disturb their chatting. Cautiously, I brought four companies through

---

1. Literally, "*ein Schnippchen zu schlagen.*" An irreverent and somewhat humorous expression.

the gap in slow, trickling movements. At the same time, we established a telephone connection to Jung's battle group, who was in turn connected with Sproesser's group. The enemy nearby noticed nothing.

Moving beyond the Romanian sentries and outposts whose front faced west we reached the northeast slope of the Piciorul, the thick network of bushes constantly coiling around us. Over to our right, rifle and machine gun fire of Jung's battle group was launched according to plan.

A very deep ravine still separated us from the main Romanian position which we wanted to stealthily approach. Climbing down through it, we crossed over multiple paths but thankfully did not run into any Romanians anywhere. Over to the right near Hill 674, Romanian artillery laid heavy fire on the position of Jung's battle group. The Romanians seemingly assumed that an attack would come from that area and wanted to prevent it.

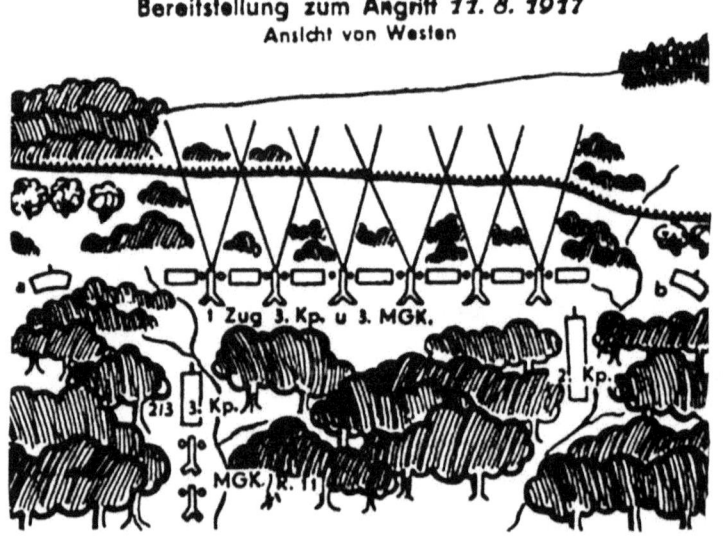

The arrangement of Rommel's troops for the attack on Aug. 11, 1917, shown from the west.

Climbing with our heavy rucksacks in the glowing August heat was monstrously strenuous. The heavy machine gun troopers carried loads of nearly 50 kilograms on their backs. It was nearly 1100 hours by the time we reached the lowest point of the ravine. Then we ascended a craggy rock slope into a light forest of fir trees on the other side. We could only move forward very slowly. The terrain created immense difficulties. Our own artillery began a fire for effect punctually at 1100 hours. It only came in very sparsely ahead of us and did not hit in the area nearby where we wanted to attack. Additionally, the fire from the 5th and 6th Companies grew stronger. The enemy replied with artillery.

Amid these sounds of battle, we climbed up the hill using every ounce of our strength. My wounded arm, rendered useless from being shot, hindered me a great deal as I climbed. My battle orderly had to help me climb in difficult spots.

Our own fire for effect began again at about 1130 hours. At the same time, the *Vizefeldwebel* going out ahead of us in the light forest was shot at and—as instructed —rapidly withdrew into the nearest concealment without shooting back. I halted the point men and secured the ascent of the companies. Then the companies marched soundlessly into a relatively tight space near a covering slope 50 meters below the point men.

During this time I was able to convey via telephone my intention to attack in half an hour to Jung's battle group. During my attempt to also report to *Major* Sproesser and to secure artillery support for the attack, the telephone connection was torn up. Probably the Romanian detachments constantly moving towards Piciorul had discovered the phone line and cut through it. The fact that our connection to Sproesser's group, the artillery and Jung's battle group broke down now in this precise moment right before our decisive attack was completely unsettling. It seemed hardly possible to get the telephone back in order, which would cost us hours of work. I had to make do with this misfortune.

We could only assume where the exact location was of the enemy position that I now wanted to attack. I thought it was in the area where Romanian sentries had shot at our advance scout. The upward bend of the hill, its carpeting of bushes and tall ferns would enable us to prepare under full cover for an attack at storming distance. It would not be possible to support our attack with machine gun fire from a higher position. Also Jung's battle group could not reach our front with their fire. There was absolutely no connection to them now. They would hopefully act on the directions I gave them according to instructions.

I took only one platoon of the 3rd Company and the whole Machine Gun Company Grau into the frontline with a width of about 100 meters. In the second line, the 2nd Company came behind the right flank and the remaining two platoons of 3rd Company with the 1st Machine Gun Company of Reserve Infantry Regiment 11 behind the left flank.

I ordered the following for the attack: At my signal, the frontline would creep silently forward through the ferns towards the enemy position up on the hill. As soon as enemy sentries or the troops occupying the position opened fire, Machine Gun Company Grau would open up on the enemy position with continuous fire from all guns and stop suddenly after about 30 seconds at a hand signal given by me. In this moment, the platoon of 3rd Company and remaining units of the detachment close behind them would break into the enemy position without yelling. Individual troops close the gap immediately from the sides. The main body breaks into the enemy's zone of resistance and, proceeding ahead in a southeasterly direction, accomplishes our first goal of taking the crest. To deceive the enemy about the location of the break-in and to lead him to scatter his defensive fire, he should be kept busy by troops throwing hand grenades from both sides of the break-in point from the first moment we open fire.

All of these preparations and conversations took place noiselessly barely 100 meters away from the enemy outpost. Because I had left *Leutnant* Haußer behind with the 5th and 6th Companies, I had to oversee all the preparations alone.

We were ready to attack barely a few minutes before 1200 hours. The Romanians had done us the favor of not disturbing us. Presently Romanian detachments in platoon strength crossed the eastern slope of the Piciorul across our path of approach. This was the peak moment for the attack. A signal of my hand unleashed it.

The detachment crept uphill. Just after a few steps, the enemy's first shot rang out from an immediately close distance. Seconds later, all the machine guns of Company Grau levied continuous fire. Hand grenades cracked right

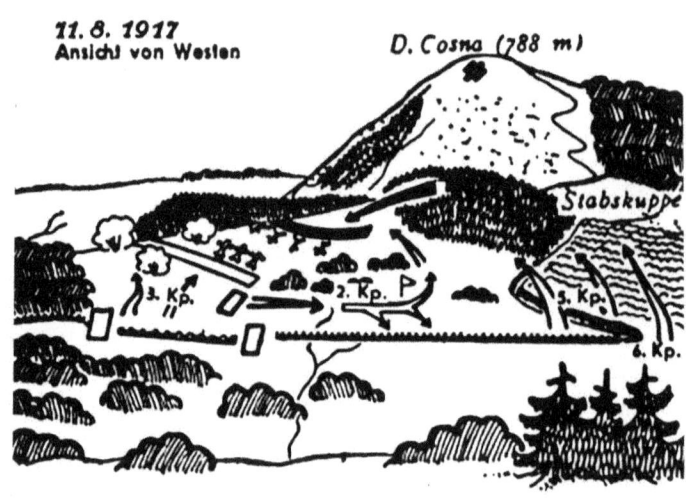

A view of Mount Cosna from the west on Aug. 11, 1917.

and left. Our own detachment lay ready to spring. The thick machine gun fire quickly forced the occupants of the enemy position into cover. Right and left, the enemy was shooting into the blue yonder. I gave the signal to halt the heavy machine guns. The mountain rangers stormed over and away uphill past the machine gunners, broke into the enemy position without sustaining any casualties, took a few dozen prisoners, sealed off the position and stormed into the enemy zone of resistance ahead on the right. Everything happened so punctually and fast that it was just like an exercise maneuver.

Soon the bushes ahead of us dwindled. As soon as we had gotten 100 meters forward, fierce machine gun fire hemmed our approach towards the slope rising to the right. The fire was coming from the mixed forest located on the highest summit, from which we were separated by a grassy meadow about 500 meters in breadth.

The fire intensified. The platoon of 3rd Company and the heavy machine guns of Company Grau took up the firefight. The remainder of 3rd Company and Machine Gun Company of the 11th Reserves spread out to the left. The enemy on the edge of the forest steadily gained more strength. Dozens of machine guns soon stood in the fight against us. A forward approach across the coverless meadow ahead was now totally unthinkable. Soon we would be struggling to hold our ground.

Enemy reserves broke out of the forest in a counterthrust. At the same time, Romanian artillery hurled shell after shell among our ranks—above all on the left flank. Frantically the mountain rangers clung to the ground. They did not want to give ground. Their rapid fire brought the enemy counterattack to a standstill. Enemy machine guns hammered with gathering intensity into our ranks.

Our own casualties quickly piled up, increasing at a shocking and horrifying rate. The situation became more perilous with each passing second. I was lying in the front line on the right flank of the 3rd Company. Next to me on the left, Albrecht's heavy machine gun platoon was locked in a heavy firefight. On my right, 2nd Company hung back covered from enemy fire in the bushes, still at my disposal. Should I also deploy them into the frontline? Could their firepower decide the battle in our favor? No! Should I then give orders to clear out? No! In that case we would have to leave our dead and wounded behind to the enemy. Then the enemy would again push out right out of their positions we had occupied, press us back into the deep ravine and then, easily, exterminate us there. The situation looked desperate, but we had to master it or else...we would remain in our places. The presence of groups of bushes down the slope on the right and the possibility to advance towards the enemy on the highest summit from the cover offered there brought me to a decision. I would deploy my last reserves in a surprise blow against the western flank of the enemy who was afflicting us so terribly. Maybe this had a chance to decide the fight in our favor. I informed those in my immediate surroundings. Then I crawled backward. A few seconds later I was with the 2nd Company, charging furiously forward towards the south. Everything depended on this. We overran a sparse force of enemies in the bushes before they even knew what happened to them. Soon we had put several hundred meters behind us. I had everyone turn east. Hopefully the remaining forces of my detachment were still holding up.

Just as I was about to lead the charge into the enemy flank, units from Jung's battle group popped up on the right behind 2nd Company. This action was according to the instructions I had given him early in the morning to attack the enemy on both sides of the mountain path. Within a few minutes, the battle was decided in our favor. The enemy had committed his full strength against 3rd Company and both machine gun companies. Therefore he could not push back the blow of three mountain ranger companies in his flanks and rear. Fleeing Romanians cleared out from the hill, leaving most of their machine guns behind on the battlefield. The exemplarily brave *Leutnant* Jung, the honored leader of his company, was on the eastern edge of the forest, 600 meters of Hill 674, when he received a deadly shot to the stomach.

The 3rd and 2nd Companies, and units of the Machine Gun Companies fired in pursuit of the enemy who was fleeing helter-skelter in complete disorder alongside the road and across the wide vale. In the meantime, just south of the mountain road, I pushed across the highest summit of the crest in pursuit of the enemy with the 5th and 6th Companies. The remaining units of Detachment Rommel received orders via courier to follow the same path as soon as possible. As the 6th Company took possession of the knoll 800 meters west of Mount Cosna's peak—known to us mostly as "the staff's knoll"—the 5th Company took more then 200 prisoners and bagged countless machine guns west and south of the mountain road. A wide ravine still separated our force from Mount Cosna.

The Romanians traveled in small herds retreating over the mountain path across the western slope. Then 6th Company's fire caught them. On the peak of Mount Cosna stood a densely packed mass of Romanian troops. It was not long before we got fierce machine gun and rifle fire from there.

Then, among others, my superb *Leutnant* Haußer received a shot through the breast.

Soon, completely exhausted, the companies arrived at "the staff's knoll" one after the other. It was no wonder since they had been marching, climbing difficult terrain and attacking nonstop since 0600 hours. The enemy, apparently sitting in well-prepared positions on the steep summit of Mount Cosna, could not be gotten the best of with such tired troops. Thus I decided to let my troops rest first, organize their groups and then take steps to attack the position on the peak of Mount Cosna.

As the 2nd Company secured itself while resting, an observation patrol of 6th Company, equipped with a telephone, reconnoitered opportunities for an approach under cover towards the Mount Cosna position. In the northeast, looking outward from "the staff's knoll," one had a view of Targu Ocna in the valley below. The distance as the crow flies was only about 4.5 kilometers. The train station at Targu Ocna was active and busy with train traffic. At about 1300 hours, the staff of Group Sproesser arrived with the group's reserves (2nd and 3rd Battalions of the Reserve Infantry Regiment) just west of "the staff's knoll." *Major* Sproesser had observed the attack of Detachment Rommel from his combat headquarters in the small oak forest and believed we had taken Mount Cosna in one swing.

Nothing was known of the actions or condition of Detachment Gößler. I reported my intentions to launch the attack against the enemy's position on the peak in an hour and asked for fire support from the machine guns of one of the two Bavarian battalions, which was to be shot outward from the "staff's knoll." I wanted to execute the attack in a similar manner to the one we launched before noon. *Major* Sprosser agreed.

At an arranged time, units of the 2nd Battalion of the Bavarian Reserve Infantry Regiment 18 opened fire against the enemy positions. At the same time I climbed east through a ravine several 100 meters north of "the staff's knoll," accompanied by the 6th, 3rd, and 5th Companies, the 3rd Machine Gun Company and the 1st Machine Gun Company of Reserve Infantry Regiment 11. Going through thick bushes parallel to the telephone line set up by 6th Company's patrol, we passed under the extraordinarily steep slope. Soon we were ascending again on the other side and reeled in the 6th Company's patrol. The ascent in the noonday heat put us under extreme strain. It took me multiple hours to ascend towards the peak with my exhausted men.

Taking similar security measures as before noon, we probed forward, making our way through lightly strewn bushes and shallow gullies, getting closer and closer to the enemy. The enemy garrison on the peak were in the meantime occupied in a lively firefight with the reserves on "the staff's knoll." The sheaves of fire from both parties zipped high above and beyond us.

Soon it became apparent that a Romanian outpost was lying about 200 meters across from the Bavarians on "the staff's knoll." Finally, using a small dip in the terrain, we closed to about 70 meters distance from the enemy on the peak. In order not to endanger us, the Bavarians had now ceased their firing at the enemy units above us. The enemy's fire also fell silent.

With the utmost caution I prepared my detachment for the attack. Two rifle platoon and six heavy machine guns would come into the frontline this time. Two companies came up behind the right and left flanks. The attack was prepared exactly as we had done in the morning: creep up close, continuous fire from heavy machine guns, hand grenades thrown from the right and left as a distraction, and then, break-in! Our preparations were not quite complete before we heard the distinct sound of rifle fire from a southeasterly direction. That had to be units from Detachment Gößler. Now I gave the signal to begin the attack immediately. After a short period of full-automatic fire, the mountain rangers broke into the peak position and swept the slope of Mount Cosna clean of the enemy in only a few minutes. The enemy was so surprised that no serious fighting developed anywhere. We stormed the peak with very few casualties of our own. We took several dozens of prisoners and some machine guns out of the well-fortified positions.

The majority of the position's occupants escaped, however. They ran head over heels down the east slope of Mount Cosna. Just as we were about to chase them, strong Romanian machine gun fire hit at us from the barren east slope. This came from positions located 500 to 600 meters east of Cosna's peak on the north-south ridge across Hill 692. These positions were exceptionally strongly fortified and furnished with wide obstacles. It proved impossible to bypass the sharp slope of Mount Cosna and descend the eastern slope during the day without strong fire support from artillery and heavy machine guns. We had to content ourselves with occupying the peak of Mount Cosna. The Romanian landscape was visible from far and wide looking out from the summit.

Soon we got contacted by the 1st Company of Detachment Gößler, who were climbing up the steep slope from the south towards the peak. Detachment Rommel dug in. The 1st Company, which I placed under my command, was located on the sharp ridge south of the mountain path; the 5th and 6th Companies were on the peak itself and on the ridges sited north of the mountain path which sloped northwest. I divided the Machine Gun Company of Reserve Infantry Regiment 11 into three companies on the front line. I kept the 2nd Company at my disposal behind the center, and the 3rd Company and 3rd Machine Gun Company behind the left flank.

About an hour after the storming of Mount Cosna, *Major* Sproesser arrived at the front with both Bavarian battalions. We now had information on Detachment Gößler.

After seizing Romanian positions on Hill 347, they ran into a very strong enemy force, and this force had attacked from the east in dense masses with the support of numerous enemy artillery batteries. Detachment Gößler had to retreat under very heavy losses and now was supposed to hold a position on the eastern slope of the rocky ravine leading towards the Cosna summit from the south. Our neighbors of the 70th Royal Hungarian Infantry Division were hanging on many kilometers away to our left opposite the Slanic valley. No communication existed with them.

In the evening hours, we watched the artillery battle north of the Slanic valley from our position on the peak of Mount Cosna, and we observed attack maneuvers of Romanian infantry in the area of Hill 722. I organized everything for the night. Among other things, I ordered for reconnaissance patrols to make contact with Detachment Gößler. I visited the individual companies in their positions to give them instructions about their tasks.

Now, however, I myself was so completely exhausted that I could no longer even compose my combat report for Group Sproesser. I dictated the report detailing the course of the battle to my new adjutant, *Leutnant* Schuster. Despite my exhaustion I could not get much rest during the night. An hour before midnight, numerous hand grenades burst into the 6th Company's position. The yells of attacking men, accompanied by the sounds of rifles and machine gun fire, resounded. Without waiting to be briefed I went with 3rd Company in a counterattack in the direction of the endangered position.

When we arrived at the scene of battle, only 6th Company was there and was clearly master of the situation. What had happened? Romanian storm troops had fallen on the company in a surprise attack but were repelled by vigilant troopers. However, some machine gunners of the Reserve Infantry Regiment's Machine Gun Company had fallen into enemy hands during the attack.

Considerations:

- The **attack plan** for Aug. 11 *developed on the basis of personal reconnaissance* in the earliest morning hours. The **normal mode of attack on either side of the mountain road**, supported by heavy machine guns and artillery, was decided against in view of the open terrain. The attack would have been recognized quite early by the enemy and probably repelled under heavy losses on our side.
- Again the Romanians had learned from the battles of previous days and had set up **combat outposts to secure their main position. By keenly observing the battlefield, even while advancing,** we recognized these just in time.
- Only the fact that I had troops with me that were accustomed to the **most rigorous battle discipline** allowed me to make the decision to probe the enemy outposts during

the daytime.
- It was extremely **difficult to calculate the time** that outflanking marches through the mountains required by this task would take. In addition to the difficulties of the terrain, there were still unexpected enemies to be found.
- **Working together with the artillery groups** during the attack came to nothing, because our wire connection was torn up at the decisive moment. The artillery would have been in a position to provide good support for Detachment Rommel's difficult attack.
- The **extremely difficult situation** after we achieved our breakthrough was **overcome with the reserve companies** we still had available. By striking **blows in the flanks and rear of the overpowering enemy force,** we turned the page quickly to our advantage. In this task, the **"map for the attack"** which had been **given early** to Detachment Jung proved to be extraordinarily valuable, because we had also lost communication with Jung at that point.
- The **fleeing Romanians were not only shot at from behind, but also units of Detachment Rommel were deployed immediately to overtake them in pursuit** towards the crest. However enemy forces in dominant positions towards the rear succeeded in bringing this pursuit to a standstill.
- As the **exhausted storm troops rested, a reconnaissance patrol ascertained means of approaching** the position on the peak of Mount Cosna via routes that provided cover. The telephone connection again proved to be properly usable. **The breakthrough into the enemy position at midday as well as the breakthrough into the peak position in the evening were successful without the support of artillery or heavy machine guns from rearward positions. Only the machine guns located in the front line of the storm troops smashed in the break-in point with their fire.** Again the fire of enemy troops occupying the position was diverted sideways by hand grenade troops. The casualties we sustained during the break-ins themselves were **exceedingly minimal**.
- The Romanians occupying rearward positions, both during the midday break-in and the capture of Mount Cosna, **took in their fleeing troops and prevented us from pursuing them.** —*Rommel*

"Better To Be A Hammer Than An Anvil"—The Combat Activity on Aug. 12, 1917

> ◊ Young commander Rommel doubles as a forward artillery observer (FAO). It's one of the many instances when he showed a willingness to "wear many hats" in battle. He demonstrated unusual flexibility and readiness to take on almost any task.

Shortly after midnight, a full moon rose. The reconnaissance patrols sent to Detachment Gößler brought back reports. Detachment Gößler was lying with its left flank some 800 meters southeast of the peak of Mount Cosna. They had sustained heavy losses and urgently needed support. The enemy was lying across from them, 500 meters ahead of their present front in a very strong position.

At about 0100 hours I reconnoitered the terrain ahead of the right half of our position accompanied by some officers from my detachment. I wanted to close the gap between Detachment Gößler and my right flank with a company before daybreak, and also to push my own position to within storming distance of the enemy positions east of Mount Cosna. However *Major* Sproesser was not in agreement. He commanded both Bavarian battalions to break through the enemy positions northeast of Mount Cosna by daybreak. Units of the mountain ranger battalion under my command would follow the Bavarians in the second line and push onward towards Nicoresti after the breakthrough.

Before the skies had become gray with dawn, shells struck from a northwesterly direction—thus from the rear to our left and hit a heavy battery. They came from the heights beyond the Slanic valley. They had little explosive effect, but they hewed trenches in the soft clay soil that were from six to seven meters in diameter and nearly three meters deep. Dirt clods fell within a 100-meter radius of them. Of course the very idea of sleeping was unthinkable. If the impacts came within a risky distance the endangered zone needed to be cleared. Soon the fire grew in intensity. Other batteries from the east and north took aim at Mount Cosna. The area surrounding the mountain peak became extraordinarily uncomfortable. Shortly before daybreak two Hungarian Honved battalions, which had also been placed under the command of *Major* Sproesser, reached Mount Cosna. One of the two attacked straightaway as they marched in without waiting for orders, heading clear over Detachment Rommel's positions and onward to the Romanian position east of Mount Cosna. It caused them to suffer heavy casualties. Afterwards the enemies' artillery groups increased their fire.

I was very happy as I led my detachment—now consisting of the 5th, 3rd and 2nd Companies, the 3rd Machine Gun Company, plus Hungarian Honved rifle company and Honved machine gun company—out of the endangered zone. Ahead of us, both Bavarian battalions had started off early on their way to carry out their task of breaking through the Romanian position northeast of Mount Cosna by daybreak. Once the breakthrough was achieved, the way across the plain would become free. Then, the Romanian mountain front south and north of the Oituz valley would collapse in a short time.

Marching in long columns of two, we crossed the western slope of Mount Cosna some 600 meters below the peak, and were often endangered by Romanian shells of various different calibers that struck without rhyme or reason to our left and right. Being on the move in the fresh and cool morning got us in good spirits again.

After marching for about half an hour through light brush on the steep slope, we reached the ridge descending from Hill 788 towards Hill 491. Tall, branchy fir trees stood on the particularly steep northern slope, with much smaller patches of adjoined fir forests below on the left. Looking through the fir trees we could see, from a birds-eye view, the Romanian positions northeast of Mount Cosna which both Bavarian battalions were supposed to break through. Worryingly, we saw improved trenches ahead of the front with broad obstacles running through them. Numerous communications trenches led across the barren ridge into the forested area on the eastern slope. Between us and the enemy position lay a basin in the earth that broadened towards the northwest. The slopes of this basin were for the most part completely overgrown with low-level brush.

The enemy positions had still not been taken. We observed units of the Bavarian battalions in a hard fight in the broad basin 1,200 to 1,500 meters north of us with the occupants of the Romanian position.

Soon a troop of wounded men from Reserve Infantry Regiment 18 passed by us. We heard from them that things really looked bad in the area ahead. They told us the battalion in the front had suddenly stumbled across the enemy position and had sustained severe casualties through rifle and machine gun fire; there were about 300 wounded. A breakthrough into the enemy position had not been achieved.

After that I organized my detachment in single files and ordered them to rest. At the same time I reported to *Major* Sproesser via telephone—a phone line had been laid down as we had marched—about the situation north of Mount Cosna. From my point of view, the exceptionally well-fortified Romanian position northeast of Mount Cosna could now only be taken with very heavy artillery support after the Bavarians' attack had miscarried. I received a promise that this would happen before noon. Because we had no artillery observers in the forward area, I requested to take over directing the artillery groups' fire myself. The visuals offered by my current position were exceptionally suited for this.

We tested the opportunity to climb down into the basin unseen by the enemy.

Unfortunately there was no path that offered complete cover. The individual groups of fir trees were spread too sparsely.

At 1130 hours, I had the first battery open fire. At the same time, Detachment Rommel descended into the basin west of the Romanian position, keeping 20 steps distance between each man. I intended to break into the enemy position northeast of Mount Cosna's peak after a short fire for effect from the artillery. The process of setting up for and shooting the artillery took a very long time. Much to our happiness, the shots of an Austrian howitzer battery landed in the Romanian position after a great deal of trouble and effort—but then, at the same time, the entire artillery received orders to fire no more shots on August 12 due to a mandatory change of position and ammunition shortages.

In the meantime, Detachment Rommel had reached the southeast section of the basin despite fierce Romanian artillery fire—the enemy had not failed to notice the descent of 700 men. The men were now lying between groups of bushes 300 meters from the enemy obstacles and could not be seen by the enemy. One man was lightly wounded during the descent. I climbed down to join the detachment. A wire for communication was rolled out along with us. The situation looked to have dim prospects of success. An attack without artillery support was just unthinkable. The enemy was waiting for us, and their obstacles and positions were strong. Yet neither did it seem possible to withdraw from the basin—during the day, up a very steep slope fully observed by Romanians equipped with machine guns and heavy artillery. My troopers could easily spring down the slope, but to climb back up the steep surface would be slow going and my men would make good targets for Romanian shells and machine guns. If it occurred to the enemy to exterminate us by firing into our area of the basin with artillery and mortars, heavy casualties would be unavoidable.

Despite the unfortunate outlook of this situation, I set my mind with determination on a single decision—to attack the Romanian position, without artillery support. I knew for a fact that my troopers could do it. Better to be a hammer than an anvil![2]

Agile and quick patrols reconnoitered the enemy obstacles and the position lying behind them. So that we would run creeping below the range of the enemy artillery I expected would fire, I took the detachment some 200 meters towards the enemy, moving through the bushes. I arranged my men in gullies and grooves in the earth to be ready to attack. On the slope above us to the right, both machine gun companies reconnoitered opportunities to provide us with fire support from higher ground. The results of the reconnaissance did not seem altogether unfavorable. Also, the enemy had seemingly

---

2. Rommel uses the phrase, *"Hammer oder Amboß sein."* This comes from a poem called, *"Geh, gehorche meinen Winken,"* by famed German Romantic poet Johann von Goethe, sometimes called the German Shakespeare. Goethe is loved for his verses about wild woods, free love, and mystical idealism. This particular poem is beautiful for its harsh elegance; it is a German marvel of steely absolutes: "You must climb or sink, You must rule and win, Or serve and lose, Suffer or triumph, Be a hammer or an anvil."

not yet noticed the prospect of an imminent attack by Detachment Rommel.

Just as I was in the very act of ordering both machine gun companies to fire at the targets they had identified, an order from *Major* Sproesser came in via telephone: "The Russians are in the Slanic valley and have broken in to the north, and within just a few moments they will be ascending towards our rear. Detachment Rommel, along with both Bavarian battalions, get back to the ridge 800 meters west of Mount Cosna immediately! The group staff is also going over there. Detachment Rommel, transmit this order to Battalions I and III of the Bavarian Reserve Infantry Regiment 18 and cover their retreat!"

What a horrible situation! The most difficult thing in my view was how I was going to get my troops out of the basin in daylight and within view of the enemy. If the enemy noticed us moving back in retreat, they would immediately try to hit us with machine gun fire and artillery fire or to pursue us. It would be next to impossible for me to prevent heavy casualties. The Russians[3] were less of a worry to me. Yet I hoped to reach the lines on the ridge before they did. If I failed to do so, my men would have to rip through these positions again with a faster, heavier thrust.

I had both Hungarian Honved companies climb the northeast slope of Mount Cosna—which was now covered by shade—under the leadership of *Leutnant* Werner (of the Württemberg Mountain Battalion). They were to reach the peak of Mount Cosna. As for the remaining four companies, I searched for the best way to personally lead them through the cover of bushes first in the direction of Hill 491, then towards "the staff's knoll." Just ahead of Hill 491, Romanian machine gun fire still managed to catch us, which resulted in several light wounds among the men.

Arriving in the area of Hill 491, I ordered the 3rd Company to occupy the lower section of ridge from Hill 788 to Hill 491 and assigned them the task of joining up with both Bavarian battalions here. I passed on the command from Group Sproesser to these battalions through an officer. Unfortunately the telephone connection with [Sproesser's] Group had been lost by this time. By a lucky coincidence I overheard a conversation at Location 491, from which I gleaned that the Group now viewed the overall situation in a much better light following the latest reports than they had half an hour ago.

Following this, I dispatched the 2nd Company on the shortest path to the ridge that descended northwards from "the staff's knoll." They were supposed to establish a firm position on the ridge some 500 meters north of the knoll, secure themselves in the direction of the Slanic valley and make observations. I marched all remaining units, except for the 3rd Company, in the direction of the knoll. I myself remained with the 3rd

---

3. Rommel clearly states here that he is not afraid of Russians. Interestingly, the 1944 U.S. Army translation, made during World War II, changed "Russians" into "Romanians" here. This swap, probably made for political reasons, falsified the original sentence.

Company. During the course of the next few hours, both Bavarian battalions were able to disengage themselves from enemy forces.

As I monitored that these situations were proceeding well in order, I climbed with the 3rd Company towards Mount Cosna. The 1st and 6th Companies were still located on Cosna's peak, which had been under continuous fire throughout the whole day and reduced to a plane of shell holes. To strengthen these troops I left the 3rd Company behind on Mount Cosna, then presented myself to the Group staff on the knoll.

First I stated my report, then I asked to be released into the hospital, because I was utterly exhausted and no longer felt that I was in a condition to lead. The bandages wrapped around my left arm had never been changed since the time I had been wounded. I was told that my papers for release were to be prepared the next morning. I laid down command of my companies and just rested in the area near the Group staff's location. The pitch-black night was warm with summer heat.

"Maelstrom"—*On the Defensive, Aug. 13–18, 1917*

◊ Something extraordinary happens: Rommel's commanding officer is at a loss for ideas and asks Rommel for advice. Young Rommel has become a de facto leader sought out by superiors. This demonstrates his natural leadership abilities. Junior officer Rommel then commands a considerably large force beyond what someone in his position ordinarily would have been tasked with—he is placed in charge of a scope of operations that exceeds his rank. Does it go to Rommel's head?

◊ Rommel's instructions for the men to pretend to have idle conversations to distract the enemy was omitted from the English translation.

◊ He maintains rigid control over his troops and places great emphasis on communication. He develops battles according to a general "game plan" but, ever the improviser, keeps forces ready in reserve nearby to allow him to "change the game." He keeps up that approach for life.

◊ When Rommel finds soldiers lagging, he personally harangues them to bring them forward, often harshly—or *"nicht freundlich,"* as he states bluntly here. Rommel would also do this in World War II.

◊ He places great emphasis on not wanting to leave behind the wounded and marks the loss of another comrade.

◊ Rommel displays South German pride in this passage when he highlights how "the Bavarians and Württembergers" held fast.

◊ While Rommel tends to avoid writing anything that sounds like a pointless complaint, he frequently alludes to his tiredness and his pained physical condition. One gains the impression he was truly facing the limits of his endurance here.

One hour before midnight, I was summoned urgently to *Major* Sproesser. I encountered a crowd of officers at his command post. *Major* Sproesser informed me that the situation looked exceptionally bad. After reports from shattered units of the 70th Honved Division (3rd Eskadron of the Austro-Hungarian Imperial and Royal Uhlans, 1st Eskadron of the Austro-Hungarian Dragoons and 1st Hungarian Honved Company), strong forces of Russians and Romanians had broken through into the Slanic valley and the area north of the 70th Honved Division—and, in the blink of an eye, these enemies had headed south and were attempting to climb the ridgeline between Mount Cosna and Ungureana.

It was now understood that Group Sproesser was already totally cut off, because no additional troops of ours existed between our rear and Ungureana, and we had to make a plan with this in mind. I was ordered to give my view on the situation.

I believed that, in this pitch-dark night, an advance of strong Romanian or Russian forces against our lines between Mount Cosna and Ungureana was extremely unlikely. We would only have to deal with that at daybreak—so, in four hours. There were five battalions at the Group's disposal, and I considered that it was not only possible to firmly hold the line with these forces but, under these circumstances, absolutely necessary to do so. Under no circumstances would I allow this territory, which we had conquered with so much skill, backbone and blood, to be relinquished without a fight merely due to alarming reports.

I suggested that the following regrouping be conducted as quickly as possible: "The mountain ranger battalion will take over the defense of Mount Cosna, the "staff's knoll," and the ridgeline until Hill 674. The remaining battalions of the group will take and hold the crest between 674 and Ungureana. All units, push reconnaissance and security elements ahead towards the Slanic valley."

For the insertion of the mountain battalion, I suggested: "Combat outposts—one of them a guard platoon strengthened with machine guns—to occupy the southern section of Mount Cosna. The field of shell craters at the summit will be left unoccupied. Reconnaissance towards the southeast and east. One platoon and one heavy machine gun platoon will occupy "the staff's knoll" with the task of preventing the enemy from occupying the bare peak of Mount Cosna. Both ridges descending northward between Mount Cosna and Hill 674 will be occupied by one company each. Reconnaissance and security towards the north. All remaining companies will be held in reserve at the disposal of the commander, nearby and to the southwest of the knoll."

Major Sproesser accepted my suggestions and advised that, since I had conquered the terrain in an attack, to also take charge of defending the sector held by the Württemberg Mountain Battalion. I could not refuse because of the seriousness of the situation, my worries about my mountain rangers and last but not least the thrill of taking on the difficult task. Group orders were issued immediately and resulted in the suggested regrouping. I had the following troops at my disposal for the defense of the Mount Cosna sector: the 1st, 2nd, 3rd, 5th and 6th Companies, the 3rd Machine Gun Company of the Württemberg Mountain Battalion and the 3rd Company of the Reserve Infantry Regiment 11 with six light machine guns.

Major Sproesser's staff returned to the oak forest 1500 meters northeast of Ungureana (at Trassenknie). I spoke with the company leaders in detail about the situation and the tasks faced by the Württemberg Mountain Battalion.

a: Gefechtsvorposten.
b: Stellung des Gegners.

Rommel's map shows German combat outposts on the high ground on the upper right and the enemy position to the center left, immediately below the incline of the hill.

Afterwards I gave the following orders: "The 3rd Company will move immediately from Mount Cosna towards the "staff's knoll." From here outward, a platoon of 3rd Company, without their rucksacks, will relieve the 1st Company on Mount Cosna; they will be strengthened by the six light machine guns of the 3rd Company, Reserve Infantry Regiment 11. They will occupy the forested southern ridge and reconnoiter the enemy position east of Mount Cosna. In the event of an enemy attack, the platoon will hold its position for as long as humanly possible; they are only to retire in the presence of a strong enemy and if threatened with encirclement, and in this case would withdraw fighting to "the staff's knoll." Later the leader will receive directions about his tasks only from me, in person.

"A different platoon of 3rd Company, as well as Albrecht's heavy machine gun platoon, with dig in on "the staff's knoll" in such a manner that their fire will dominate the crater field on Mount Cosna and its western slope. Their task is to prevent the enemy from crossing the bare section of Mount Cosna during the day and threatening our combat outposts on our left flank.

"The 2nd Company will occupy the small knoll (later to be named "Russian knoll"[4] 600 meters north of "the staff's knoll," perform reconnaissance and implement security facing the Slanic valley, and maintain communication during the night with combat

---
4. *Russenkuppe.*

outposts on Mount Cosna via patrols. They will also deceive the enemy and divert his artillery fire by lighting enormous campfires on the northwest slope of Mount Cosna and chattering in idle conversations during the whole night.

"The 5th Company, strengthened by the heavy machine gun platoon, will occupy the knoll 800 meters northeast of Hill 674 and set themselves to setting up defense on all sides there. They will reconnoiter, implement security towards the Slanic valley and maintain contact with the 2nd Company and neighboring troops positioned in the area of Hill 674, such as at Piciorul. To deceive the enemy and divert his artillery fire, the company will light huge campfires in the basin 800 meters northwest of "the staff's knoll" and will chatter throughout the whole night.

"A platoon of 3rd Company, Aldinger's machine gun platoon, the 1st and 6th Companies of the Württemberg Mountain Battalion and the 3rd Company of Reserve Infantry Regiment 11 will move to a place to remain at my disposal; this will be in the area between "the staff's knoll" and the descending slope 400 meters southwest of it. Security and reconnaissance in the direction of Grozesti. More detailed orders will follow when I arrive there personally.

"The detachment's combat command post will be 50 meters west of "the staff's knoll." A signal communication platoon will lay wire communications between the combat outposts, as well as to the 2nd and 5th Companies."

As the leaders repeated their orders to me, a bustling surge of movement began all around us. Bavarians and Honveds pulled back. Soon afterward movement stirred in the companies of the Württemberg Mountain Battalion. There could be no thought of sleeping tonight. Here and now, the details organizing individual locations and positions needed to be set out for everyone. After about three hours the companies stood in their places. The campfires on Mount Cosna and in the basin northwest of the knoll burned. Communication between the various different units was established. While the deployed units worked vigorously, the companies in reserve at my disposal rested. Reports from scouts during the night contained nothing concerning.

I had requisitioned to my staff *Leutnant* Schuster as my adjutant and *Leutnant* Werner as special-missions staff officer. Just before 0500 hours, a few artillery observers came forward. With them (the Hungarian *Oberleutnant* Zeidler was in this group, among others), I set out on the way to the combat outposts on Mount Cosna. The sun rose right over the horizon just as we arrived at the position of Platoon Allgäuer (of the 3rd Company) ahead. Platoon Allgäuer had, in accordance with orders, positioned itself on the steep, thin ridge leading south from Cosna's peak with their left flank on the edge of the forest 200 meters south of Hill 788.

The Romanian position lay misted with haze on a bare ridge about 100 meters below and 700 meters away from their front. You could see what appeared to be a large number of Romanian troops there judging from the glinting of their steel helmets. There were no shots to be heard for far and wide. Our men, unable to rest during the night, now slept in freshly dug foxholes. Only the sentries peered sharply at the enemy, observing them. The slope ahead of the platoon's position dropped sharply towards the east and was barely covered with scrub brush. On this thin ridge itself, as well as on the west side of the crest, were forests of tall trees with hardly any undergrowth.

Map of the action on Aug. 18, 1917.

As I discussed normal barrage fire and destruction fire with the artillery observers, various sentries reported: "The Romanians are abandoning their positions in skirmishing lines and are climbing towards the peak of Mount Cosna." Shortly after that, heavy Romanian machine gun fire was fired at the crest. At the same time, heavy shells, coming from a southeasterly direction, burst in the area of "the staff's knoll." I demanded, via telephone, destruction fire on the Romanian positions east of Cosna's peak, which an increasing number of enemy troops were climbing towards. At the same time, a report arrived stating: "Strong enemy forces are located very close ahead of the line of our combat outposts and are also climbing from the right of there towards the crest."

The cracking of countless hand grenades, as well as fierce carbine fire and machine gun fire all around confirmed the report. It came back to bite us that close-in security on the craggy eastern slope had been lacking.

Using the telephone, I commanded the reserve platoon of the 3rd Company and Aldinger's machine gun platoon to run ahead at double-time to strengthen the combat outposts. Then I requested normal barrage from the group. After that I went to go look around the front lines. Below on the right, the Romanians had already achieved a significant lead towards the crest and were shooting into the flank of our combat outposts. All attacks at the front had so far been repelled. Our artillery fire now struck amid the numerous reinforcements of Romanian troops on the bare slope. On the left, outside of the forest, our units located on "the staff's knoll" used fierce machine gun fire and carbine fire to prevent strong Romanian forces from crossing the peak and northwest slope of Mount Cosna and protect the left flank of the combat outpost.

I gave the order to *Vizefeldwebel* Allgäuer to hold the position [where we were] at all costs until the arrival of reinforcements. Then I personally ran back in order to bring these reinforcements forward as quickly as possible. Heavy shells struck "the staff's knoll" continuously. I met the two platoons there as they were just about to launch a breakout attack. I moved forward with them at double speed. Meanwhile the sounds of battle had become noticeably louder there since I had left. Hopefully Allgäuer was still holding on!

In the dip between "the staff's knoll" and Mount Cosna, we came across numerous light machine gun squads of the 3rd Company of Reserve Infantry Regiment 11, who were supposed to have been supporting Allgäuer's platoon. Apparently things had gotten too hot for them up front. I did not handle them in a friendly manner, and brought them up front again.

We ran straight into Allgäuer's entire platoon coming towards us east of the dip. Allgäuer reported that the Romanians had pushed through onto the slope in large masses, and additionally the strong enemy fire from the right had forced him and his men to give up their position. Because I was not reconciled to the idea of giving up Mount Cosna to the enemy at such an easy price, I deployed the forces available to me in a counterthrust. *Leutnant* Aldinger took two heavy machine guns into position in the forest on the right and opened fire on the ridge that had been previously occupied by Allgäuer's platoon.

At the same time, I climbed through dense clusters of bushes towards the ridgeline. As soon as we arrived up top, we dove in swinging and dusted our totally shocked enemies right off the hill and down towards the east. We also took the small hill below to our right back into our possession.

However, the Romanian is tenaciously tough and doesn't let go easily. We could clearly hear the verbal orders of the enemy commanders [shouting] from the domed hill below us. Soon, bitter hand grenade battles began in many places. The hill that the enemy was climbing up was so steep that our hand grenades would not burst at the distance of 30 to 40 meters where the approaching Romanians were getting ready to storm us from, but instead exploded much farther down the hill. If we wanted to shoot the enemies with our carbines, we would have to get close enough to expose our chests while aiming. At this close distance, things were exceptionally unfavorable for us. Our casualties piled up. Dr. Lenz had a great deal of work to do in our front line.

Our infantrymen fought with exemplary courage. Numerous wounded men kept fighting in the front line after they were bandaged up. Just as soon as Romanian storm troops seized some place on the ridge, another group of our mountain rangers would repel them back again in a counterstroke. This hard battle, so rife with casualties, lasted for many long hours. Ammunitions and hand grenades gradually became scarce.

The enemy artillery fire aimed at "the staff's knoll" increased. The telephone connection between the combat outpost position and the knoll was blown up. If I wanted to hold the combat outpost position longer, the moment was now critical to bring over additional forces, ammunition and hand grenades. In order to get this in motion quickly (without a telephone connection), I handed over command of the 3rd Company to *Leutnant* Stellrecht with instructions to hold out and went as fast as I could to "the staff's knoll."

There I was met with the following situation: The platoon of the 3rd Company and Albrecht's heavy machine gun platoon had nearly shot all of their ammunition at the enemy, who was menacing the left flank of our combat outposts in the crater field on Mount Cosna. The companies still remaining at my disposal (the 1st and 6th of the Württemberg Mountain Battalion, plus the 3rd of the Reserve Infantry Regiment) had occupied the southern slope of "the staff's knoll" on their own initiative, because patrols had reported a strong enemy force climbing towards there from the ravines of Grozesti.

Just as I was preparing to deploy units of these companies, a report came in stating that strong Romanian forces had been spotted advancing from the south as well as the north towards the saddle between "the staff's knoll" and Mount Cosna, and that our combat outposts on Mount Cosna had already cleared out in order to retreat to the knoll. Within the next few minutes—during which time I still had not even a single man at my disposal—the din of battle grew worrisomely nearer to the knoll.

Harried severely by the numerically superior and pugnacious enemy, the infantrymen of the 3rd Company fell back to the knoll. In the center of their group, they carried their dead and wounded comrades—including, among others, the dead *Leutnant* Hummel. None of these casualties should be allowed to fall into enemy hands!

Hand grenades and machine gun ammunition had run out in the front. Carbine ammunition had become scarce. Right and left, the enemy threatened to encircle us. The lack of ammunition and hand grenades made it extraordinarily difficult for our forces on the knoll to bring the onslaught of the Romanian masses to a standstill. Our heavy machine gunners had to defend their positions with pistols and hand grenades. The two couriers of my subordinate staff were deployed to endangered positions. Soon, a violent fight was burning across the entire front. At the same time, I discovered a large number of Romanians in the forested area of the cauldron land formation 600 meters northwest of the knoll. Via telephone, I described to the 2nd and 5th Companies the new danger threatening their flanks and rear.

A violent struggle was underway in all areas of our sector. It was impossible to even think of withdrawing our forces. What would happen if we totally ran out of ammunition on the knoll? Then that imposing hill would fall into enemy hands. Then all units of my detachment would fall under the most dire oppression. Then our entire line of defense would utterly collapse. But we were not about to permit that to happen!

A group photo of the 3rd Machine Gun Company of the Württemberg Mountain Battalion.

There was still a telephone connection to Sproesser's Group. I described to them the immense crisis which, as of this moment, we were still managing to hold off, and I implored them to send over additional forces, weapons for close quarters fighting and ammunition with utmost urgency and as soon as possible.

Then I survived a really hair-raising half hour. In the very last seconds, help arrived from the 12th and 11th Companies of the Bavarian Reserve Infantry Regiment 18 and a heavy machine gun platoon. I deployed the 12th Company with the heavy machine gun platoon on the knoll. I kept the 11th Company at my disposal on the slope 300 meters west of the knoll, to which I now also relocated my detachment's command post. Looking outward from this location, I had an extremely good view overlooking the whole battlefield.

I had the reserve company resupply the front lines with ammunition and hand grenades. When they weren't exchanging fire with enemies nearby, the troops were working furiously with their spades. This work became extremely trying on the knoll and the ridge west of it due to enemy machine gun fire aimed down from higher positions on Mount Cosna. I pulled Aldinger's heavy machine gun platoon from the front line and planted it in the zone of resistance in the area of my detachment command post. I also had depots set up for ammunition and weapons for close-quarter fighting. The supplies were arranged and organized.

The fighting around "the staff's knoll" and "the Russian knoll" lasted for hours without reprieve. The enemy constantly hurled new forces against our thin lines. Romanian artillery fire was consolidated on the slope just west of "the staff's knoll," tying down our movements to our front lines and shattering the telephone lines.

Yet the Bavarians and Württembergers at the front stood fast. During the course of the battle, our artillery shot a well-aimed barrage ahead of our threatened positions. Our shells thinned the rows of Romanians assembled en masse in assault jump-off positions.

Heavy extermination fire from multiple batteries was prepared to combat the strong enemy forces pressing constantly into the cauldron 800 meters northwest of "the staff's knoll." This fire was prepared in such a manner that it could be unleashed within only a few minutes upon request. Our teamwork with the artillery groups was good, but I still lacked artillery observers in the front line and a direct telephone communication with their command post.

By about midday, mountains of dead and wounded Romanians were piling up ahead of "the staff's knoll." However, among our own, the 12th Company of the 18th Reserve Infantry Regiment had suffered much, and their ranks had to be replenished with units from the 11th Company. Later, additional units from the 11th Company had to fill gaps in the ranks of the 2nd Mountain Company.

The defense of "the staff's knoll" and the "Russian knoll" was now oriented in such a manner that the front line was not densely occupied. By contrast, strong assault groups lay under cover close behind the terrain sectors most threatened by the enemy. These troops had the task of repelling any enemies that pressed through in immediate counterthrusts. This troop disposition proved itself especially useful in this terrain. After midday, the 10th Company of the 18th Infantry arrived as reinforcements. I had them make an approach trench from the "staff's knoll" to the detachment command post. Now the Romanians directed the point of concentration of their attack on the "Russian knoll." There, Hügel's platoon was nested in old Romanian positions and was set up to defend the area from all sides. The platoon was hard-pressed by countless masses of numerically superior enemy troops from the east and north. The enemy sought persistently through hand grenade battles to gain those positions, which had already been fortified through weeks-long work. The enemy's attempts to pounce on Hügel's platoon from the west were constantly foiled by Aldinger's heavy machine gun platoon shooting outward from the detachment command post. Thus the 2nd Company bravely held their positions.

The battle churned on nearly unceasingly throughout the late afternoon. I had to order for resupplies of ammunition and hand grenades to be taken up to the frontlines for the third time. Through the drifting plumes of smoke of our own shells (our defensive fire consisted of shells up to 30.5 cm caliber), we saw new Romanian troops climbing towards us up the slopes of Mount Cosna again and again. As the 2nd Company reported to me that their numbers were melting away to such an extent that they would have to give up the Russian knoll, I sent the remaining units of the 11th Company of the 18th Infantry to support them. At the same time, I ordered two heavy machine gun platoons to prepare destruction fire to be aimed at the Russian knoll. As this was happening, I commanded the 2nd Company to now quickly evacuate the knoll. We waited for the enemy forces to storm the bare knoll in a dense mass. In the same blink of an eye, the destruction fire of both machine gun platoons struck among the enemy and mowed them down like ripe wheat. The survivors speedily cleared out from the dangerous knoll. Shortly afterward the strengthened 2nd Company was back in possession of the knoll and now had peace.

Sometime later, the Romanian forces that we had already observed hours ago in the cauldron 800 meters northwest of "the staff's knoll" began moving uphill toward the south. Destruction fire from our artillery was now called for and drove our opponents back again into the low-lying forests. The rifle and machine gun fire of the 2nd, 12th, 10th and 5th Companies as well as from three heavy machine gun platoons which we had prepared to welcome this enemy was thus no longer needed.

During the fight, reports came rushing back from the front lines. My adjutant and special-missions staff officer bore an enormous degree of responsibility for speedily requesting artillery barrages, resupplying the troops with ammunition, close combat weapons and food, and for making situation reports to Group Sproesser. Telephone connections to the especially imperiled spots along the battlefront as well as to Sproesser's group were now laid in double lines and were kept in working order under the guard of tireless communications men. Their task was extremely dangerous due to the practically ceaseless searching and traversing fire of machine guns and artillery.

Despite extremely heavy casualties, the Romanians continued their attacks as night fell, yet without gaining even a footstep of ground. As the din of battle faded away in the night, we heard everywhere the wailing and cries of the wounded ahead of the front. However, our stretcher bearers were shot at while attempting to carry away some of the poor men, and had to turn back empty-handed.

I considered that the enemy was repeating his attack of Aug. 14 with the use of stronger artillery and fresh infantry forces. We could not allow ourselves to suffer such acute casualties as we had on Aug. 13. Therefore I spent the short hours of the night attending to strengthening the positions with meticulous detail and redirected the defense [approach] of the various positions. I instructed the company commanders and platoon leaders, the majority of whom had little experience in such battles, as to the main line of resistance in the terrain and ordered the types of construction of the defensive installations. The field of fire facing the front had to be opened up in various locations during the night. Furthermore, the disposition of the rifle and heavy machine gun nests had to be reviewed, since the enemy was in a position to seize them from the high elevation of their positions on Mount Cosna. I entrusted the extensive work at "the staff's knoll" to the 233rd Pioneer [Engineer] Company which was brought over and placed under my command shortly before darkness fell.

By just about midnight, all units across our broad sector had been briefed and were totally absorbed in hard work. Exhausted, I reached the detachment's command post. A warm bite to eat refreshed me again. I could not think of resting tonight for the time being. Wounded men had to be cared for. Ammunition and hand grenades still had to be restocked with the companies at the front line and in the depots before daybreak. Meals had to be arranged for individual companies. The communication patrol had to lay a double [phone] line connection to the artillery command. And then I still had to prepare the combat report for Aug. 13 for Group Sproesser.

As we finally finished all of this work at about 0400 hours, I tried to sleep. However it was so cold that I preferred to oversee the night's work with *Leutnant* Werner as dawn broke. My feet were swollen due to my continuous wearing of my boots— I had not been able to take them off for five days. Also I'd found no opportunity until now to change the bandage on my left arm, or to change my bloody uniform jacket which was now hanging from me, or likewise to change my trousers which were just as bloody. I felt completely sapped of all my strength. Yet I found the idea of retiring to the hospital unthinkable right now, only due to the weight of responsibility lying upon me.

On Aug. 14, a Honved infantry company with a light machine gun arrived and was placed at my disposal. I sent them to relieve the 1st and 3rd Companies and took both these companies into reserve just west of my command post. The 11th Company of the 18th Infantry was in position on "the staff's knoll" while the 12th had taken over the position on both sides of the mountain path. I left the 10th Company of the 18th Infantry in the position they had occupied during the night in the forest 300 meters west of the "Russian knoll." They had secured their position to the north and northwest towards the Slanic valley.

The fight could begin. We were armed and ready. During the entire morning, the Romanian artillery shot very lively fire at our positions near "the staff's knoll," mountain path and "Russian knoll"—however without doing much damage. All sectors now worked hard to improve the fortifications of their positions which they had deepened with trenches during the night. A strong Romanian attack across the whole front was repelled effortlessly at about noon. The 2nd Company on the "Russian knoll" suffered heavily under the fire of a Romanian battery situated in an open position about 1,500 meters away. Since we had no artillery observer in that sector—all requests to the artillery were made via telephone to the artillery headquarters in the oak forest—it was impossible for us to silence this battery. The enemy strengthened his positions on the western slope of Mount Cosna. The enemy wounded continued wailing in the area ahead of our front. Our own casualties on Aug. 14 were few. Also Aug. 15 ran its course with relative quiet. During these days I had two artists duplicate and create gridlines on a map of Mount Cosna and surrounding terrain with a 1:5,000 centimeter scale that I myself had drawn. Copies were given to Group Sproesser, artillery commanders and artillery observers. Additional copies of these sketches were made by the artillery so that every battery had a sketch.

Artillery located in forest and mountain areas unable to see geographical landmarks around them indicated on a [regular] map could easily direct fire on points requested by the infantry with the aid of a grid map or sketch. For example, I could transmit to the artillery: "Requesting normal barrage in Quadrants 65 and 66." If the requested fire fell outside of this area, it was enough to tell them: "The normal fire we requested in Quadrants 65 and 66 is actually falling in Quadrants 74 and 75," in order to bring the fire quickly into the desired area. Also, it made writing combat reports within our own units and to Sproesser's Group considerably easier. For example: "Romanian battery was recognized in 234a."

During the night from Aug. 15–16, the *Minenwerfer* mortar company under *Leutnant* Wöhler arrived. During the night reconnaissance of enemy positions continued. In the gray hours of early morning, the *Minenwerfer* launchers were assembled. *Hauptmann* Gößler came forward. He was supposed to relieve me, since I'd had no rest during the day or night for an entire week. Command remained in my hands.

In the afternoon, the 4th Company also arrived as reinforcements. The strength of my force was thus increased to 16 ½ companies—a greater number of troops than a regiment.

On our right, the Reserve Infantry Regiment 11 was in contact with us. On the left we were hanging as if in midair. The brigade was most anxious that a continuous front should also be established here. However, the troops we had available were not enough for this. The steep, forested cliffs of the Slanic valley guzzled up an untold number of troops for its defense. On the evening of Aug. 16, the atmosphere unburdened itself of a heavy lightning storm following oppressive humidity. The thunder clapped with booming echoes through the mountain peaks. The rain rattled down from the low-hanging clouds. Close to the west of the detachment's command post, covered Romanian positions provided the detachment's reserves and the staff with shelter from this maelstrom. However, these positions soon got full of water and had to be evacuated.

Absolutely soaked to the skin, we lay out in the open. Lightning flashed all around us. Suddenly, bursting shells of all calibers became louder than the rolling of the thunder. Ahead at the front, very heavy rifle and machine gun fire was leveled at us. Hand grenades burst! There was no doubt the Romanians were hoping to surprise us during the storm. Would the front hold? Or had it already been broken through? The rain lashed against our faces so hard that we could only see a couple meters ahead. Should I wait to receive a report? No—I needed to act! Within a few minutes I was lying with the 6th Company with fixed bayonets just west of "the staff's knoll," the fighting's main hotspot, ready to launch a counterattack. Our normal barrage utterly raked and churned the terrain into which the Romanian masses were storming. A combat telephone line connected me with my staff and thus with all locations within our sector.

The Romanian attack failed in all locations. The tumult of battle in the pouring rain finally came to an end that night. After suffering a huge forfeit in dead and wounded, the enemy cleared out from the foreground of our positions.

Upon returning to my combat command post after the battle concluded, I found that the place where we used to pitch our tents had been completely ploughed up by heavy shells. Under these circumstances, I moved the command post 250 meters to the right. Still wearing our soaked clothes, we dried ourselves in the heat of a fire kindled by Romanian prisoners. We were in excellent spirits!

Considerations:

- The task given to the Württemberg Mountain Battalion on Aug. 13 **to defend** sections of **Mount Cosna** and the elevated terrain in its immediate west **was exceptionally difficult**. With both flanks exposed, the battalion not only had to reckon with **strong enemy attacks** in the front, but also **in both flanks.** The **exceedingly furrowed terrain**, overgrown with dense woods, on both sides of the bare crest facilitated the **enemy's approach** to within **storming distance** of us. Furthermore, the Romanian artillery was located in a half-circle formation around the battalion's position.
- Under these circumstances, a **deep distribution** of defense and **the retention of strong reserves were put in place.** Active **battle reconnaissance** to the south, east and north before daybreak was necessary in order **to determine where the enemy intended to attack.** Additionally, the blind areas of foreground just ahead of our positions had to be continuously **watched very closely.** In areas where that did not take place, for example by the combat outposts, unpleasant surprises were unavoidable.
- The **battle of the combat outposts** turned out to be very difficult. Indeed they had a wide field of fire from the sharp spine of Mount Cosna, but they just could not reach the foreground right in front of their position, which was a domed bluff overgrown with scrub brush. Also they neglected to keep this terrain under close surveillance. The Romanians had already prepared since before daybreak to attack right at this spot. Their attack came as a complete surprise to the combat outposts.
- The **left flank of the combat outposts** was protected by machine gun fire and rifle fire from "the staff's knoll" aimed at the bare terrain around the peak and the very lightly forested western slope of Mount Cosna. Only when the ammunition ran out on "the

staff's knoll" was the enemy able to set foot on Mount Cosna.
- **It was possible, with quickly dispatched protective fire of a heavy machine gun platoon,** to recapture the lines that the combat outposts had already cleared out from without incurring any casualties on our side. Both fire and movement of the storming troop were used best here in **unison**.
- The battles of the combat outposts and the battle for "the staff's knoll" demonstrated just **how fast that ammunition and close quarter combat weapons are needed at the epicenter of a battle**, and how critical the situation can become because of this. Resupply must be initiated very early in such locations (especially in the mountains). To accomplish this, a battalion must have a reserve of ammunition and close quarter fighting weapons on hand. A battalion station responsible for the resupply must keep continuously informed about the latest developments about available ammunition in the front lines and must lead the resupply efforts there. The resupply during the course of the battle on Aug. 13 worked out well.
- In the difficult battle on Aug. 13, **reserves** were urgently needed. Without them the position would have been impossible to hold. Losses in the main battlefield had to constantly be replenished through reserves. Also, elements of the reserves were the ones who brought the resupply of ammunition and close combat weapons to the front line. A reserve company had to dig a communications trench from the battalion command post to "the staff's knoll"—the nucleus of the battle—during the battle. Without it, resupply would have incurred heavy casualties due to enemy machine gun fire from the high elevation of their Mount Cosna position.
- The Württemberg Mountain Battalion was already positioned **in depth** throughout the main battleground **at the beginning of the defensive battle.** The 5th Company, 2nd Company and the forces located on "the staff's knoll" **could support each other on all sides with fire.** During the battle, reserves added to the depth of forces at the battle's focal points ("the staff's knoll" and "Russian knoll"). It would have been a mistake to deploy them all into nests at the front line. The heaviest casualties were recorded there and these numbers became higher still when the garrison forces were increased. A *line* can be easily broken.
- The teamwork with the artillery was supremely satisfactory on Aug. 13. Still, an artillery liaison detachment with the battalion or an artillery observer sent forward into the battalion's sector could have reaped even more fruitful results. The grid sketches we produced during the days of defense really demonstrated their value. Today, they correspond to target-designation grids or coordinate scales. —*Rommel*

"Into Our Own Artillery Fire"—*The 2nd Storming of Mount Cosna on Aug. 19, 1917*

> ◊ Rommel's demonstration to Friedel with the binoculars was removed from the 1944 English translation.
> ◊ As the battles he describes got harder and heavier, so does Rommel's use of slang. Like most infantrymen he uses rough slang and colloquial expressions when describing aggressive actions, i.e. "a piece of cake to rip up the enemy position..." It seems Rommel's writing wasn't much different from how he spoke.
> ◊ Rommel enjoys chasing enemies and cannot resist a pursuit. Whenever he describes chasing opponents it is always with a strong sense of pride. He comes across as disparaging towards enemies who run away and very pleased with himself for running after them.
> ◊ He makes a somewhat reckless decision to lead his men through a German artillery barrage. Until this point, his actions are usually undertaken with some degree of forethought. This split-second choice to lead his men through shelling seems foolhardy—particularly since the Romanians had little chance to regroup anyway. Rommel's sleep deprivation may have played a part. It is a medical fact that excessive sleep deprivation interferes with cognitive abilities. When one considers Rommel was running on sheer adrenaline and had not slept for nearly a week, his apparent recklessness becomes more understandable.
> ◊ He pushes his energy beyond normal limits, as we have already seen, followed by a "crash period" when his body simply gives out on him. Rommel is used to getting away with it, but he has pushed himself too far this time. How hard his "crash" will be when it catches up to him will become apparent.

After hard battles, our neighbors on the left (70th Honved Division) managed during these days to come forward north of the Slanic valley. Plans were made to carry the attack onward in a broad front on both sides of the Oituz and Slanic valleys on Aug. 18th. Once again, Mount Cosna was supposed to be taken by storm, and the positions east of it were also to be taken on the same day. Our leadership hoped thus to achieve a breakthrough.

Group Madlung (Reserve Infantry Regiment 22) on the right and Group Sproesser (Württemberg Mountain Battalion and 1st Battalion, 18th Infantry) on the left were deployed against the massif of Mount Cosna. I was charged on Aug. 17 with preparing all units in the sector of Group Sproesser which would be deployed in the front line for their attack. In addition, I had to brief regimental and battalion commanders in Group Madlung's sector about the terrain they would be attacking. I was on my feet without pause from the earliest morning hours until the evening. As I returned to my command post, the Romanians attacked the Piciorul in the wake of a strong artillery barrage, coming from the Slanic valley—thus from the left and rear of our positions. I knew that units of the Bavarian Reserve Infantry Regiment 18 were in position over there. Judging from the noises of battle, the attack was gaining ground fast. The flanks and rear of my force seemed threatened and I had to fear that we would be cut off at the ridgeline from Sproesser's group. To prevent this from happening, I rushed at a running pace with a portion of my reserves (2 rifle companies and 1 machine gun company) into the area of 674 and placed them there in concealment in groups of bushes, under cover and ready to make a counterattack. Meanwhile a telephone connection was laid from my command post [to the group]. I learned from the group that the Bavarians at Piciorul had brought the attackers to a standstill. From that point on my reserves were not deployed.

Before daybreak, the attack troops of Group Sprosser prepared themselves for attack in the cauldron northwest of "the staff's knoll." A new arrangement of units came into play. I led the following companies which would be sent out into the very front line: the 1st, 4th, and 5th Companies, 2nd and 3rd Machine Gun Companies, an army storm troop and an engineer platoon. *Hauptmann* Gößler was to follow in the second line with the 2nd and 6th Companies and the 1st Machine Gun Company. Group Sproesser however retained direct control over the 1st Battalion of Infantry Regiment 18.

My detachment placed itself in clumps of bushes and patches of forest just west of "the Russian knoll," while the remaining units of Group Sproesser were ready further west. The enemy we were to attack had set up a continuous position on the ridge leading from Mount Cosna's peak northwest in the direction of Hill 491 and had put up obstacles in front of it. By sharp reconnaissance through binoculars, you could recognize parts of the positions and the obstacles visible between bushes.

According to the orders of the division, these positions were to be taken after an hour-long bombardment by our artillery. After an additional preparatory artillery bombardment of about an hour, the positions 800 meters east of Mount Cosna's peak, which was exceptionally strongly fortified, was also to be taken. We had been lying before this position on Aug. 12. I made up my mind to break into the enemy position on Mount Cosna during the artillery barrage, carve out a piece of this position [for my troops] and then direct our artillery fire at the second Romanian position and lead the attack against that one.

On Aug. 19 we again had absolutely gorgeous summer weather. In the early morning hours all forces in the Mount Cosna sector rested. The assault troops lay under the cover of bushes. At about 0600 hours, I sent *Vizefeldwebel* Friedel (of the 5th Company) with 10 men and a telephone troop after briefing them about my attack plan, giving them the following task: "Reconnaissance Patrol Friedel is to climb from 'the Russian knoll' into the ravine to the east, proceeding into the hollow over there (drawn on the terrain in a map) opposite the planned break-in point; they will go along a covered path using bushes and depressions in the earth for concealment. Once there they will reconnoiter the obstacle which is located in front of the enemy position. They will take wire cutters with them. The telephone troop must maintain continuous communication with the detachment's combat command post, even during their advance."

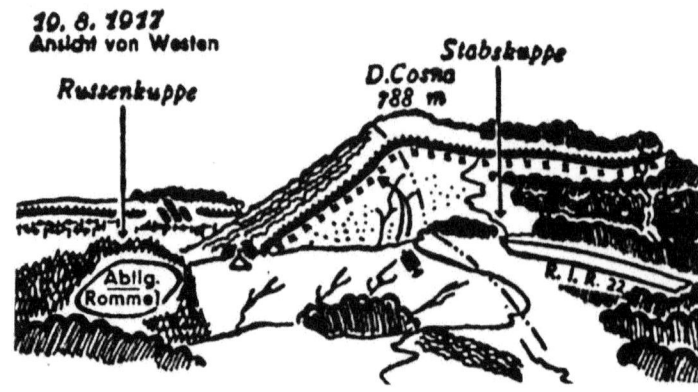

The "Russian knoll" is shown left and the "staff's knoll" to the right across from it, on a small bluff next to Mount Cosna. The sketch depicts the situation on Aug. 19, 1917 with a view from the west.

Magnifying the view of the lenses of my binoculars, I used them to show Friedel the point of our planned break-in and our anticipated way of approaching it.

Half an hour later, I saw Friedel's patrol climbing the western slope of Mount Cosna. In the meantime I had discovered the presence of Romanian trench sentries in the area of our planned breakthrough. The telephone connection to Friedel was working, so I could tell Friedel about everything that was going on in the enemy position above him. I could also tell him exactly how far he was from the enemy position at any given time and thus lead him to the planned break-in point in this manner. Friedel and his men arrived quickly at the enemy obstacle.

But just then the Romanian sentries in the trenches got restless—it seemed they had somehow noticed or overheard the patrol. I immediately withdrew Friedel 200 meters away from the wire obstacle and had *Leutnant* Wöhler's *Minenwerfer* company shoot at the break-in point from our rear.

Soon the shots were falling close behind the enemy sentries. Some of these sentries took full cover. Others cleared sideways out of the endangered position. As Wöhler's company now conducted harassing fire, I had Friedel's troop cut a narrow alley through the enemy wire only 50 meters from the bursts of our own shells. By all outward appearances they carried out their work briskly and without any disturbance.

At about 1100 hours, the artillery preparatory fire was supposed to begin. At about 0900 hours, I climbed with the detachment through the path that Friedel had blazed.

Telephone communication wire had already been set up robustly along the route. The hill leading from "the Russian knoll" into the ravine to the east was now lying in the bare sun; the bushes did not offer enough concealment from the enemy position on the opposite hill. Soon, the Romanians recognized our movement. Despite the great gaps in distance between my individual troopers and our accelerated movement, Romanian machine gun fire now fired at us caused several light casualties. However by contrast, on the domed western hill of Mount Cosna, the ascent was carried out under complete concealment from enemy fire, and by all outward appearances, was also unnoticed by our opponents.

Arrows show Rommel's forces on the move in this map of his break-in to the Mount Cosna position.

When I reached Friedel's patrol with the tip of my column, I found that Friedel had already cut through the enemy wire with only a few wires remaining. During my detachment's advance, *Leutnant* Wöhler, who had been left behind on "the Russian knoll" to observe, had kept me continuously informed about all activity in the enemy position. From time to time, he had at my request shot numerous mortars as harassing fire.

Some 50 meters below the break-in point, I had the companies close ranks. Meanwhile I reconnoitered the possibilities for preparing and launching an attack from an area closer to the break-in point. It was now about 1030 hours and Detachment Gößler now came marching into the nearest hollow on our right. The 1st Battalion, 18th Infantry was already climbing at this point. It was just before the beginning of our planned artillery barrage of the enemy's position; I had to hurry up with my preparations.

To deceive, divert and pin down the enemy garrison above the break-in point, I deployed the entire 2nd Machine Gun Company and a platoon of the of the 5th Company. These units crawled in concealment into position and would unleash their fire at my command. Their left flank was lying just above the gap in the barbed wire. A few seconds after they opened fire, Friedel's storm troops would penetrate the enemy position through the gap in the wire and secure the break-in point on both sides. I myself would follow close behind the storm troops with the remainder of the 5th Company, the heavy machine gun platoon of *Leutnant* Leuze and all remaining units for my detachment.

After achieving the breakthrough I wanted to go full speed straight ahead with the 5th Company without worrying about what was going on to the right or left; I'd keep going until I broke through to the nearest ridge in a northeasterly direction, followed by the 3rd Machine Gun Company, the 1st Company, 4th Company, army storm troop and engineer platoon. Leuze's heavy machine gun platoon received the task of sweeping powerful machine gun fire right and left into the enemy position once they had penetrated the break-in point. I kept all remaining units at my disposal. The gunners who had opened fire to deceive the enemy were to come after us into the captured position as soon as the situation allowed. I agreed with *Hauptmann* Gößler that he and his forces would also follow close behind me. The 1st Battalion, 18th Infantry, with units of Detachment Rommel, were ordered to roll up the enemy position on Mount Cosna from the break-in-point outwards in the direction of Hill 491, with remaining units remaining at the disposal of Sproesser's group.

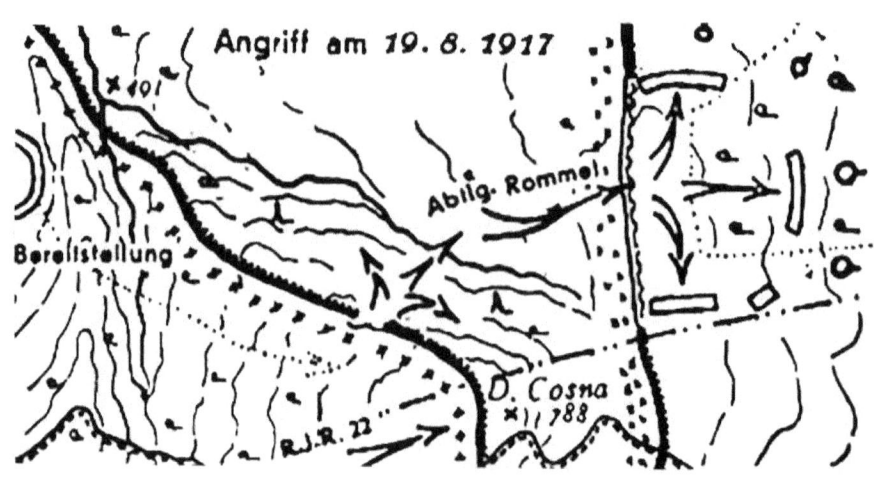

The attack on Aug. 19, 1917. Rommel's forces move to the right and penetrate enemy-held areas, starting from the attack preparation area, shown on the left,

Detachment Rommel and individual units prepared for action in the space most suited for the rapid development of the attack, which was somewhat above the break-in point. Our preparations had not quite finished before our own artillery began to bombard Mount Cosna. Columns of dirt as high as towers, springing up from 21-centimeter and 30.5 centimeter shells, sprouted from the earth ahead of us. Thick clods of dirt and tangled shrubs came hailing down. The hearts of the mountain soldiers sang at seeing this powerful help from our comrades-in-arms.

The break-in point itself, called Quad 14, remained free from artillery fire as per agreement. Our own *Minenwerfer* mortars had already prepared this area well enough. Five minutes after the artillery fire began, I signaled my detachment to attack.

The gunners above us let loose. A few seconds later, Friedel's storm troops rushed through the avenue in the barbed wire into the enemy position. The front elements of my detachment got moving. The sharp boom of hand grenades in our immediate area overpowered the din of battle to above on our right. A few lunges through dust and smoke put us standing in enemy trenches.

Friedel's storm troops rendered courageous work here. Yet the gallant *Vizefeldwebel* Friedel himself fell—fighting at the head of his flock of men, while storming the enemy, killed by the bullet of a pistol wielded by a Romanian cavalry captain. At this, the mountain rangers hurled themselves even more wildly at the enemy and overpowered the trench garrison in hand-to-hand combat. The cavalry captain and 10 men were taken prisoner. Then the storm troops divided to the right and left to seal off the area.

I took the enemy trenches with the first troops of the detachment. The garrison of the enemy trench above to the right prepared to defend themselves against a direct frontal assault by us which they supposed would come from this direction. Since we had already broken through, these enemy units could not see in certain areas due to bushes and in other areas due to dust clouds. They also could not see company after company running in columns through the break-in point.

A riotous commotion ensued. Hand grenades cracked all around. Rifle and machine gun fire clattered right and left through the bushes. Heavy shells burst in the immediate distance. The storm troops had carved about 40 meters out of the enemy position and had sealed it off on both sides. It would have been a piece of cake to rip apart the enemy position uphill, but I stuck firmly to my plan and left this task to the rear units.

Soon the 5th Company was already streaming, in accord with their original orders, northeast through the bushes to the nearest ridge. Shortly afterwards, *Leutnant* Leuze and his heavy machine guns opened fire from our newly secured spot inside the enemy position at the enemy garrison occupants both above and below the hill. I could now advance untroubled with the 5th Company into the enemy defense's zone of resistance. The adjutant reported the breakthrough we achieved to Sproesser's group via telephone and requested that heavy caliber artillery fire be spread in a sweeping manner on the positions east of Mount Cosna.

The detachment surprised Romanian reserves in the enemy's main resistance zone and took more than 100 men prisoners. The rest fled. While we chased after them, some of our own 30.5 centimeter shells hit on the left and right of us. The shells tore craters in the clay soil. Entire companies found safe shelter in these craters. We stopped and held our breath for a few seconds, but no harm came to anybody. Onward we went.

As we reached the ridge 400 meters northeast of our break-in point, our second attack goal was the nearest Romanian position, which was about 600 meters away and lay far below us just like on Aug. 12. German shells struck the hollow ahead of us, causing multiple Romanian companies to flee their positions in disorder.

With utmost speed I had a heavy machine gun platoon open fire at the retreating enemy and commanded the rest of the detachment to climb down into the basin and pursue the enemy retreating there. Using the telephone—the phone line had been brought up next to me during the advance—I requested strong artillery fire into Quads 76, 75, 74, 73, 72, 62, 52 and 42. Staying true to my original plan, I wanted to storm the second Romanian position as well after a short fire for effect from the artillery. Things turned out differently!

Giving the brief instructions and speaking on the telephone took several minutes. Soon the first German shells began hitting below us in the basin. The Romanians rushed back into their new position along a narrow footpath through the bushes. Now the machine gun fire of multiple heavy machine guns began striking between them. The effect on them was overwhelming at such a close distance. So, wouldn't it be possible now to exploit the enemy's panic and overpower the second position with a brisk pursuit? To be sure, we would then be going right into our own artillery fire, but we had already escaped the impact of one of our own 30.5 centimeter shells without being harmed. Things ahead of us couldn't be any worse than that.

So, downhill we went as fast as our legs could carry us. Shells of a howitzer battery struck right and left of us in the basin. Our heavy machine gun fire got closer and closer to the enemy, who was now squeezing through tight corridors in their obstacles to get back into their position. Soon I was close on the enemy's heels along with the foremost units of my detachment. In the heat of the fight, we did not worry about the German shells striking right, left and behind us. The enemy ahead fled head over heels. Surely they had no idea just how close we were to them, because not a single shot from their side checked our forward rush. A large number of dead and wounded Romanians now lay right and left of us. Our heavy machine gun spread fire to the left. We hurried through obstacles and shortly afterwards were standing in the enemy position. After a short exchange of bullets and hand grenades, the enemy garrison on the right and left was running away.

As my companies arrived on the scene, I quickly sent them in the following directions: the 1st Company to the east, the 5th Company to the north, and the 4th Company to the south. Every company was to roll up 150 meters of the enemy position, then halt the main body, occupy the conquered terrain and only sent assault troops out further in the previous direction they were going to harry the enemy further and to conduct reconnaissance.

Just after a few minutes I received the report that our set goal had been accomplished overall. On the right by the 4th Company, the enemy garrison showed itself to be really tenacious and even tried to regain their lost place in the position by a counterthrust. All in vain—because mountain infantry troopers, once they have taken something, never allow it to be torn away from them.

Romanians were withdrawing on the east and north and even the enemy batteries set up right behind the position rapidly cleared off the field. Rearward on the right, the enemy still held on Mount Cosna within Group Madlung's attack sector. Over on the right the enemy now occupied the second position. After the first counterthrusts against us failed, the enemy restricted himself to holding this position. A huge gap yawned in the enemy's defense ahead of us and to the left. If all our available reserves were thrown at it now, we would surely achieve breakthrough.

Telephone connection with Group Sproesser had already been set up. My communication troopers were fabulous and hadn't held back from coming with my storm troops! Soon I briefed the group about the situation at the front, asked for all available reserves to be sent and for artillery fire to be adjusted towards the second enemy position in Sproesser's sector. It was now 1145 hours and I learned that Group Madlung on the right had still not achieved its objective of taking the enemy position on Mount Cosna. I promised to send over Detachment Gößler and the 1st Battalion, 18th Infantry immediately.

Now I managed as well as I could with the forces available to me. I had to reckon with the probability of a counterattack from the direction of Mount Cosna and from the south. The engineer platoon was deployed to seal off the positions in the 4th Company's sector. They widened their front eastward until reaching a knoll with a small wooded area on it. From there outward, a heavy machine gun platoon gave continuous fire at enemy batteries near Nicoresti (across a distance of over 2,500 meters), causing them to draw their teams of horses together and clear out from their positions at a gallop. To the east, the 1st Company remained hot on the heels of the enemy as they fled downhill through light forests. To the north, the army storm troop was deployed to roll up the enemy position beyond the line that the 5th Company had reached. Everything went ahead briskly. In the same direction, Targu Ocna lay nearby within our reach (three kilometers). Our artillery now began laying heavy fire on that location. An endless stream of trains stood at the train station there, and numerous columns were parked in the vicinity. We could be there in half and hour and choke off that wide valley from which many units on the Romanian front were being supplied.

I waited with total impatience for the arrival of Detachment Gößler and the 1st Battalion, 18th Infantry. According to reports from Sproesser's group, both should have already been on their way a very long time ago. As the hours drained by in minutes, nobody came. There was continuous fighting going on in our neighboring sector on the right for the occupation of Mount Cosna. In the meantime, the number of prisoners taken by Detachment Rommel had increased to 500 and several dozen Romanian machine guns were captured. More than two hours after our successful storm attack on the enemy's second position had flown by. By now the Romanians in the north had recovered from their fright and were pushing the army storm troop back. At the same time Romanian artillery groups in the area of Satul Nou[5] were firing many hundreds of shells against the 4th Company. Thankfully most of the shots passed too far to reach us and burst on the northeast slope of Mount Cosna without inflicting any damage on us. The enemy in the south would risk no counterattack, but forced us to make careful use of our positions and our communications trenches through extremely heavy machine gun fire. From time to time, hand grenade tussles would occur in the 4th Company's area, but these brought the enemy no additional advantage.

As Detachment Gößler arrived at about 1600 hours (4 and a half hours after our successful attack), a powerful Romanian counterattack from the north forced the deployment of our 6th Company in the gap between the 1st and 5th Companies. Without reserves, it was now unthinkable for us to make a forward thrust east into the valley. The enemy attack from the north was beaten off with hand-to-hand combat.

At about 1830 hours, Group Sproesser reported that Group Madlung had taken Mount Cosna (the southern sector) and was now advancing east through the ravine to attack the enemy's second position.

Shortly before darkness fell, we observed rearward movements of strong Romanian infantry groups near Nicoresti and Satul Nou. At the same time, multiple platoons drove out from Targu Ocna, one after the other, in an easterly direction. Then we finally established contact with the Reserve Infantry Regiment 22, whose left flank had taken the Romanian positions near Hill 692.

Because I hoped that a breakthrough would come on the following day, I occupied an outpost position extending far to the east with my detachment. Reconnaissance patrols were sent forward as far as Nicoresti. In the north however, very strong enemy forces lay across from the 6th and 5th Companies.

Until midnight the usual concerns of providing meals for the troops, replenishing ammunition and composing my combat report kept me on my feet. Then I lay down to sleep in a tent with *Hauptmann* Gößler.

---

5. Present-day Moldova.

Considerations:
- The **attack** on Aug. 19, 1917 a**gainst two fortified Romanian positions secured with barbed wire**, with about 800 meters between them, presented the Württemberg Mountain Battalion with a new type of task. These positions were supposed to be

taken **after artillery preparations which lasted respectively an hour.** The mountain troopers had, while the artillery bombardment against the first position was still underway, broken through both positions while sustaining minimal casualties and torn a gap of about 600 meters in the second position, taking over 500 Romanians prisoners. Thus the path for a breakthrough into the east was free, because it was hardly imaginable that the Romanians had created more than three fortified and occupied positions in the lowlands east of Mount Cosna.

- Unfortunately t**his tremendous success could not be exploited**, because the prepared r**eserves were not fast enough to follow into the second position and were too weak**.
- The **terrain required a type of attack that deviated from the usual sorts**. Once we succeeded in pressing through into the enemy position just below the peak of Mount Cosna, it then became easy to rip open the enemy position on the steep slope declining toward the northwest—especially since our attack here could be supported by heavy machine gun fire from "the Russian knoll."
- For Detachment Rommel, deployed in the very front line, it was a question of leading in the **blow as quickly and as deeply as possible into the enemy's defensive zone.** The troops could not split apart while ripping up the first enemy position. Also, while tearing open the second position, the troops concentrated their strength of force together, to retain it for striking the additional blow expected with the arrival of the reserves.
- The **teamwork with artillery,** *Minenwerfer* **mortar units** and **heavy machine gun troops** was prepared in advance with the utmost attention to detail. The *Minenwerfer* held the enemy at our break-in point pinned down even before the planned artillery fire, and enabled Friedel's storm troop to cut through the enemy's barbed wire obstacles. The **artillery** held the first enemy position down as Detachment Rommel broke through. A machine gun company and a platoon of the 5th Company assailed the enemy above the break-in point with fire and held him down as Friedel's storm troops and the remaining units of Detachment Rommel broke in.
- Ahead of the very **heavy German preparatory fire** on the first enemy position, the strong Romanian troops located behind it quickly withdrew to their second position. **Detachment Rommel thrust straight into this rearward enemy movement**, making use of the favorability of the situation. We broke **into the position** following close **behind the fleeing enemy**, who was being harassed by artillery and especially heavy machine guns. **Thus the mountain rangers accepted the risk of going through the German artillery fire**, which could not be reoriented at such short notice.—*Rommel*

"For The Rest Of My Life"—*Again on Active Defense*

> ◊ Rommel has practically knocked himself out. He is forced to relinquish command due to illness. Almost the very same thing will happen to him much later in North Africa.
> ◊ His touching sentence about his lifelong attachment to his troops emphasizes "happiness" along with "pride." Rommel makes it clear he derives gladness from his memories. This sentence sheds light on the personal affection Rommel felt for the men he commanded.
> ◊ He mentions going to the beaches on the Baltic Sea. One wonders if Rommel is visiting Lucie during this time, but—in contrast to his mother and siblings, who he refers to specifically earlier on—Rommel makes no mention of his wife.
> ◊ After describing his heartfelt attachment to his men, Rommel ends the chapter with a brisk, informal remark about getting back "up to snuff"—very typical Rommel.

Already at 0300 hours on Aug. 20 the enemy opened the fight against the Mount Cosna massif with a heavy surprise fire from numerous batteries. A vast number of heavy shells struck in close proximity to my command post and next to the reserve located in the area. It forced us to clear out the endangered area and seek cover in the basin 800 meters north of Hill 788. The enemy fire became stronger and stronger. It was primarily concentrated on the position we had taken east of Mount Cosna, where the Romanians estimated us to be. I was very happy that none of my detachment had dug in there, because the fire soon rendered this position into a heap of rubble.

At about 0700 hours, a strong enemy force advanced from the east against the 1st Company, who occupied deeply entrenched forward outposts. Soon the basin near Nicoresti was also brimming with Romanians. At the same time reports of enemy preparations for an attack came from the 6th Company, secured towards the northwest. Now there was no longer any doubt. The Romanians wanted to tear the land we had conquered on Aug. 19 away from us again. It was now really high time to turn our attention toward defense.

A continuous line had to be drawn up in the creviced and forested terrain. The unsupported north flank especially needed protection. I declined to occupy the Romanian position, which had been under heavy fire that whole morning. The Romanians had zeroed in very well on that position and were familiar with it in every detail. For us to fight a defensive battle there would cost us heavy casualties.

Despite the overload of work and the very short time that remained for us to do it before we came into a direct collision with a strong enemy force, I moved a forward slope position eastwards into the forest. I quickly gave the necessary commands across the terrain. As the deployments were underway, the 1st Company made a fighting retreat in the fact of a strong enemy force and the companies dug in. Shovels were able to penetrate to great depths in the soft clay soil. My reserve helped the companies in the front line to fortify their position and construct communications trenches. The position had already been dug as deep as a man's height as the outpost troops, harried by the enemy, retreated into it. The impending attack of strong enemy forces was easily repelled at a short distance. Afterwards the Romanians also seized their spades and dug themselves in 50 meters ahead of our position.

The Romanian artillery searched in vain for our forward slope position in the woods. They could not lay fire on our new position without endangering their own troops and confined themselves to firing on the previous Romanian position up on the hill.

The fight on the eastern front (the 4th and 1st Companies) thus didn't worry me too much. Things looked different in the north and northwest. There was a great big gap over there. Our connection to the left (with the 1st Battalion, Bavarian Reserve Infantry Regiment 18) lay on the northeast slop of Mount Cosna on the ridge ascending towards the peak from Point 491. Thus a Romanian force coming in from the north succeeded in climbing into the basin to our rear. The 3rd Company, kept in reserve until now, had to close the gap between the left flank of the 5th Company and the 1st Battalion of the 18th. They made a very tough stand against a vastly numerically superior enemy in a bushy terrain that offered little visibility, but they held out.

Hour by hour the fight intensified. During the course of the day, the enemy stormed our positions following long artillery bombardments at least 20 times; the enemy had surrounded us in a half-circle formation. The few reserves left available to me had to be constantly sent from one hotspot of the fight to another. The enemy artillery fire hacked apart the ridges we held. But the mountain rangers didn't waver. Compared to the casualties suffered by the enemy, our own losses—20 men altogether—were relatively few.

I was so utterly exhausted, certainly due to my crushing amount of activity of the past days, that I could now only give orders while lying down. In the afternoon I had a high fever and began hallucinating the most fantastic things. Now I was finally no longer able to lead. In the evening I handed command over to *Hauptmann* Gößler and discussed the most important things with him. As darkness fell, I wandered along the mountain path across Mount Cosna back to Group Sproesser's command post, located 400 meters southwest of "the staff's knoll."

The Württemberg Mountain Battalion held its positions against all Romanian attacks until Aug. 25. Then it was relieved by Reserve Infantry Regiment 11 and moved behind the front as a divisional reserve. The battles for Mount Cosna had made colossal demands on the young troops. During the two and a half weeks, 500 men fell out of action and

Rommel is shown interacting warmly with his fellow mountain rangers during World War I. Being part of this unit was a source of great pride to him.

Romanian soil blanketed [the graves of] 60 courageous mountain rangers.

Although the offensive's set goal of "shattering the southern flank of the eastern front" was not reached, the mountain ranger troops had continuously performed in an exemplary manner in response to every task set before them against tenacious, brave and supremely well-armed opponents.

To have been the leader of such troops has and continues to fill me with happiness and pride for the rest of my life. After the hard days on Mount Cosna, a few weeks of leave on the beaches of the Baltic Sea brought me back up to snuff again.

Considerations:

- **In the defense on Aug. 20, 1917,** the **main battle line** was set **in the thick forest area on the forward slope** so as to negate the effect of the very strong Romanian artillery fire we expected. Doing so proved its value many times over, because during the course of the battle, the enemy did not manage to hit this well-camouflaged main line with artillery fire.
- As the [garrisons of] **combat outposts made a fighting retreat** from the battlefield, the outpost positions were dug further in. At the same time, reserve companies were deployed to draw concealed **connective trenches** to the front line. These trenches proved important **for resupply of every type** and for **transporting the wounded** away in the midst of **enemy fire,** enabling us to carry these things out with minimal or no casualties. Afterwards the reserves dug in at their assigned places.
- **The defensive battle** on Aug. 20 necessitated the **deployment of reserves to focal points that often changed. Wherever danger** threatened, the reserves had to occupy the main battle area in **deep distribution**. Strengthening the front line with reserves was avoided whenever possible.—*Rommel*

5

The Tolmein Offensive 1917

"Unfavorable Circumstances"—*Deployment and Preparations for the 12th Battle of the Isonzo*

Rommel's troops are pictured on a summit during a mountain training exercise in World War I.

In the wonderful Carinthia region[1], where the Württemberg Mountain Battalion had been transported via a detour through Macedonia, I resumed command of my detachment at the beginning of October. Reserves had arrived to replace our losses at Mount Cosna. The fighting strength of the rifle companies was noticeably increased as they were now armed with light machine guns. The short time of peace was put to use for detailed training with the new weapons. We didn't know what the Army Command[2] had planned for us. Perhaps the Isonzo front?

The main goal of Italian operations since the beginning of the war in May 1915 had been to take Trieste. During the course of two years of war, 10 battles had slowly but steadily pushed the Austrians back along the lower stretch of the Isonzo river. During 6th Isonzo battle in August 1916, the Italians had managed to set foot on the east bank

---

1. In Austria.
2. Written as *Heeresleitung*.

of the Isonzo near Görz[3] and to take the town.

The 11th Isonzo battle which began in August 1917 was launched from Cadorna and carried off in the style of the battles on the Western Front: supported by 5,000 artillery guns, the infantry consisting of 50 divisions attacked along the narrow area between Görz and the sea. In a violent grappling struggle, the courageous troops of Austria countered the initial gains of the Italians, but in the second part of the battle, the Italians crossed the midsection of the Isonzo and took the high plateau of Bainzissa[4]. By mobilizing their last strength, our allies succeeded here in bringing the attack to a standstill. The violent battle lasted until the beginning of September; then things became quieter.

The Italians' prospects for a new battle looked especially promising from the new land they had taken east of the river's midsection; this terrain dominated the southeast far and wide. Their end goal of Trieste now moved within close reach. The Austrians no longer felt themselves able to handle the expected new attack. Thus they asked for German help. Despite the monstrously immense need for troops in battles on the Western Front (Flanders and Verdun), the German Supreme Army Command[5] sent an army consisting of seven battle-hardened divisions.

A joint offensive of [German and Austrian] allies was supposed to bring about the desired relief on the upper Isonzo front. As the goal of this operation, it was thought that the Italians were to be pushed back

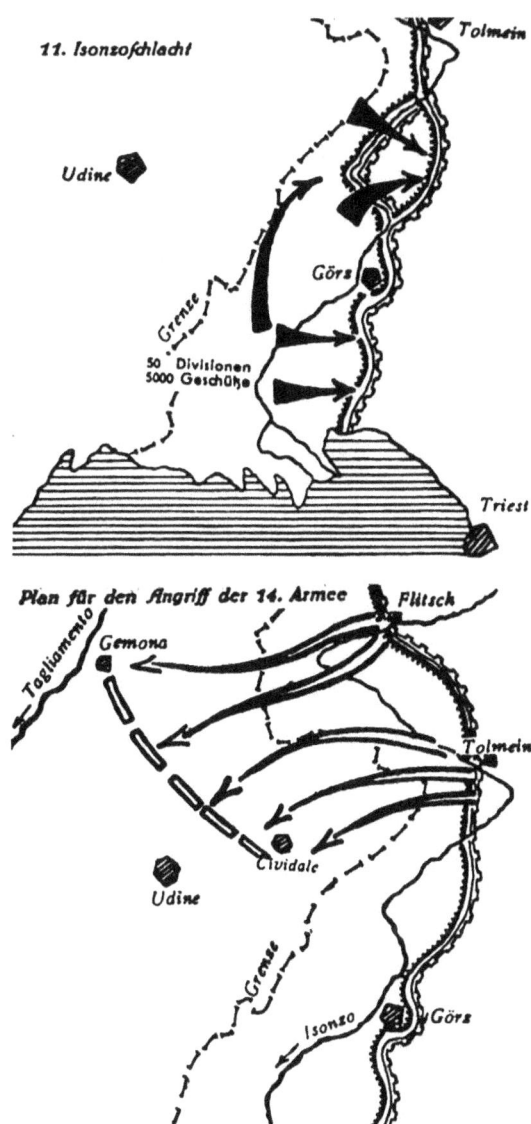

Top: Rommel's depiction of the 11th Battle of the Isonzo. Bottom: The plan for the attack of the 14th Army.

---

3. Present-day Gorizia.
4. Today Banjsice.
5. *Oberste Heeresleitung,* or OHL.

over the Austrian border, and as far back as across the Tagliamento wherever possible.

The Württemberg Mountain Battalion arrived as part of the newly established 14th Army and was attached to the *Alpenkorps*[6]. On Oct. 18 we went from the assembly area in the vicinity of Krainburg[7] straight into our advance to the front. *Major* Sproesser's marching group, consisting of the Württemberg Mountain Battalion and the Württemberg Mountain Howitzer Detachment No. 4, made the way in pitch-dark nights and sometimes through pouring rain across Bischoflak[8], Salilog[9], and Podbordo up until Kneza, which we reached on Oct. 21. During this march, the troops had to reach their designated marching destination before daybreak and be hunkered down in tight and uncomfortable shelter to guard against possible air reconnaissance before dawn each day. These night marches made steep demands on the poorly nourished troops.

My detachment consisted of three mountain companies and one machine gun company. I marched on foot with the detachment staff at the head of the long column. Kneza lay eight kilometers east of the battlefront near Tolmein. On the afternoon of Oct. 21, *Major* Sproesser and the detachment leaders reconnoitered the allotted preparation area for the attack: the northern slope of the Buzenika mountain (Hill 509), located 1,500 meters south of Tolmein, which declined sharply towards the Isonzo.

Very powerful and fierce Italian harassing fire from countless batteries on formidably high positions burst far and wide behind the front. It seemed like the Italians had a surplus of ammunition. The preparation of 11 companies of the strong battalion was difficult in the terrain assigned to us. Only the edges of some screes and a few tight, exceptionally steep water canals falling towards the Isonzo offered opportunities for preparation on the slope, which was otherwise hardly possible to walk across. It was frightening that the enemy, from his very high positions on the peak of Mount Mrzli Vrh (Hill 1360) northwest of Tolmein, could clearly see the entire northern slope of Buzenika from an almost flanking perspective. We also had to reckon with the possibility of heavy rockslides in addition to artillery bombardment on the steep slope. The battalion was supposed to be ready and in position to attack in about 30 hours. Was that going to proceed well?

We had to resign ourselves to all these unfavorable circumstances. There were no other possibilities. The mass of the assembled troops being prepared for attack in the basins of Tolmein was too large. We returned to the battalion under fierce Italian artillery harassing fire, aimed at us especially in the narrows near St. Luzia[10] and Baza di Modreja.

We were noticeably less informed about the plan for the whole operation than the <u>Czech traitor</u> who ran over to the Italians that day with a collection of orders and maps

---

6. Elite mountain ranger unit of the German Army.
7. Present-day Kranj.
8. Today Skofja Loka.
9. Zali Log.
10. Present-day Most na Soci.

for the offensive north of Tolmein.

During the night of Oct. 22–23, the battalion moved into the assembly position. Enormous searchlights from the Italian positions lit up our approach route on the hills of Kolovrat and Jeza. Strong artillery fire struck between us time and time again. The strong, glaring light beams from the searchlights forced us to lie totally still for minutes on end. As soon as the lights flitted in another direction, we rushed through areas imperiled by shells. During the advance we all had the impression that we had come within range of an extraordinarily active and well-armed enemy.

We had to leave the pack animals behind on the eastern slope of the Buzenika mountain. Heavily laden with machine guns and ammunition, Detachment Rommel reached its designated scree shortly after midnight following an arduous climb. We stripped off our burdens. Relieved of our loads, everyone rejoiced at having passed through all the dangers unscathed[11]. Yet rest was unthinkable. The few night hours had to be used most diligently for digging in and camouflage. Hurriedly I directed the companies to their areas.

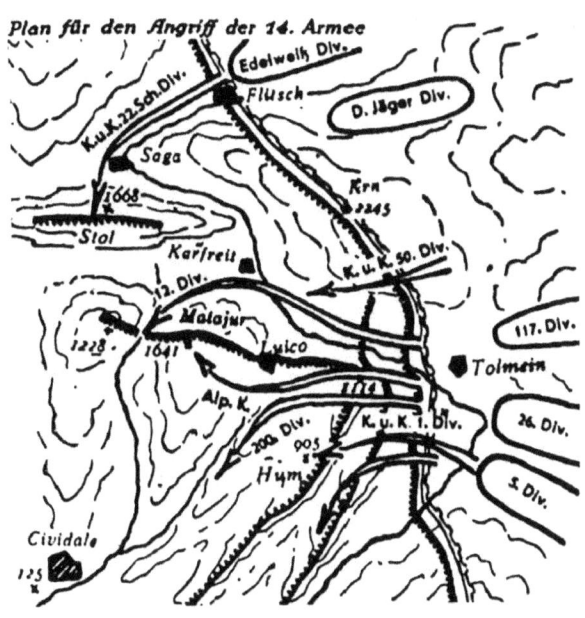

A more detailed view of the 14th Army's attack plan.

On the western slope, the staff and two companies nested themselves on a scree about 20 to 40 meters across on either side of a path screened on the northwest side and leading to the front line. The remaining two companies held a narrow canal assigned to them 200 meters to the east. Everyone worked feverishly—officers as well as men. As daylight came, the slope was deathly quiet. The troopers caught up on their missed sleep in foxholes covered with brush and twigs.

But the peaceful quiet did not last. Some heavy Italian shells struck the slope above us. Rockslides tumbled past us, down to the Isonzo valley. There was not much sleep to be gotten now. Had the enemy already recognized our preparations and zeroed in on us? Heavy artillery fire on this slope, steep as a tilted roof, would surely have a devastating effect! Minutes later the fire quieted again. One quarter of an hour later, shells again struck another location in our immediate area. Then we had peace for awhile.

---

11. Another informal and slightly humorous expression from Rommel. The word he uses is *ungerupft*, literally meaning "with no feathers plucked."

Men of Rommel's regiment line up for chow at a field kitchen. The field kitchens often had difficulty reaching the troops.

The Italian artillery now concentrated its main effort at the Isonzo valley. Throughout the day, we observed the violent effect of the heavy calibers against installations, positions and supply routes in Tolmein below. By contrast, our own artillery only seldom fired a shot.

For me, the day dragged on with infinite slowness as I worried, terrified, about the wellbeing of the men entrusted to me.

By traveling a few steps ahead to the west along the screened path, one could clearly see the enemy's foremost position in the valley. It crossed the Isonzo 2.5 kilometers west of Tolmein and then ran south of the Isonzo close to the east towards and past St. Daniel to the eastern edge of Woltschach[12]. The positions and above all the wire obstacles seemed to have been built exceedingly well. There wasn't much to see of the remaining enemy positions in the murky weather.

The second enemy position allegedly crossed the Isonzo in the area of Selisce, nine kilometers northwest of Tolmein, and ran south of the Isonzo over the Hevnik toward Jeza. The Italians had laid the third well-fortified position on the hills south of the Isonzo in a line from Matajur (Hill 1643) over Mrzli Vrh (Hill 1356), Golobi, Kuk (Hill 1243), Hill 1192, Hill 1114 (at which point the line turned sharply southwest across Clabuzzaro) to Mount Hum. This line was recognized through aerial photographs. Isolated points of

---

12. Volce.

support also occurred in the terrain between the individual positions.

The forces of the 14th Army were prepared in this manner: Group Kraus (consisting of the Imperial Austrian 22nd Rifle Division, the *Edelweiß* Division, the Imperial Austrian 55th Division and the German *Jäger* Division) would attack the focal point on the Stol by crossing Saga. Group Stein (consisting of the 12th Infantry Division, the *Alpenkorps,* and the 117th Infantry Division) had already been prepared near Tolmein and in the bridgehead position south of Tolmein; this group would lead the main attack. The 12th Infantry Division would break through on both sides of the Isonzo into the valley to Karfreit. The *Alpenkorps* was charged with capturing the high positions south of the Isonzo—above all, Hill 1114, Kuk, and Matajur. Finally, Group Berrer (consisting of the 200th Infantry Division and the 26th Infantry Division) would be deployed south to Cividale across Jeza and St. Martino. Still farther south, Group Scotti (consisting of the Imperial Austrian 1st Division and the 5th Infantry Division) was to take the positions south of Jeza and, still further, Globocak and Mount Hum. The Royal Bavarian Infantry Life Guards Regiment and the Austrians' 1st *Jäger* Regiment relieved the *Alpenkorps* in the frontline. The goal of the 1st *Jäger* Regiment was to attack the hills west of Woltschach, Knoll 732, and Hill 1114 from the southeast. The Württemberg Mountain Battalion had the task of protecting the right flank of the Life Guards, taking the enemy batteries near Foni and following the Life Guards to Mount Matajur.

Around the evening time of Oct. 23, the weather was overcast and misty. As twilight set in, pack animals carrying meals arrived in the preparation area. As soon as everyone had satisfied their hunger, each man sought out his foxhole to catch up on as much sleep as possible before the days of attacking ahead. After midnight, a soft rain came trickling down and forced men to tuck their heads under their shelter halves. It was attack weather!

Considerations:

- Just moving into position and preparing for the battle of attack on Tolmein placed **hefty demands on the troops**. In strenuous night marches, mostly during the pouring rain, they successfully overcame the Karawank mountain range, altogether about 100 kilometers as the crow flies. During the day, the troops lay concealed from planes in the tightest shelters. Nourishment was scarce and monotonous, but the mood was exemplary despite it all. The troops had learned during three years of war to bear hardships without losing their resilience
- During the advance into the preparation area in the night of Oct. 22–23, a reserve of belted ammunition was brought forward with the machine gun company and units of the mountain troop companies. **The battles at Mount Cosna had clearly proved the**

> - Because we had to reckon with strong enemy surprise fire in the preparation area, the troops **dug** in during the night and carefully **camouflaged** the newly created installations before daybreak.
> - It was impossible to **feed the troops** in the preparation area during the daytime. Pack animals were only able to bring up warm food to the troops after darkness fell. —*Rommel*

"To Try My Luck"—The 1st Day of the Attack: Hevnik, Hill 114

◊ Rommel and his men are selected to be the advance guard of the entire battalion on a dangerous march—a prestigious duty. Rommel relishes it and is later upset when he is not allowed to serve in the advance guard; a change of duty perhaps owing to brash behavior which will become apparent.

◊ Freak accidents arise during war in the mountains. Rommel survives a terrifying ordeal when an enormous boulder rolls down and almost smashes him. He is wounded when his foot is crushed by a rockslide. Never at any point has Rommel, despite being wounded, leaned on his comrades literally or figuratively unless he is totally incapacitated—until now. Describing his pain level as "excruciating," a rare thing for him to admit, Rommel cannot walk and accepts the help of two men. However he carries on with his duties.

◊ While the Bavarians stop for a rest, Rommel drags his men up a mountain peak. He draws attention to this. This belief in driving troops hard would stay with Rommel, known for being a hard taskmaster.

◊ Rommel's pursuit of nonstop action and habitual scorning of basic human needs did not always work out well, as he ran himself and his troops to the brink of exhaustion.

◊ He gets into a dispute with the commander of the Life Guards, who is jockeying for authority. He is backed by his commanding officer Maj. Sproesser, who has been showing more faith in Rommel, and indeed leaning on him for advice. Like the mythical Icarus, however, Rommel will eventually get his wings burned by taking Sproesser's goodwill for granted.

In the dark rainy night of Oct. 24, 1917, our own hitherto silent artillery began preparatory fire. Soon the barrels of more than 1,000 artillery guns were rattling on both sides of Tolmein. A noise of ceaseless bursts and cracking explosions droned from enemy territory. It echoed mightily back from the mountains, like the heaviest kind of lightning storm. We were astonished as we saw and heard this colossal event.

Italian searchlights tried in vain to pierce through the rain. The enemy fire for destruction that we had anticipated and feared in the areas around Tolmein did not materialize. Only a few enemy batteries countered the German fire. That was very comforting. Again we sought out our shelter and, half-asleep, heard the weakening of our own artillery fire.

By daybreak our own firing swelled up again to a powerful degree. Below near St. Daniel, heavy mine shells were now smashing enemy positions and obstacles to pieces. Sometimes you could see nothing more of the enemy installations at all due to the dense smoke and fumes. The tornado of fire launched by our artillery and *Minenwerfer* steadily increased in strength. The enemy's counteraction seemed only to be very weak.

Shortly after daybreak, the Württemberg Mountain Battalion deployed from the preparation area in a march toward the front. Now it began to rain harder and thus our view was severely obstructed. Following Sproesser's staff as they hurried ahead, Detachment Rommel climbed down into a scree declining towards the Izonso. Arriving below, we drew up above the steep bank of the Isonzo close behind the right flank of the Royal Bavarian Infantry Life Guards.

A few shells struck on both sides of the long marching column without inflicting any damage. The column stopped close behind the front line. Soon the rain pressed down all the way through our clothing, soaking us to the skin. Freezing, every man longed for the attack to begin. But the minutes stretched by only slowly.

In the last quarter of an hour before the storm attack was to begin, the fire climaxed to a monumental ferocity. A shot bursting from a vortex of wind enveloped an enemy position lying only a few hundred meters ahead of us in smoke and clouds. Gray swathes of smoke drifted above the valley; the results of the hours-long bombardment. Low-hanging rain clouds encased the mountain peaks of Hevnik and Kolovrat.

Shortly before 0800 hours the storming troops ahead of us left their own positions and approached nearer to the enemy. The [enemy] defenders, amid the whirl of artillery fire, neither saw this nor defended themselves against it. We also used the new space made available to us [left by the troops who had moved up] to prepare ourselves to storm forward.

0800 hours came! The artillery and mine fire jumped towards the enemy. The Life Guards ahead of us threw themselves into a storming attack. Following closely on their right flank, we moved half to the right and took the enemy positions around St. Daniel. What remained of the garrisons emerged from these ruins with their hands up and hurried towards us with faces distorted with fear. We hurried forward across the broad plain that still separated us from the north slope of the Hevnik [mountain]. While machine gun fire from the eastern foothills of Hevnik hampered the advance here and there, our

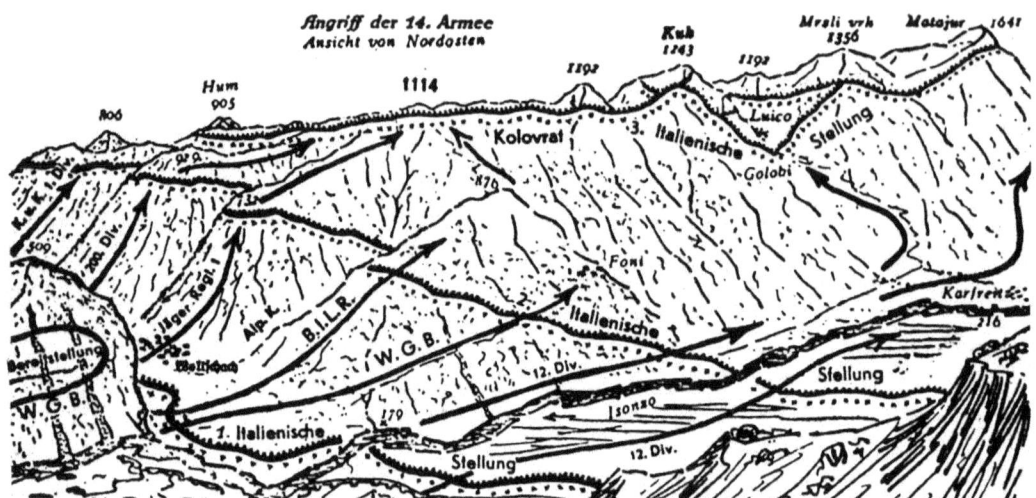

The attack of the 14th Army. View from the northeast. The Württemberg Mountain Battalion is shown moving up the center of the mountain range from the lower left, with the B.I.LR. indicating the Bavarian forces moving towards the eastern slope in parallel with them.

attack across the open ground remained in flux.

As the "Lifeys"[13] strove for the eastern slope of Hevnik, our goal was the northeast slope. *Major* Sproesser had already hurried ahead over there with his staff. The infantrymen with their heavy rucksacks, machine guns or ammunition couldn't get forward so quickly.

After [we] reached the area of Hill 179 the rising, forested slope of Hevnik protected us against enemy fire from the hills on the left.

The entirety of Detachment Rommel had now reached the cover of the slope. On orders from *Major* Sproesser, the detachment mustered as the advance guard of the Württemberg Mountain Battalion on march along a footpath from the northern slope of Hevnik towards Foni. A group of the 1st Company under *Vizefeldwebel* Seitzer served as the point men. With a distance of 150 meters between them, they were followed by a platoon of the 1st Machine Gun Company, the Detachment staff, the 1st Company, 2nd Company and remainder of the 1st Machine Gun Company. With *Leutnant* Streicher, my new adjutant, I positioned myself a few meters behind the point of the column.

The footpath we were using to ascend towards Foni was narrow and thickly overgrown with connecting bushes. From all outward appearances, the enemy had hardly used this path. The slope on both sides of the path was very steep and densely forested. Autumn leaves were still hanging on the trees. You could only see ahead a few meters through the thick undergrowth. Opportunities to view into the valley rarely presented themselves. A few deeply dug ditches led to the Isonzo.

<u>The sound</u> of heavy German shells came dully from the valley and also rearward from

---
13. Rommel's slang reference to the Life Guards.

the left from the direction where we supposed the Life Guard Regiment was. The slope ahead of us lay eerily quiet. Somewhere on this hill we were supposed to clash with the enemy. Once that happened no artillery of ours could help us in this dense mountain forest. We would be totally on our own. With exceeding wariness and caution the point of our column crept forward, halting often and listening into the forest before creeping forward again. But all caution was of no use here. The enemy had been on the lookout and was waiting for us. Suddenly, one kilometer east of Hill 824,

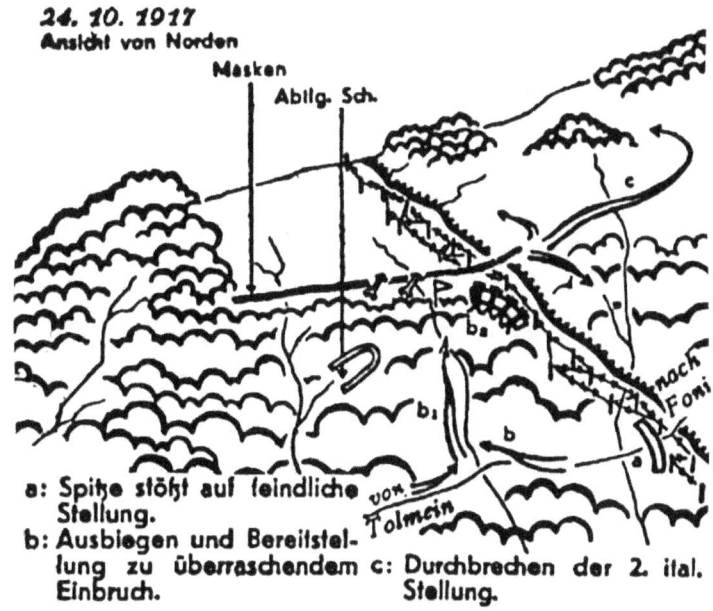

A view from the north of action on Oct. 24, 1917. Starting from the lower right: (a) the point men run into an enemy position; (b) the troops turn and move into position for a surprise attack; (c) the breakthrough into the 2nd Italian position.

the point men were shot at with machine guns at close range. I received the report: "Enemy ahead of us in a fortified position behind wire obstacles. Five point men are wounded."

An attack without artillery support on both sides of the path on this slope, which was as steep as a rooftop, through the dense underbrush and across obstacles against an extremely watchful and well-entrenched enemy seemed fruitless to me, or at least unattainable without sustaining severe casualties. Thus I decided to try my luck in another place.

Those who had previously served as the point of our column remained near the enemy. I formed another group from the 1st Company as a new point and ordered them to ascend south in a stone canal some 200 meters ahead of the enemy position in the hope of bypassing the enemy above and to the left. I sent a report to *Major* Sproesser.

The ascent proved to be really difficult. About 40 meters behind the new point, *Leutnant* Streicher and I climbed into the canal. The squad of a heavy machine gun followed close behind us on foot, carrying their weapon divided into pieces on their shoulders.

At this moment, a massive boulder came speeding down from above, zigzagging straight towards us below. The canal was only three meters wide. It would be hard to fall back, and a retreat was no longer possible. Within a split second it was clear to us—whoever got hit by that boulder would be smashed to a pulp. We all pressed ourselves against the left wall of the canal. The enormous rock zigzagged between us, then zigzagged down into the valley without leaving a scratch on any of us.

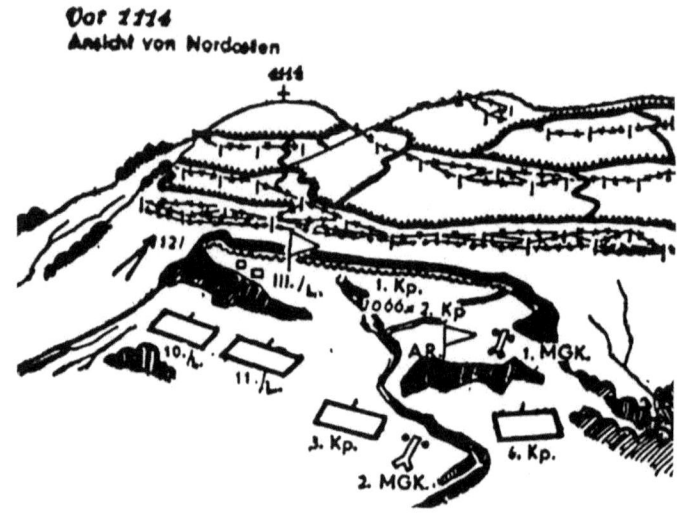

In front of Hill 1114. View from the northeast.

To our relief our assumption that the Italians had rolled the stone down at us was proved false. The rock had loosened under the steps of the point men.

As we climbed further, a rockslide tore the heel strap on my right boot off and severely crushed my foot. For the next half hour I could only move forward under excruciating pain with the help of two men.

At last the steep canal lay behind us. In driving rain and soaked wet to the skin, we climbed uphill through thick underbrush, cautiously watching and listening all around us. The forest ahead of us grew sparser. Using the map I determined that we had to be 800 meters east of Hill 824. With extreme caution we stalked closer to the tree line. Here were discovered a concealed path leading downhill in an easterly direction. Behind it we determined the existence of a continuous enemy position well-fortified with wire obstacles on the bare, rising slope stretching uphill in the direction of Leihce peak. This enemy position was also as quiet as the grave and no German artillery fire had been directed at it up until this point. My decision stood firm: We'd make a surprise break-in following a short surprise fire with numerous heavy machine guns and with our left flank stretched along the tree line. The situation reminded me vividly of the situations preceding the assaults on Mount Cosna on Aug. 12 and 19, 1917.

Protected by one of the heavy machine gun platoons brought into concealed position in the bushes, I positioned the detachment in a small basin 60 meters ahead of the enemy obstacles in the forest and prepared it to attack. Thanks to their exemplary combat discipline the mountain ranger troops accomplished this whole movement in the pouring rain without any noise whatsoever. The sounds of battle in the Isonzo valley resounded far and wide. The "Lifeys" seemed to be doing battle a bit closer to the left on the reverse of the crest. A deep, silent stillness prevailed all around us and also across the surface of the meadow. Now and again you could see a few people moving inside and to the rear of the enemy position—a sign that the enemy ahead of us still had no idea that we were nearby. Now isolated German shells began to fall 600 meters on the left to our rear. Judging by its course, the enemy position located ahead of us must be connected with the enemy position on both sides of the path to Foni, which we had come up against three quarters of an hour ago.

Thus we would probably best focus on the second Italian position. To succeed in approaching up close without making any noise was impossible in the dense underbrush. The detachment was standing ready. Should I begin the attack? We faced 60 meters of undergrowth, then obstacles! Even if the enemy only partially manned his sentry posts, I couldn't bet on an easy victory here.

The exceptionally camouflaged footpath on the tree line beside us on the left brought new thoughts to me. This path had probably been used up until this point for access to the Italian frontline at St. Daniel, or the garrison of the position on the eastern slope of Hevnik, or the artillery observation posts located there. No Italians had used the path throughout the time we were lying here. The path was full of turns. The screen on the southern side blended so well going uphill towards the Italian position that from there outward it would hardly be possible to recognize whether friend or foe was moving on the path. Using this path, you could get into the enemy position in half a minute, if the enemy didn't stop you. If we seized the position fast, we could perhaps apprehend the enemy garrison right and left of the path before things came to a shootout. This was a task for stouthearted men! If the enemy put up a defense, I could immediately unleash the attack of the entire Detachment Rommel with protective fire from the Machine Gun Company.

I gave *Gefreiter* Kiefner (of 2nd Company), who had a giant physique, the difficult task of advancing on the camouflaged path with eight men as if they were Italians returning from the front, to break into the enemy position by surprise and, if at all possible without shooting or throwing hand grenades, to apprehend the enemy occupants on both sides of the path. In case things came to a fight, covering fire and support from the whole Detachment Rommel was assured to them. Kiefner understood me and sought out his comrades. A few minutes later, Kiefner's patrol moved along the concealed path towards the enemy. The sound of their steady footsteps faded away. Would they succeed? In suspense we listened, ready to lunge, and accordingly prepared to release continuous fire. A single shot would unleash the storm attack of three companies. Long harrowing minutes continued to pass by. There was no sound apart from the rain in the forest. Then footsteps came towards us, and a soldier reported in a quiet voice: "Patrol Troop Kiefner has cleared an enemy dugout. Seventeen Italians captured. One machine gun confiscated. Enemy in the position is completely unaware."

Then I led the entire Detachment Rommel in the following sequence—the 2nd Company, then the 1st Company, and then the 1st Machine Gun Company following the path into the enemy position. Detachment Schiellein (consisting of the 3rd and 6th Companies and the 2nd Machine Gun Company), which had placed itself under my command shortly before Kiefner's successful break-in, followed as well. Using assault detachments, I widened the break-in point on both sides by about 50 meters without making any noise. During this action, numerous dozens of Italians taking refuge from the rain in their dugouts were captured by the clever and agile mountain rangers.

The enemy on the slope above continued not to recognize the movement of the six companies thanks to the thick screen. I now faced the decision of whether to roll up the enemy position uphill or downhill toward the valley, or to break through in the direction of Hevnik's peak. I chose the last option. If we got the peak first, then it would be easy to roll up all the Italian positions on the slopes. The farther we penetrated the enemy's zone, the less aware the enemy occupants would expect our coming, and the fighting would thus be easier. I wasn't worried about support on the right and left. Six companies of Württemberg mountain rangers could protect their own flanks by themselves. Furthermore, our attack orders stated: "The day's goal is to push forward continuously to the west with no limits on space or time, knowing that you will have strong reserves next to and behind you."

The 1st Machine Gun Company was echeloned farther forward because I wanted, in the event of a collision with the enemy, to have rapid firepower at my disposal. The machine gun crews, bearing loads of up to 80 pounds, determined the tempo of the climb. The tremendous physical performance accomplished by these troopers can only be understood by someone who has carried a similar load while climbing off the path in the high mountains under the same weather conditions.

In constantly streaming rain, we crept forwards in a kilometer-long from bush to bush, climbing in basins and canals under cover and seizing position after position. Nowhere did it come to fighting. For the most part we approached the enemy installations from the rear. Whoever among the enemy didn't surrender immediately at our sudden surprising appearance fled head over heels into the deep forest below, leaving behind their weapons. We did not shoot after these fleeing enemies in order not to alert the enemy garrison still occupying the position uphill above us.

Yet again we were endangered during this advance by our own strong artillery fire. We could not give off flare signals because these would alert the enemy garrison to our presence. One man of the detachment was wounded in a rockslide caused by a heavy German shell. Among other things we pressed into the position of an Italian 21-centimeter gun battery that had been under fire from gas shells. The crew had vanished without a trace. Mountains of shells lay close beside the enormous guns. The dugouts and ammunition storage chambers, blasted into the rock face, were undamaged. Hardly 100 meters above, we spotted a medium-heavy battery as we passed by; its guns were completely secured against fire in rock chambers with only very narrow openings to shoot out from. Also here the gun crew had vanished without a trace.

At about 1100 hours we reached the crest running east from the Hevnik peak and established contact with units of the 3rd Battalion of "Lifeys." We climbed the length of the crestline towards Hevnik's peak with them for awhile, us and them both climbing at the same elevation. The peak was now under heavy German fire. As the "Lifeys" laid up for a rest, waiting for a barrage from our artillery, I turned with my companies towards the northern slope of Hevnik. From there we scaled the Hevnik peak (Hill 876) at about 1200 hours without encountering any resisting enemies. All around us, scattered Italians kept showing up in great numbers; some of them were captured.

The rain had stopped. Movement came in the clouds hanging thickly above us. For awhile we had a clear view of Hill 1114 and the Kolovrat ridge. From there, heavy Italian fire now came bursting onto the Hevnik peak. Apparently we mountain rangers had been recognized by Italian artillery observers ahead of Hill 1114. To avoid unnecessary casualties, I withdrew both detachments of the Württemberg Mountain Battalion in a northerly direction out of the endangered zone, and had them clear out enemy artillery nests – in accord with orders given to the Württemberg Mountain Battalion—between the Hevnik peak and Foni. Together with reconnaissance patrols, I wiped out [the enemy from] the southern slope of the knoll forming the Hevnik peak and the Nahrad saddle (Hill 807, 300 meters southwest of Hevnik peak).

The number of enemy guns we captured and marked with chalk climbed to 17; among them were 12 heavy caliber guns. Tins of Italian food and readily prepared food from their field kitchens sated our powerful hunger.

As units of the Life Guards arrived at the Nahrad saddle at about 1530 hours, I assembled both of my detachments and moved over there as well. Half an hour later, the 3rd Battalion of the Life Guards climbed with three companies towards the camouflaged mule track leading over Hill 1066 (Nahrad) towards Hill 1114. Bearing in mind the orders given to the Württemberg Mountain Battalion to protect the right flank of the Life Guards Regiment, I followed with the six mountain ranger companies in tightly closed ranks in the following order: Detachment Rommel, Detachment Schiellein.

With *Leutnant* Streicher, I marched at the head of my column. The weather had brightened up. The Kolovrat ridge, Hill 1114 and the mountain range from Hill 1114 towards Jeza were sharply outlined. No enemy hampered our ascent at the outset. As we got nearer to the rocky crest of Hill 1066 (Nahrad) at about 1700 hours, shots fell near the Life Guards company marching furthest in front. Afterwards two companies of the Life Guards went into cover below the rocks and east of the path. I had Detachment Rommel march into cover to the right of the path to the same level as the Life Guards companies in the second line did, and I reconnoitered the situation in the area of Hill 1066 with *Leutnant* Streicher.

Here we came across units from the 12th Life Guards engaged in a firefight with a strong enemy force. The enemy was positioned in towering, multi-storied fortifications sited across from one another and fortified with very tall and strong wire obstacles. These were on the summit 500 meters northwest of Hill 1114 and also on Hill 1114 itself. Also additional Italians were position right of the path at the same elevation as the right flank of the 12th Life Guards.

I brought the 1st Company under *Leutnant* Triebig forward with utmost speed and sent them to clear the enemy from the positions right of the path in the area southwest of Hill 1066. The companies went about this task quickly and with great dexterity. The positions were taken without casualties on our side, with seven Italian officers and 150 men made prisoner. Meanwhile, the 2nd Company and the 1st Machine Gun Company had cleared out enemy position areas, shelters and observation posts west of Hill 1066 at my prompting. Detachment Schiellein moved into the area 100 meters northwest of Hill 1066, just below the rocky peak we had cleared off, and placed itself at my disposal.

I went with *Leutnant* Streicher to the right flank of the 12th Life Guards, in order to have a look at the situation ahead of Hill 1114 for myself while it was still daylight, to establish contact with the 3rd Life Guards Battalion and to keep myself abreast of their future intentions. We met many officers of the 3rd Life Guards Battalion in the front line 50 meters south of Hill 1066. They showed us a patrol who were just about to attempt creeping up to the nearest enemy position in the saddle between Hill 1114 and the summit 500 meters northwest of it, by going through a basin that led in that direction.

The prospects for this patrol hardly looked favorable. The enemy seemed totally undaunted and sprayed the bare grassy slops ahead of his positions from time to time with machine gun fire from different directions. By all outer appearances the enemy garrison there was really on their toes and not disposed to give up any part of their position.

Leutnant Streicher and I, in addition to officers of the 3rd Life Guards Battalion, were of the unanimous opinion that any immensely strong and dominating mountain position held by the enemy on either Hill 1114 or the summit 500 meters northwest of it (which was between 1,120–1,130 meters high), which had hitherto been untouched by our artillery fire and was defended by a strong and tenacious enemy force, could only be taken by us alongside comprehensive artillery support.

Using the binoculars, I observed the unique details of the enemy positions for a long time. As I did so, repetitive machine gun fire from the direction of Hill 1114 forced me into cover. At last darkness fell. The 1st Company's attempts to seize additional parts of enemy positions on the Hill 500 meters northwest of Hill 1114 failed. The Württemberg Mountain Battalion's units settled in for the night. The 1st and 2nd Companies received orders to conduct combat reconnaissance during the night. An Italian artillery observation post served as Detachment Rommel's command post. It was situated close behind the 1st Company. Here I discussed prospects for continuing the attack on Hill 1114 and the Kolovrat ridge with *Leutnant* Streicher and individual officers of the 3rd Life Guards Battalion. The 10th and 11th Companies of the Life Guards had as yet not been deployed. Nothing was known of the results of the 12th Company of the Life Guards attack against Hill 1114.

At about 1900 hours, I was summoned from the command post of the 3rd Life Guards Battalion, which was located in a dugout about 100 meters from Detachment Rommel, to the leader of the Life Guards Regiment who had just now arrived—*Major* Graf Bothmer. I reported the mission of the six mountain ranger companies under my command. *Major* Bothmer immediately demanded them to be placed under his command. I took the liberty of objecting that until this point I took orders from *Major* Sproesser, who to the best of my knowledge was senior in authority to the leader of the Life Guards, and that this matter would have to be taken up with him as soon as he arrived here.

At this, Bothmer forbid my detachment from performing any action to the west or directly against Hill 1114, because he claimed this matter was the business of the Life Guards. He left it to the units of the Württemberg Mountain Battalion to either occupy Hill 1114 on Oct. 25 after the Life Guards had taken it or to follow the Life Guards westward in the second line. I answered him that I would report to my commander about this. Then I was dismissed.[14] I was not happy as I returned to my command post. Fighting in the second line was in no way acceptable to us mountain rangers, and I started trying to devise ways and means of obtaining total freedom of action for my fighting force again. It was clear to me that the opportunity would only come after the arrival of *Major* Sproesser.

At about 2100 hours, *Leutnant* Autenrieth, the mess officer of the Württemberg Mountain Battalion, arrived in Detachment Rommel's command post. He had come over to us after visiting the 12th Life Guards Company and then the 3rd Life Guards Battalion command post, where he had been present during a discussion of the Oct. 25 attack being planned against the mountain positions on Kolovrat with artillery support. I learned from him

---

14. Rommel adds as a footnote: "See amendments of the Reichsarchiv to the Volumes Isonzo I and II from Spring 1930."

that *Major* Sproesser had continued the attack against Foni with Detachment Wahrenberger and had broken through there shortly after darkness fell. Furthermore, *Leutnant* Autenrieth reported that the 12th Division had done very well in the Isonzo valley. I

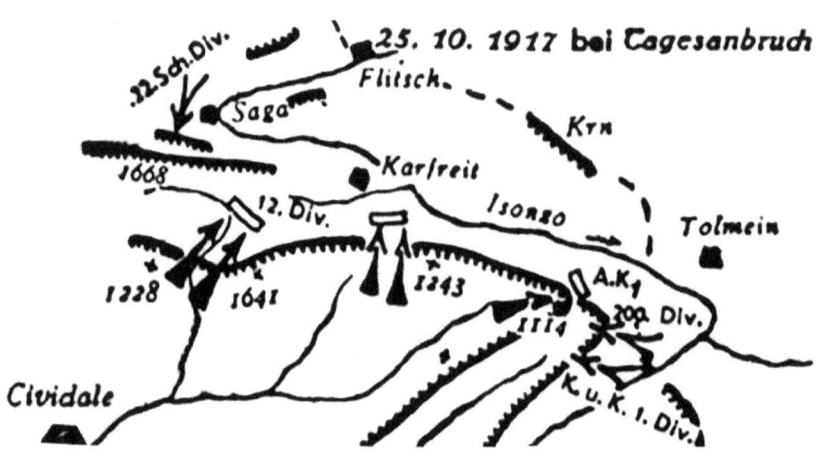

The area around Cividale and Hill 1114 on Oct. 25, 1917 at daybreak.

put him in the picture about the situation here ahead of Hill 1114 as well as the relationship with the Life Guards. I charged him with reporting this as fast as possible to *Major* Sproesser and to ask Sproesser to still come, with or without Detachment Wahrenberger, to Hill 1066 before daybreak, and thus arrange freedom of movement for Detachment Rommel again. *Leutnant* Autenrieth joyfully took on this truly difficult task and carried it out immediately in the pitch dark night and across a landscape that had by no means been cleared of the enemy.

The night of Oct. 24–25, 1917 was uncomfortable, to say the least, for the units of the Württemberg Mountain Battalion remaining at Hill 1066. We were in wet clothes and the wind was really cold. Night raids made by companies deployed towards the front brought in dozens of prisoners who had been seized from the area in front of the enemy obstacles. Moreover, none of the raids managed to bypass the obstacles to approach the enemy's frontal position. The Italian sentries proved themselves to be extremely watchful in all areas and were quick at hand with grenades and machine gun fire.

The 3rd Life Guards Battalion informed us that reserve companies north of Hill 1066 were deployed in the late evening hours on the northeast slope on Hill 1114 to the left, but until this point had not linked up with the *Jäger* Regiment I attacking over Hill 732. We learned nothing of the capture of Hill 1114's peak by the company of *Leutnant* [Ferdinand] Schörner[15] of the 12th Life Guards.

Half-asleep on a hard pallet, I contemplated how to carry out the attack. A frontal assault? Without precise support from the artillery, which could come into action on the first morning hours of Oct. 25 at the earliest, it seemed impossible to me to carry the attack through from the positions we had reached so far against the formidable enemy

---

15. The same Ferdinand Schörner who would later become a Nazi Party member and Field Marshal.

installation system on Kolovrat. Furthermore, the Life Guards did not want the frontline participation of the Württemberg Mountain Battalion in such an attack.

However, if deciding against the extremely time-consuming set up of artillery support, the question then arose of managing a surprise break-in to a section of the 3rd Italian position, hitherto not attacked by us, located west or southwest of Hill 1114 and several thousands of meters away from the battle's focal point. West of Hill 1114, the 3rd Italian position crowned the barren, stepped knolls of the Kolovrat ridge rising towards Kuk. An successful breakthrough west of Hill 1114 would inevitably impact enemy positions located at lower elevations around Hill 1114. Also, this area offered good prospects for the adventurous officers and men of the Württemberg Mountain Battalion.

Southwest of Hill 1114, the enemy positions were lower than those on Hill 1114 itself. A breakthrough there would hardly improve the situation on Hill 1114. That was inconsiderable for the Württemberg Mountain Battalion, positioned to the right of the Life Guards Regiment. But had the commander of the Life Guards not expressly forbidden my detachment to undertake any activity against the enemy to the west?

Aside from a short hand grenade brawl, the night of Oct. 25 passed quietly. Patrols dispatched towards the enemy positions in the early morning hours were driven off by attentive Italian sentries, just like the patrols sent before midnight had been. Detachment Rommel was given no update from the 3rd Life Guards Battalion in the early morning as to whether the situation had changed during the night.

It was still pitch black when, at about 0500 hours, *Major* Sproesser walked into my command post. Following close on his heels came the rest of the Württemberg Mountain Battalion (4th Company, 3rd Machine Gun Company and the Communications Company). I put him in the picture about the situation here at Hill 1114, the attitude of the Life Guards towards us and my plan of attack. To carry out my plan, I requested command of four rifle companies and two machine gun companies.

Major Sproesser agreed with the plan to attack the 3rd Italian position. However he only gave me two rifle companies and one machine gun company and told me that he could give me additional forces if my plan were successful.

As I gave orders to my new group for departure, *Major* Sproesser argued with the commander of the Life Guards, who arrived at my command post.

Considerations:
- **The 1st Italian position by St. Daniel consisted of a continuous trench** in the front line with numerous shelters and dugouts in addition to strong wire obstacles. **Single machine gun nests and strongpoints** were sited **in the zone of resistance organized in depth** between the 1st and 2nd positions. The camouflage of the front line was insufficient in contrast to the positions of the zone of resistance, which could hardly be recognized.

The **Italian front line was battered to ruins** by the **German preparation fire** and its **garrison** nearly **driven out**. The **few enemy machine gun nests,** and so forth, **located in the zone of resistance** that the preparatory fire had not smashed to pieces **were not capable of bringing** the incoming **attack** on a wide front **to a standstill.** If the Italians had arranged numerous machine gun nests in the zone of resistance between the 1st and 2nd positions, perhaps the German attack could have been held off. To destroy a position with a deep main defensive area as we see today via a barrage would require a really monumental artillery effect.

- The **point** of Detachment Rommel lost five men during a run-in with the second Italian position on the narrow path of a steep wooded hill. Had the distance apart per man been greater, losses would have been fewer. In Romania, the Cossacks rode as the advance guard with 200 meters distance between each man in open terrain. Whenever the first man came across something, the nearest man reported it. An infantry point must do similarly. The leader of the point men must **fight against the herd instinct to close together in a dense mass**.

- As the Italian occupants of the 2nd position on the road to Foni proved themselves to be very vigilant, the enemies occupying the same position some 800 meters southeast demonstrated that they did not have enough sentries. It is not enough to just have attentive sentries in a position, but also **the area ahead** must be **continuously watched** through regular sweeps, especially in bad weather and when the terrain is covered with vegetation and many crevices.

- Situation of the assault battle on Oct. 25 at daybreak: Group Kraus, attacking in the Flitsch basin, had just reached Saga on the evening of Oct. 24 in a thrust into the valley. They marched on the morning of Oct. 25 from there [Saga] to attack Stol, which was at 1,668 meters elevation. In the Isonzo valley, the 12th Division—advantaged by the rainy, cloudy weather which interfered with enemy artillery fire from the mountains into the valley – pushed forward as far as the Natisone valley near Creda and Robic, via a route from Idersko through Karfreit on Oct. 24. Group Eichholz (two battalions, 1st Artillery Platoon) was diverted from Luico Pass. On the morning of Oct. 25, weak units of the 12th Division (Company Schnieber) climbed the northern foothills of the Matajur massif as Group Eichholz remained near Golobi in a fierce battle against vastly superior Italian forces. Among the *Alpenkorps*, the Bavarian Infantry Life Guards Regiment and the Württemberg Mountain Battalion were fighting for the corner post of the 3rd Italian position near Hill 1114. Although Schoerner's company (of the 12th Life Guards) had the position on the peak of Hill 1114 firmly in hand, the Italians held onto their positions all around it with the utmost fierce tenacity and tried to win back areas of positions their had already lost through counterattacks. The *Jäger* Regiment 1 was still fighting over the 2nd Italian position in the area of Hill 732. The 200th Division had taken Jexa with *Jäger* Regiment 3. *Jäger* Regiment 4 fought for the 2nd Italian position west of Hill 497. Group Scotti had taken the 1st and 2nd Italian position with the Imperial Austrian 1st Division and had reached the line stretching across Ostry—Kras—Pusno—Srednje—Avska.

- Overall summary: The 3rd Italian position on the towering heights south of the Isonzo (across Matajur, Mrzli Vrh, Golobi, Kuk—Hill 1192—Hill 1114, la Cima, Mount Hum) was still, with the exception of small pieces of positions on Hill 1114, firmly in the hands of the Italians. Their garrisons were fresh for the fight; they had no lack of reserves. Up to this point their position had not suffered much under German artillery fire. —Rommel

"Stouthearted Warriors"—The 2nd Day of the Attack, Oct. 15, 1917. Surprising Breach into the Kolovrat Position

◊ Rommel, as usual, kicks a hornet's nest after he decides to keep pressing forward. Before long, a superior enemy force is bearing down on him and threatening to wipe out his forces. Again he leads from the front; his staff and communications personnel are practically in the front line.

◊ The 1944 U.S. Army translation omits two sentences about Rommel's worry for his comrades and him personally rushing forward to come to their assistance.

◊ Young Rommel becomes extremely emotional about his troops in a bad situation. He regularly refers to his troops as his "little flock." When some of his comrades are threatened, he throws his entire force into action against superior enemies to try to rescue them. It's an all-or-nothing battle— Rommel would rather risk himself and his whole force than leave his men to die.

◊ Rommel personally intervenes to save some unruly Italian officers taken prisoner from being killed by his own men. In this case, Rommel was willing to take a stand against fellow Germans to save the lives of enemy POWs.

◊ Rommel pens an especially heartfelt tribute to his men who died. He describes the men as "warriors" and with poetic flair writes they have "left behind" their young lives, in the same way a person would describe having to take leave of something, instead of using phrases such as "to die" or "to be killed."

At first daybreak on Oct. 25, 1917, I left the western part of the rocky peak near Hill 1066 accompanied by the 2nd Company and the 1st Machine Gun Company. We climbed down in a northwesterly direction along a steep, narrow gully towards a thick network of bushes some 50 meters below. However, the highly observant enemy recognized this movement and caused a few casualties among us with machine gun fire; these were light wounds. Soon all of us had reached the cover of the bushes. Here Detachment Rommel ran into the 3rd Company. A heavy firefight was breaking out above us on Hill 1114.

Before we marched off in advance, the company commanders were briefed about the plans for the undertaking. I wanted to move 200 to 400 meters west below the enemy's Kolovrat position on the steep northern slope, settle in some 2000 meters from the chaos of battle on Hill 1114 and then search for, and use, an opportunity to launch a surprise attack on the enemy's 3rd position from favorable terrain. It was especially important that all of our movements remain totally invisible from the perspective of the Italian position.

The 2nd Company under *Leutnant* Ludwig pressed forward as our point, with I myself personally leading them with hand signals. My detachment staff (adjutant, numerous couriers, telephone troop) came along 30 meters behind the point. The telephone troop lay down a wire connection to *Major* Sproesser's command post on Hill 1066 as we advanced. Following in single file an additional 50 meters behind were the 2nd Company, the 1st Machine Gun Company, and 3rd Company.

Moving along like this was especially invigorating after spending a cold night in wet clothing. We had Italian tinned rations in place of our morning coffee. Behind us on the left by Hills 1066 and 1114, the sounds of battle swelled ever closer the more light it became. We distanced ourselves from this battle noise. Silently my detachment crept from bush to bush, from hill to hill. At first, terrain formations and vegetation allowed us to move barely 200 meters below and alongside the enemy position. Then obstacles became visible on the bare knolls of the lengthy Kolovrat ridge and forced us to take detours towards the valley which consumed both our time and our strength. From these obstacles, and possibly even from areas ahead of them, the eyes of countless enemy sentries spied the hillsides we were moving across. If even one of them spotted us, he would sound the alarm—and then the entire objective of my undertaking would become extremely questionable, if not made impossible altogether.

I always ordered a halt whenever there proved to be a need for personal reconnaissance of avenues of approach. So much depended on finding the right way. With extreme caution, we crossed through multiple deep gorges, then again went forward over a grassy hill. We didn't only have to watch out from the area above on our left—we also had to make sure that the column was out of the enemy's sight on from ahead and from the rear. We could only guess at what the enemy's perspective war from the steep height above. Their continuous and highly placed obstacles hinted at strong positions. Because groups of bushes thinned out the higher up one climbed, it was only possible for us to make a concealed approach using narrow gullies; these furrows were numerous on the slope.

We had been underway for over an hour and had already put about 2,000 miles as the crow flies between us and Hill 1066. Since leaving there we had not been shot at by the enemy anywhere. The sound of really fierce machine gun fire kept coming from the direction of Hill 1114. Were the Life Guards attacking?

A view of the action on Oct. 25, 1917 from the north.

The knolls of the Kolovrat ridge and their fortifications lay above in the morning sun. It promised to be a wonderful, warm autumn day. A deep silence was all around us. About 200 meters below the enemy obstacles, the point men crept past isolated groups of bushes into a basin ahead. I was reflecting on exactly how and where I could cross the bare, sharp ridge barely 100 meters ahead of us when I heard a faint sound from behind me. Looking back, I saw some mountain rangers of 2nd Company ducking under a large clump of bushes below the path used by our point men.

What was going on? The troopers at the head of 2nd Company had discovered Italians sleeping in some bushes on the hill below. Within the next few minutes they had eliminated an Italian outguard of 40 men with two machine guns. There were no shots fired during this time, and no loud words were spoken. But a couple of enemy sentries fled down into the valley as fast as their legs could carry them. Thankfully they forgot, in their agitated state, to alert the garrison above them with shots or yelling. I made sure that nobody on our side fired any shots after them.

These sentries had probably been charged with guarding the garrison on the Kolovrat ridge from surprise attacks from the Isonzo valley. Probably the enemy had more weak outguards in the bushes which were several hundred meters below. These sentries were probably only expecting us to come out from the Isonzo valley. They had never considered that we could come from the east from the direction of Hill 1066.

After the enemy's main security force ahead of their position had been soundlessly thrown out, the surprise break-in to the Kolovrat position that I planned here had especially good prospects for success. Our avenues of approach to the obstacles above were favorable. Above all, the deepest part of the basin in which my point men were staying for the moment, was totally invisible from the ridge positions above it on either side. I decided to stake my break-in attack from here.

The prisoners were taken to the rear of the column. I had the point climb from the basin until up to 100 meters from the enemy obstacles. The uppermost part of the obstacle pickets were just coming into view. The point took over security duty while the detachment prepared to attack. With utmost caution I drew the individual companies into the basin one after another and got them ready to attack, positioning them beside one another yet still concealed from the enemy. The space was narrow and they were massed together quite heavily. I quickly instructed the leaders about my objectives. After that, the troops prepared with the utmost caution. All of them did so, including the ones right behind the point men, which meant all of this occurred about 100 meters from the enemy obstacles. The hills were very steep and had high arcs.

Nothing moved in the enemy position ahead of us. Far over to the left, the sounds of battle became increasingly loud. My adjutant, *Leutnant* Streicher, volunteered to reconnoiter the obstacles ahead to judge their strength and look for any gaps, and also to cut gaps into the wire if necessary. I gave him five men of 2nd Company and one light machine gun. The patrol were instructed only to use firearms in the case of extreme necessity.

Streicher crawled uphill with his men. *Leutnant* Ludwig sent a few troopers forward to keep in contact with the patrol. In the meantime the telephone troop had established contact with *Major* Sproesser's command post at Hill 1066. I reported the course of events in the undertaking until that point and my decision to break in soon and by surprise to the enemy position on Kolovrat, about 800 meters east of Hill 1192. Additionally, in the event that we succeeded, I requested that reinforcements be placed under my command and be sent over as quickly as possible. The reinforcements were promised.

Major Sproesser had been observing our entire advance from his command post using binoculars. He imparted to me that the situation ahead of Hill 1114 had changed insofar as the Life Guards had attacked the Italians with strong forces. Under these circumstances it had been impossible for artillery to support an attack by the Life Guards.

At the very moment I set down the telephone and was in motion of tearing into a piece of Italian white bread, a short report came in from the front from Streicher's patrol via the communication people: "Patrol group has broken in. Artillery seized. Prisoners taken."

Above in the enemy position there was only a complete silence. Not a single shot rang out. I now with utmost haste set my decision in motion to break in with my entire detachment following the same path as the patrol. The resourceful man can tear quick success from every second of delay. Exerting all their strength, all the men of Detachment Rommel climbed uphill from the steep basin in the minutes that immediately followed. In the next few seconds, the enemy obstacles had been reached and climbed past. Afterwards we crossed over the enemy position. The long barrels of a heavy Italian battery rose ahead of us. Around them, Streicher's men cleared out individual shelters. Several dozens of captured Italians stood in the immediate vicinity of the artillery guns. *Leutnant* Streicher reported that he had surprised the gun crew while they were doing laundry.[16]

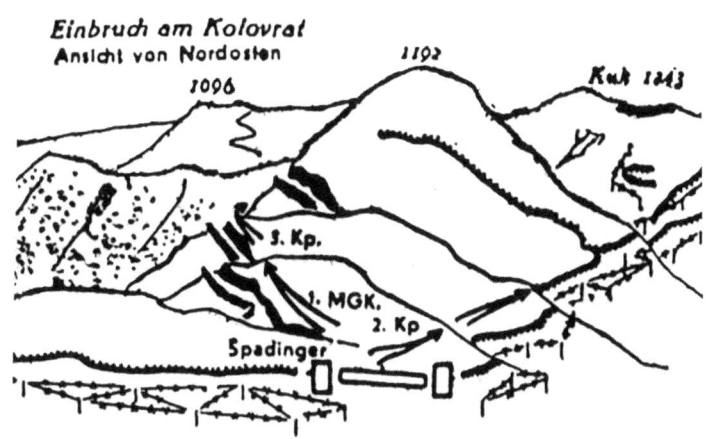

Rommel's map of his breakthrough into the Kolovrat position, shown from the northeast.

We stood in a narrow saddle land formation. On the bare knolls on both sides were numerous earthworks and connecting trenches leading towards the strongly fortified continuous position located on the northern slope. On the southern slope of the saddle, only 100 meters from the position on the northern slope, ran the high road leading from Luico, Kuk—Hill 1114, which was well-camouflaged from sight from the ground as well as from the air.

A third of Detachment Rommel was now in the saddle formation. The soldiers gasped after the mighty exertion of climbing in assault up the steep hill. It seemed the enemy garrison in the Kolovrat position had still not noticed our break-in. Were they still asleep?

Judging by the number of prisoners taken in the saddle formation, which was only about 50 meters wide, the position had to be strongly manned. Seconds would now decide our fate. I ordered: "Detachment Rommel will close up towards the east and penetrate further to the west. *Vizefeldwebel* Spadinger, with the 1st Machine Gun Group of the 2nd Company, is to seal off the enemy position on the northern slope towards the east, block the high road in the same direction and thus cover the rear of Detachment Rommel as it advances west. *Leutnant* Ludwig will go west and tear open the enemy position on the northern slope. Shooting is to be avoided wherever possible. I will go with

---

16. Rommel's sense of humor here. He seems to have been both slightly disdainful and amused by the idea of the enemy gun crew being vulnerable while distracted with laundry.

the 3rd Company and the 1st Machine Gun Company on the high road to the west. *Leutnant* Streicher will take over security in this undertaking with his patrol. Get going as quickly as possible!"

All units of the detachment jumped to their tasks with spunk and great circumspection. Under the vigorous *Leutnant* Ludwig, storm troops of the 2nd Company swooped into the enemy position, going from shelter to shelter, from sentry post to sentry post. The majority of the enemy garrison were found still in their shelters. A single mountain ranger was enough to oversee the lining up, disarming and marching away of enemy occupants of a shelter.[17]

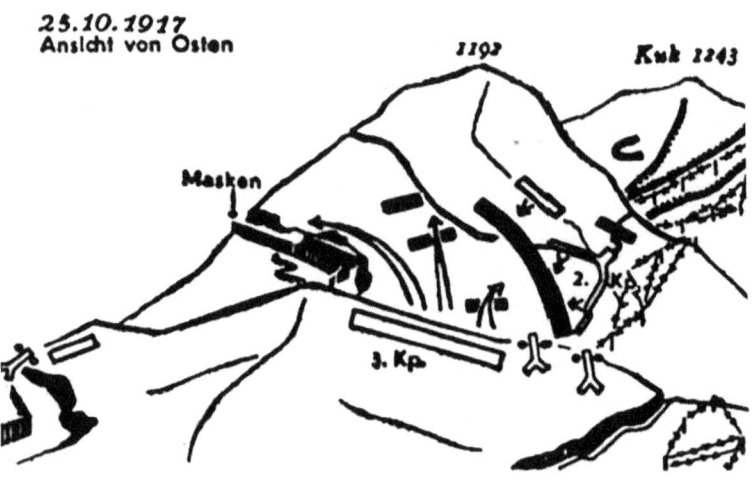

A view of the 3rd Company's maneuvers from the east.

In the enemy sentry positions, the solitary sentries kept watch facing the valley. The Isonzo valley, lying in the beams of the morning sun with its mighty 2,000-meter mountains in the background, radiated a picture of bewitching beauty.

As the troopers of the 2nd Company suddenly came up standing behind the sentries as if they had jumped out from the ground, the sentries went limp with horror and showed as much neglect for signaling alarm as their comrades stationed ahead of the position had half an hour before. The number of prisoners quickly grew into the hundreds in this manner. The main body of the detachment also got far ahead on the high road. It was lucky for us that we were masked from the enemy on the heights east and west. Multiple gun batteries positioned in blasted openings in the rock face on our right were captured. Our sudden appearance in the morning silence far from the sounds of battle near Hill 1114 also surprised the enemy garrisons here out of their wits.

As the goal of the advance down the road, I had intended to possibly surprise nearby reserves. Yet I also wanted to quickly grab by the scruff of the neck the enemy on the northern slope, who would probably mount some resistance against the 2nd Company.

---

17. Some boasting from Rommel. He points out that a "single mountain ranger" could manage a garrison of Italians. In doing so he shows his pride—at being German and a mountain soldier—but his actions demonstrate that, although he viewed Germans as better fighters, he did not hold disdain for Italians as people.

Events took a different course! About 10 to 15 minutes had passed since our penetration into the Kolovrat position. The lead units of the 3rd Company were approaching the saddle formation 300 meters east of Hill 1192. It would probably come to a battle over there. Patrol Troop Streicher, which had just reached the saddle formation 300 meters east of Hill 1192, came under machine gun fire from the southern slope of Hill 1192 and was soon hard-pressed by Italian infantry storming across the high road from the hill's southeastern slope. The patrol troop retreated from the northern slope of Hill 1192.

Heavy machine gun fire from the direction of Hill 1192 brought the advance of the 3rd Company and the 1st Machine Gun Company on the road to a standstill. Soon the units of the Machine Gun Company were deployed. Only they could assert themselves against the overpowering enemy force. To attack Hill 1192 from the roadside across the steep, coverless slopes of the Kolovrat ridge would be supremely difficult, because now enemy machine gun fire was striking at the left side of the road from between the screen of the mountains.

Within a few seconds the sounds of battle swelled up violently ahead of us, especially to our right where I estimated the 2nd Company to be. Hand grenades exploded amid the distinctive sound of fierce gunfire from my mountain rangers' carbines. They all seemed to be in the firing line down to the last man.

I couldn't see anything. You couldn't signal [to them] from the bare knoll to the right of the street, otherwise enemy machine guns from Hill 1192 would shoot at you. Could the 2nd Company hold out against the enemy? They only had 80 carbines and six light machine guns! If they were overpowered, the enemy would quickly win back his lost positions on the northern slope and free all the prisoners we had taken. I could tell that the enemy facing us was very strong from listening to the violent firefight. A few minutes had been just enough to turn the whole situation to our disadvantage and make things quite gravely serious for us. It seemed most urgent to me to block the road towards the west and to come to the aid of the harried 2nd Company as quickly as possible.

The shortest way north over the bare summits was blocked on both the east and west by numerous enemy machine guns. An attack on both sides of the road to the west against Hill 1192 would get caught by similar enemy fire and therefore had no prospects of breaking through. I came up with a different solution.

A machine gun platoon was already in a firefight with the enemy on Hill 1192 and some troopers from the 3rd Company were tasked with blocking the road towards the west. I rushed with the rest of the 3rd Company and the Machine Gun Company as fast as I could along the road back towards the saddle 800 meters east of Hill 1992. Cover blocked the enemy east and west from observing this movement and targeting us with fire. The enemy's occasional broad sweeping of camouflaged areas with fire did not do much

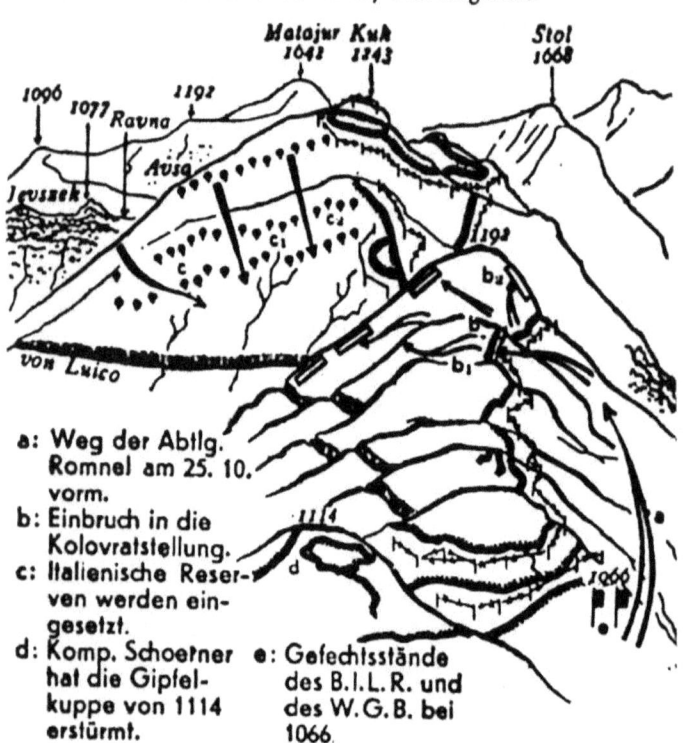

Enemy reserves shown being deployed to Kuk. Starting at the lower right: a) the path taken before noon by Detachment Rommel on Oct. 25; b) the breakthrough into the Kolovrat position; c) Italian reserves are sent in (over the hill in the top left); d) Schörner's company has stormed the peak of Hill 1114; e) the command posts of the Bavarian Life Guards and the Württemberg Mountain Battalion on Hill 1066.

to hinder our movement. We reached the saddle formation.

Here the hardy Spadinger with his eight men held the Italian garrison to the east at bay. I had an additional two groups hop over to strengthen him. Then we went running over into the Italian position on the northern slope that had been cleared out by the 2nd Company and again towards the west. Two mountain rangers stood guard over some 1,000 Italian prisoners between the position and the wire obstacles 150 meters west of the saddle. I called out to them to move the prisoners to the hill below the wire obstacles and left it to them to carry out this task. They pulled it off! The sweeping Italian machine gun fire from east and west coming over the hills helped to get the prisoners moving faster.

A few 100 meters ahead of us, the sounds of battle near the 2nd Company swelled to a very fierce pitch. Hand grenades cracked, machine guns fired ceaselessly, carbines gave rapid fire. I demanded speed to the utmost extreme from the companies following me. Our help must not arrive too late.

I hurried forward with a few battle orderlies on my staff. I surveyed the situation from a knoll 350 meters east of Hill 1192. The 2nd Company held a few trench sections on the northeast slope. They were encircled from the west, south and east by enemy forces five times their strength—an entire Italian reserve battalion. The foremost enemy units lay hard-pressed about 50 meters across from them. In the rear of the 2nd Company stood the vast and tall Italian obstacles, which made a retreat on the northern slope impossible. The troopers defended themselves desperately against the violent mass of the enemy. If the enemy chose to storm forward despite the firing, the little flock would surely be crushed to death.

Should I bring the troopers arriving behind us into the firefight right away? No! It quickly became clear to me that the 2nd Company could only have an exit cut out for them by a surprise thrust with my remaining detachment into the flank and rear of the enemy. Then the close quarters battle against vastly superior enemy troops would be decided by the victory or defeat of the mountain rangers.

Breathlessly the foremost units of the 3rd Company gasped through the deep trenches, followed by the first gunners of the Machine Gun Company with their disassembled machine gun. Their leaders were briefed with a few words about what was going on and what they were supposed to do. We went towards the left and out of the trench from a shallow basin. The men of the 3rd Company were lying here concealed from the nearby enemy ahead and ready to storm. In the basin to the right, a heavy machine gun crew in concealment brought their gun into working order at a flying speed and reported that it was ready to fire. The crew of an additional heavy machine gun came gasping over to us; on the left the majority of the 3rd Company had reached the basin and were lying ready to spring.

I could no longer wait for the second heavy machine gun to be ready to fire. The densely massed enemy 100 meters ahead of us, fired up by the orders of their officers, was preparing to storm the trapped men of the 2nd Company. I gave the signal to attack to the 3rd Company and the 1st Machine Gun Company. As the first heavy machine gun thrust out from its previous cover and began striking the enemy with continuous fire and the second heavy machine gun shortly pitched in with fire afterwards, the mountain rangers rushed left upon the flank and rear of the enemy with wild determination. The surprise blow in the flank and rear hit home. The Italians stopped their rush against the 2nd Company and tried to wheel to face the 3rd Company. But now the 2nd Company threw themselves into action and stormed the enemy from the right.

Seized on two sides and pressed into a tight space, the enemy cast aside his weapons. Only the Italian officers defended themselves right up until we were only a few meters from them, using pistols. Then they were physically overpowered. It was necessary for me to personally intervene to save them from the wrath of the mountain rangers.

A whole battalion with 12 officers and over 500 men were captured in the saddle 300 meters northeast of Hill 1192. Thus the total number of our prisoners taken from the Kolovrat position increased to 1,500 men. We took the peak and the southern slope of Hill 1192 and captured there yet another heavy Italian battery.

Our tremendous joy about our hard-earned successes was dimmed by very painful casualties we had sustained. Apart from numerous wounded, two especially stouthearted warriors had been forced to leave behind their young lives amid close-quarter fighting—*Gefreite* Kiefner (of 2nd Company), who had distinguished himself so very exceptionally the day before on Hevnik as an assault troop leader, and *Vizefeldwebel* Kneule (of 3rd Company).

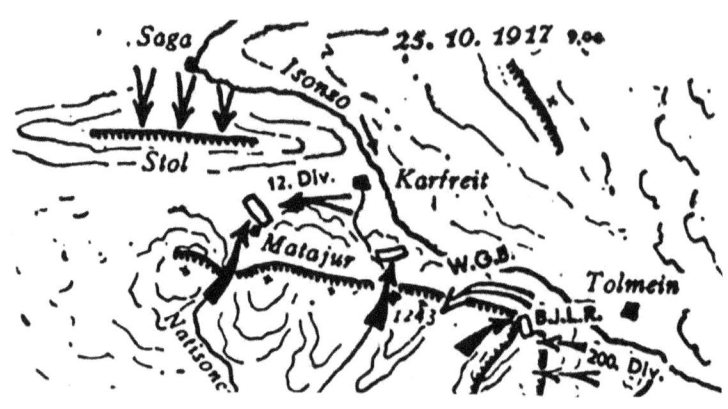

A map showing the overall situation on Oct. 25, 1917.

At about 0915 hours, Detachment Rommel was in absolute possession of a 800-meter-wide section of the Kolovrat position, including Hill 1192 and 800 meters east of it. Thus a wide breach had been punched through the enemy's main position. I had to reckon with additional attempts from the enemy to win back what they had lost. Well, the Italians could go ahead and come! We mountain rangers were unaccustomed to give back what we had fought for and seized in hard battle.

From the west, southeast and east the enemy swept the hills we occupied with machine guns. Also Italian artillery groups on Mount Hum and to the west were not unaware of the breakthrough onto Kolovrat and the fighting around Hill 1192. Their heavy shells forced us to quickly seek the cover offered to us by the northern slope.

With my available forces I could not even think of immediately continuing the attack. It would be enough to hold what we had conquered until the arrival of reinforcements. The 2nd Company and half of the Machine Gun Company occupied Hill 1192 facing west. Spadinger sealed our position off further east in the saddle 800 meters east of Hill 1992. I kept the 3rd Company and half of the Machine Gun Company at my disposal in the positions we had seized on the northeast slope of Hill 1192.

Then I oriented myself about the situation facing us on all sides from the peak of Hill 1192. It seemed on first glance that the most endangered front was west in the direction of Kuk (Hill 1243). Apart from dozens of machine guns shooting at us from positions which for the most part were at higher elevations and similarly arranged to levels of a multistoried building, strong reserves were visible on the highest summit and on the southeast slope. Soon, multiple waves of firing lines moved towards us across the broad eastern slope of Kuk. I estimated their strength to be from one to two battalions.

To the south, Mount Hum was crawling with movement just like an anthill. There, a mighty mass of enemy artillery were firing. Columns of trucks busily transporting troops filled the high road from Cividale across Mount Hum in both directions. On both sides of the street, groups of the enemy in closed formation streamed towards the battlefront. In the east, one could see above the entire Kolovrat ridge, which declined gradually until Hill 1114. Strong massing of the enemy was distinctly recognizable on the south and southwest slopes of Hill 1114. The Italians there seemed to be attacking. Italian reserves from Crai were being taken there in long truck columns and unloaded on the western slope of Hill 1114. Also you could see enemy forces advancing from the east towards us along the high road and over the knolls above. From all outer appearances the enemy wanted to hit us from two sides at the same time.

Considerations:

- **The surprise break-in into the Kolovrat position** on Oct. 15, 1917 was successful because the Italians in the foreground of their 3rd position were **not keeping watch sharply enough**—a mistake that the Romanians continuously made at Mount Cosna.
- Also the **garrison of the position** itself was **unprepared for battle**. Only two kilometers from the focal point at Hill 1114, they thought they were removed from any danger. Thus the mountain rangers initially had an easy game.
- The **counterthrust of the Italian reserve battalion,** undertaken with great energy, was brought to **a standstill under fire from the 2nd Company.** However, **it would surely have led to the eradication** of the 2nd Company, had the strongly massed Italian battalion been hit in the flank and rear at the decisive moment.
- **It would have been a mistake to lead** this attack **with too few forces** or **only to limit it to hitting the flank with fire**.
- After the successful break-in to the Kolovrat position (Oct. 25, 1917, at 0915 hours) the battle situation was as follows:
- **Group Kraus** attacked outward from Saga with the Kaiser-Schützenregiment 1 in three columns towards the Stol Line (from Hills 1668–1450).
- **Group Stein.** The 12th Division stood with the Infantry Regiment 63 as in the previous evening near Robic and Creda and fended off enemy advance parties. Company Schnieber reported that they were 100 meters north of the peak of Mount Matajur. (Probably they meant the Mount della Colonna.)
- Group Eichholz was attacked by superior Italian forces from the Luico Pass but resisted this enemy in a tenacious grappling match and held positions north of Golobi.
- As part of the *Alpenkorps*, Detachment Rommel achieved the breakthrough into the Kolovrat position from Hill 1192 up to 800 meters east. The majority of the

> Württemberg Mountain Battalion was on the march from Hill 1066 to Hill 1192.
> - The Life Guards Regiment held the positions around Hill 1114 they reached on the 24th against heavy Italian attacks. *Jäger* Regiment 1 had taken Hill 732 and was advancing towards Slemenkapelle.
> - With the 200th Division, *Jäger*regiment 3 had taken Hill 942 west of Jeza.
> - **Group Scotti:** The 7th Mountain Brigade with the Imperial Austrian 1st Division attacked Globocak. —*Rommel*

"That Was Fun!"—The Attack on Kuk, the Valley of Luico-Savogna Is Cut Off and the Luico Pass Opened

> ◊ By this time Rommel has developed a sense of himself as part of an elite force. He draws attention to his skills and those of his men in glowing terms. He also points out the bewilderment of his enemies, whom he seems to view as complacent and incompetent.
> ◊ He seems to enjoy spooking opponents with stealth, speed and surprise. Although he has opportunities to massacre enemies after catching them off guard, Rommel demonstrates a unique preference for capturing his startled opponents rather than killing them. Other soldiers would arguably have not bothered to take so many prisoners. Although he comes across as arrogant sometimes, Rommel consistently chooses to spare lives.
> ◊ Rommel describes himself as excited to watch troop movements through binoculars. Binoculars would become a key part of Rommel's image during World War II. He was rarely seen without them.
> ◊ In a humorous anecdote he describes his exploits capturing all the traffic going both ways on a road, and trying to hide his prisoners, horses and mules in between ransacking vehicles. He seems to have enjoyed the experience and also remembering it, describing this caper literally as "fun."
> ◊ In a strange foreshadowing of future warfare, he equips a vehicle with a heavy machine gun and rides off to explore. In this he nearly resembles the fighters of the British Special Air Service (SAS) who later fought against him in North Africa during World War II.

Totally against my expectations, the enemy brought his multiple-wave advance across the eastern slope of Kuk to a halt in position. Was he just intending to box us in, or was he pausing to prepare for an attack? The first explanation made most sense, because the enemy riflemen began digging in under the protection of the positions on Kuk's northern slope in three lines on the eastern slopes. An attack from these forces, supported by numerous machine guns from positions at higher elevation, would really have given me serious worry. The enemy's switch to defense and the resulting pause was exceedingly welcome from my point of view. I also knew that *Major* Sproesser was on his way with the majority of the Württemberg Mountain Battalion.

As soon as additional forces arrived at Hill 1192, I wanted to attack the enemy on Kuk. We should allow him as little time as possible to keep digging; otherwise he would anchor himself so deep into the ground that it would be difficult to toss him out. It would depend decisively on using the time to make detailed preparations for the planned attack.

To guarantee the element of surprise, I let the enemy continue his digging work without disturbing him with fire. It looked like he wasn't able to penetrate the rocky earth at any great speed. I reported via telephone via Hill 1066 to the *Alpenkorps*—since the staff of the Württemberg Mountain Battalion was already on the move towards Hill 1192—about the success up to this point and my intent to attack Kuk as soon as reinforcements arrived at Hill 1192. Additionally I detailed my attack plan against Kuk to the *Hauptmann* in the general staff of the *Alpenkorps* and requested the support of two heavy batteries for the attack. My wish was granted immediately. In a few minutes I was connected with the fire control officer of an artillery group near Tolmein. With him I reached an agreement for fire for effect from both heavy batteries from 1115 hours to 1145 hours on the broad eastern slope of Kuk and the positions on the northeast slope. Thus the artillery support for the attack was firmly secured. I very much expected rockslides due to the effects of heavy shells in the stony terrain.

Now things depended on setting up infantry fire support. To this end I positioned the light machine gun of the 2nd Company and the entire 1st Machine Gun Company on the northern and southern slopes of Hill 1192 under concealment from the enemy on Kuk. The task of both gun crews was to support the attack on Kuk, which at first would be carried out with very weak storm troops, by pinning down the enemy. I pointed out target areas for the individual weapons.

At about 1030 hours, *Major* Sproesser arrived with the 4th and 6th Companies, just as the 2nd and 3rd Machine Gun Companies did, in the saddle just east of Hill 1192. I reported to the commander about the situation and the attack preparations I had already made against Kuk and requested the support of the necessary troops for the attack. After he surveyed the enemy situation from Hill 1192 outward, *Major* Sproesser placed the 6th Company under the command of *Leutnant* Hohl with the task of rolling up the enemy positions on the Kolovrat ridge in the direction of Hill 1114. My attack plan against Kuk found approval. Except for the 2nd and 3rd Machine Gun Companies and the 1st Machine Gun Company, the 4th Company, as well as the 2nd and 3rd Machine Gun Companies, were placed under my command. I rapidly finalized the preparations of my forces for the attack.

All gun crews (of six light machine guns, the 2nd Company and the 1st Machine Gun Company) under *Leutnant* Ludwig lay concealed on the north and south slopes of Hill 1992 ready to launch surprise fire on the enemy on Kuk at 1100 hours. An assault troop of the 2nd Company at two groups' strength held themselves in positions on the northern slope of Hill 1192, while an assault troop of the 3rd Company of the same strength was positioned at the same elevation in the battery positions on the southern slope. Both assault troops were ready to storm. Their task was to storm as soon as firing began and to take the saddle between Kuk and Hill 1192 under strong support from the artillery and machine guns. Then they were advance as far as possible against the Kuk garrison lining the position on the northern slope and the basins along the southern slope. I wanted to probe the enemy position with these assault troops. The 3rd and 4th Companies as well as the 2nd and 3rd Machine Gun Companies lay concealed and at my disposal in the saddle just east of Hill 1192. I wanted to deploy them on the northern or southern slope following the results achieved by the assault troops.

Shortly before the attack began, the first of the Life Guards Regiment began arriving in the saddle east of Hill 1192. Previously the 2nd Battalion of the Life Guards had tried, after waiting in vain for artillery support, to attack the Kolovrat ridge positions from Hill 1114. However very heavy defensive fire from the Italian position 500 meters northwest of Hill 1114 hindered every movement forward. Afterwards the Life Guards had followed the path blazed by the Württemberg Mountain Battalion on the northern slopes of the Kolovrat ridge below the enemy positions between Hill 1114 and the saddle 800 meters east of Hill 1192, which were continuously and tenaciously defended by strong enemy forces. Arriving here the Life Guards encountered the 1,500 prisoners taken by Detachment Rommel, who were being moved off by only a few mountain rangers.

Punctually at 1115 hours the first heavy shells roared over from the barrels of Tolmein and hit right in the middle of the newly formed Italian lines on the eastern slope of Kuk. Rockslides tumbled down to the valley. A good premiere[18] for the attack! Now the machine gunners on Hill 1192 began their work and the assault troops on the northern and southern slopes of the summit started moving. Filled with suspense, I followed their advance using my binoculars.

Our own artillery fire was laid on excellently as in previous times. Shell after shell hit amid the Italian lines. Now the fire of our machine gun crews concentrated on the location where the 3rd Company's assault troop would come next. Soon the assault troop had reached a hand grenade throw's distance from the foremost enemy skirmishing line. Individual mountain rangers used handkerchiefs to wave to the enemy occupants, who

---

18. The word Rommel uses here would be used to describe a concert opening or event kick-off.

were nearly totally exposed in the open to our fire. It worked! The first few enemies crumbled away from the position and ran over to us.

Then came the moment to deploy my main assault force, which was four companies strong. Quickly I gave the following orders to the assembled company commanders:

"Southern assault troop will climb Kuk, take prisoners. Detachment Rommel then attacks with its four companies on the southeast slope of Kuk. The 3rd Machine Gun Company, 4th Company, 3rd Company and the 2nd Machine Gun Company will follow the detachment staff at a hurried pace on the camouflaged high road. The gun crews on Hill 1192 will give especially heavy covering fire as long as the situation allows."

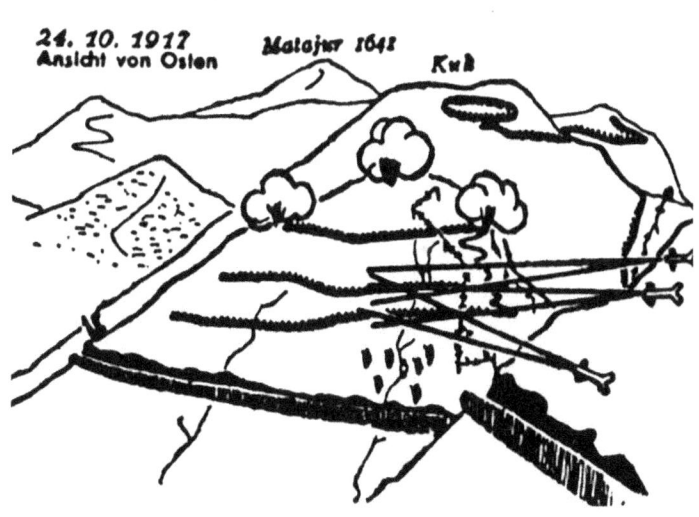

This map gives a perspective of Mount Matajur in the distance as events were unfolding on Kuk on Oct. 24.

We stormed forward in single file along the screened path. Had the enemy on Kuk been more observant, he would certainly have seen this movement from his position at higher elevation. By all outward appearances, however, our gun crews on Hill 1192 and the hand grenade battle on the northern slope was drawing the enemy's full attention. With both sides using up a great deal of ammunition, the machine gun duel roared over and away above us. Only a few stray shots hit the high road. Under these circumstances we reached the saddle between Hill 1192 and Kuk quickly. The place was currently lying in a corner that was covered from the fire of the Italians on Kuk. The detachment came running after us in a long column of twos.

In the meantime, the number of prisoners taken by the 3rd Company's assault troop on the slope above had increased to some hundred. A report came from the rear that units of the Life Guards were joining the forward assault of Detachment Rommel along the high road. Therefore I knew that I had a troop strength at well above regimental strength with a marching depth (columns) spanning a distance of two to three kilometers behind me. Under these circumstances, should I not set our goal even farther?

On the eastern slope of Kuk, the enemy was firmly nailed down in place by heavy machine gun fire and artillery fire in the blink of an eye. Any prisoners who came crumbling away from the enemy side under this bombardments were collected by the 3rd Company's assault troop.

Wasn't the camouflaged high road, which encircled the southern Kuk slope with its entire garrison, a tempting prospect for an encircling advance? The vision of choking off the garrison of Kuk floated before my eyes. Of course I would have to reckon with fighting additional strong reserves on the southern slope and also with the fact that the defenders below us on the steep mountain could storm upwards towards us with considerable force.

But on the other hand I knew that there was no task too difficult for my mountain rangers, who had proved themselves so exceptionally in so many fights. So I chose the forward thrust without any long deliberation. The attack was in motion.

I set the Ravna area as our target—a tiny mountain village on the southwest slope of Kuk. I hurried at a running pace with the forward elements of the detachment ahead down the street. Wheezing under the weight of their heavy machine gun, which they had been carrying on their shoulders since the beginning of the offensive, the troopers

Rommel's troops calmly bypass bewildered enemies, whom he describes as "flabbergasted people" who froze on sight of his forces.

of Company Grau stormed together with their exemplary leader close behind the few riflemen in the point. They all knew that everything depended on wrenching the utmost physical demands from their bodies.

The continuously well-camouflaged high road sank in the direction of Ravna. It had been blown up nearly straight into the barren bluff of Kuk. The enemy garrison on the hill above could not see what was happening on the road below. Their entire attention was continually fixed on the battle with Hill 1192. On the other hand, our own field of vision was also very restricted on the street. The many winding turns left only about 50–100 meters in clear view. Towards the right it was bordered by vertically upright rocks, while to the left and also beyond the slopes ahead of us the areas were screened from view. At the moment this was extremely advantageous to us.

At the shortest distance, sometimes, when straight paths made sharp turns at only a few meters, we ran into completely clueless enemies standing in the street or coming towards us. We were next to them or running past them before they could grab their weapons. It was enough for us to make signals to these flabbergasted people and point in an eastward direction in order to get the Italians weaponless and send them marching alongside our column towards Hill 1192. All of them were petrified by our sudden appearance.

We stormed past batteries, road trains, and enemy infantry groups in closed formation. Nothing held us back. Nowhere was there shooting. Rearwards to the right, above on the slope, the firefight between Hill 1192 and the Kuk garrison continued to play out. A few stray bullets chirped high and away over us. The Italians on Kuk were probably still expecting the Germans to come over from Hill 1192 eventually and make the usual infantry attack in broad formation. Shortly before Ravna the screens on the left side of the street cleared up. The view became wider. We now could see to the right, uphill, isolated long stretches of bushes in rows on the otherwise steep and barren hill. What if Italian reserved were behind or even inside these bushes? The first houses of Ravna lay 300 meters ahead of us. To the left below on the steep rock slope stood single farm buildings. Behind these lay the forested knoll of Hill 1077. Once again we increased our speed as much as we physically could and reached Ravna without being shot at by anyone.

It was noon! The sun burned hotly upon the southern slope. It was no wonder that the Ravna garrison, who supposed themselves far away from any shooting, discovered us for the first time when we were actually storming amongst and away past the few houses and huts. Terrified out of their wits, the Italians scattered in every direction and fled head over heels into the valleys of Luico and Topolo. Their pack animals ran around masterless in the open country.

To our great astonishment no shots were fired at us from any direction. The southern slope of Kuk lay still and silent as if dead. The reserves reputed to by there seemed to already have been deployed against Hill 1192. The last units of Ravna's garrison vanished over the small knoll just west of the village in the direction of Luico. These ones had probably been in charge of the pack animals. Following hot on their heels, I reached the knoll with the foremost riflemen in my detachment and now had a wonderful view of everything that lay to the west.

To the right and below lay the mountain village of Luico in the saddle between Kuk and Mrzli peak. The village, in addition to large military cantonments around it, was filled with many groups of Italian soldiers. The majority of them were milling around peacefully in and around Luico, as they usually did behind the frontlines. Active supply could been observed taking place on the road between Luico and Savogna in both directions. Among other things a heavy battery was being pulled by horses from Luico away toward the house. Loud sounds of battle sounded from the area north of Luico. I estimated the units of the 12th Division to be fighting there. (It was Group Eichholz, three battalions strong. They fought in a counterattack against strong Italian forces, which was planned over Iderslo on Karfreit and was supposed to join up in the flanks and rear with the 12th Division which had pushed ahead into the valley north of Matajur.) Beyond Luico, the Matajur road drew uphill in many winding curves across the partially forested eastern slopes of the Mrzli peak and Mount Cragonza. We observed little traffic on it. Near Avsa and Perati, Italian artillery groups stood in a firefight against units of the 12th division near Golobi.

The companies of my detachment came behind me, marching at quick pace. Their enthusiasm could not be allowed to diminish at Ravna. Instead this offensive spirit needed to be directed quickly in the decisive direction. There was no time left for long deliberations. I quickly considered the following three possibilities for action:

Climbing the southern slopes of Kuk and pushing out the Kuk garrison. This garrison stood with its mass facing east battling against elements of the Württemberg Mountain Battalion. Other units fought the 12th Division, facing north. I no longer regarded this force as a dangerous opponent. I could leave them to the rearward units of the Württemberg Mountain Battalion or the Life Guards. To me their fate looked already sealed.

Attack the enemy forces near Luico and open the Luico pass for the 12th Division. This attack would have good prospects for success. My two machine gun companies could give great fire support from positions at high elevation. The possible avenues of approach towards the massed enemy force around Luico were promising. However the attack couldn't result in the annihilation or capture of enemy forces around Luico, because the furrowed and forested terrain on the eastern slope of Mrzli peak offered the enemy opportunities to clear the pass without losing too much of his force. I also declined this attack possibility and decided upon....

Choking off the enemy forces around Luico by sealing off the valley between Luico and Savogna and the Matajur road on Mount Cragonza (Hill 1096). The forested slopes on both sides of the Luico—Savogna valley were convenient for this undertaking. We could be in the valley near Polava before the enemy forces around Luico noticed anything of how close we were. Then, as rearward units of the *Alpenkorps* pressed forward against Luico from the east, the bottled-in enemy would hardly be able to escape total destruction or capture.

But was my detachment dispersed? I could not clearly see how it appeared in this regard due to the camouflaged high road along the southern slope of Kuk. The men had probably loosened up their contact with one another due to the rapid tempo of our

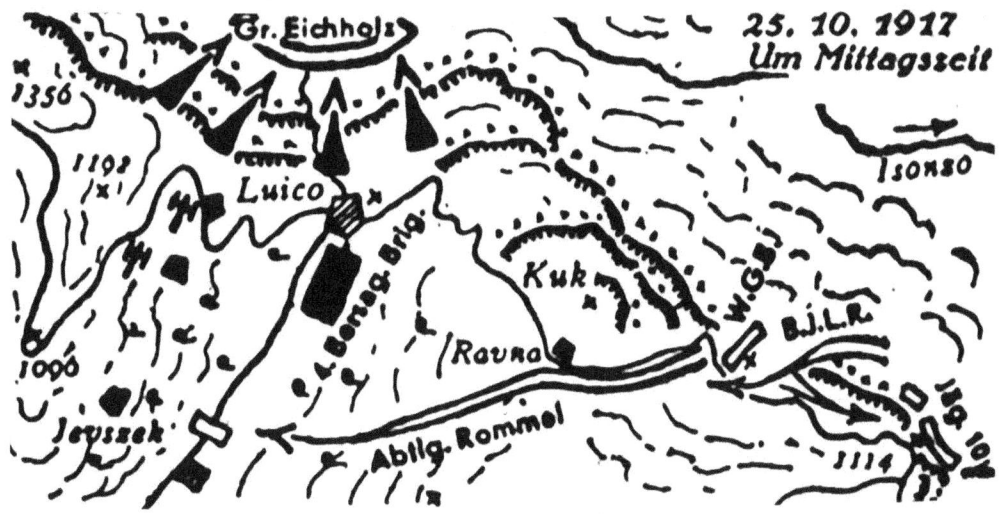

Rommel shows the advance of his forces towards the valley on Oct. 25 around noon.

advance. But I could not wait. Every second was too costly and decisive!

Going out from Ravna I turned sharply southwest, accompanied by the foremost units of the detachment, and made for the Luico—Savogna valley across the forested western slope of Hill 1077. I sent couriers back to Ravna with instructions to send all companies of the detachment to follow us in the directions of Polava.

We captured eggs and bunches of grapes from the baskets of pack animals we caught as we rushed past. Onward we went, fast as the wind! I could not discern whether the knoll of Hill 976 over on our left was still occupied by enemy forces so I left it alone. I didn't want to get bogged down.

Just like a few hours earlier on the Kolovrat ridge, I chose a path through clumps of bushes and small patches of forest. We couldn't see any enemy either towards Luico nor Hill 976. We descended downhill across soft meadows, taking wide and careful steps. We wanted to capture the heavy battery being transported from Luico towards Savogna at all costs. Soon we neared closer to the base of the valley.

The point of Detachment Rommel reached the valley two kilometers southwest of Luico at around 1230 hours. As the point men—among them *Leutnant* Grau, *Leutnant* Streicher, *Leutnant* Wahrenberger and myself—popped up out of the bushes 100 meters east of the road leading to the western side, the oblivious Italian soldiers walking and bicycling along the road got a heart-stopping shock. They had not imagined an enemy here in the peaceful valley three kilometers behind the battlefront near Golobi. They fled as fast as they could into the bushes alongside the road, probably fearing that we were going to shoot them in the next second. Nothing was further from our minds than this.

We reached the road. We nested ourselves in a position where the road made two

Rommel (shown close-up in the inset on the right) chats with fellow soldiers handling supplies and pack animals during World War I.

sharp turns. Multiple phone lines belonging to the enemy were quickly severed. The newly arriving 4th Company and 3rd Machine Gun Company were positioned in bushes and hedgerows on the slopes on either side of the valley in such a manner that they dominated the valley far and wide to the north and south with firepower.

Unfortunately it became clear that our connection with the remaining companies of my detachment was severed while we were crossing over from Ravna across the western slope of Hill 1077. That hit me hard. Without at least two to three additional companies, I couldn't even think of carrying out the advance I planned toward Cragonza to block off the Matajur road. I sent *Leutnant* Walz off with the task of bringing the remaining companies up and then reporting to *Major* Sproesser on event up until now and about what we planned to do.

In the meantime, to our complete astonishment, Italian traffic resumed on the Luico—Savogna road. Individual soldiers and vehicles came obliviously towards us from the north and south. They were politely taken into custody by a few mountain rangers when they reached our hiding place near the sharp curves in the road. That was fun![19]

There was no shooting during this. We placed great emphasis on not breaking the cars' speed as they made the sharp turns on the road. As individual mountain rangers dealt with the drivers and their passengers, others grabbed the reins of the horses or mules and led the teams over to a parking place I designated. Soon we were having trouble stashing away everything that was streaming towards us from both sides. In order to make

19. Lit. *Das macht Spaß!*

room, the wagons had to be unhitched and crammed next to each other. The prisoners, horses and mules were stowed away in a ravine just below our roadblock. Soon we counted over 100 prisoners and 50 vehicles. Business was booming!

For us famished warriors, the contents of the various vehicles presented tasty morsels such as we had never dreamed of. We unpacked chocolate, eggs, jam, grapes, wine, and white bread and divided it amongst ourselves. The brave troopers in the frontline on the slopes on either side of us were treated first. Soon all the struggles and strains of the past hours were forgotten. The mood—only three kilometers behind the enemy front! —was marvelous.

Rommel's drawing of his troops moving cross-country.

This paradise-like situation was disturbed by the alarm call of a sentry. An Italian car was approaching from the south at top speed. Rapidly we shoved a wagon sideways across the street to make an ambush. However, contrary to my express orders, a machine gunner who thought all opportunities for loot were lost opened fire at 50 meters. The car screeched to a halt in a cloud of dust. The driver and three officers jumped out and surrendered themselves, with the exception of one of the officers who managed to reach the bushes below the street and thus escaped. A fourth officer lay fatally shot in the car.

They appeared to be the officers of a high-ranking staff based in Savogna who, unsettled by the break in their telephone lines to the front, wanted to personally report on the conditions of the battle. The car itself proved to be in sound condition. Its former driver moved it into the parking place with the other vehicles we had captured.

Perhaps an hour had gone by since we had blocked the road. There was no trace of the remaining units of my detachment. There were no sounds of heavy fighting to be heard either in the direction of Luico or of Kuk. Hopefully the enemy front had not closed up again behind us. In this case we would have to break back out to our own lines.

A new report from the sentry on the eastern side of the valley turned our attention north. A very, very large enemy infantry marching column was supposed to be coming from the direction of Luico. Marching along as in peacetime—without security—thinking of themselves as far behind the frontline, the lead of the column soon was coming towards us. Alarm! Everybody prepare for battle!

In a few minutes it would probably soon come to a fight: 150 mountain rangers against an overpowering force. But our position was strong and our machine guns commanded the valley far and wide. The closer I let the enemy approach against the roadblock, the less he would be able to spread out his superior forces and bring them into attack. Therefore I told the troopers to hold fire and release only at my direction with my signal whistle.

The start of the enemy column was now marching towards our roadblock from about 300 meters distance. To avoid unnecessary bloodshed, I sent *Offizierstellvertreter* Stahl as negotiator to the enemy, wearing a white armband. He was supposed to make the enemy aware of our forces on the hills on both sides, and order the enemy to cast down their weapons without fighting. As he hurried to the enemy column, *Leutnant* Grau, *Leutnant* Wahrenberger, *Leutnant* Streicher and myself went ahead around the bend in the road. We wanted to lend weight to Stahl's words by waving with handkerchiefs.

Presently Stahl reached the head of the enemy column. Officers stumbled forward, ripped his pistol and binoculars away from him, which, in his hurry, he had not set aside, and took him prisoner. Hardly any words were spoken. Our waving availed nothing. The Italian officers let their forward groups shoot at us. We disappeared quickly around the corner. Then my whistle unleashed a hail of bullets at the enemy column—still standing in the middle of the road—and the swept the street empty in just a few seconds. As the enemy took full cover, Stahl managed to free himself and hurried back over to our side.

Because we had to be very sparing with ammunition, I had everyone hold and adjust their fire for a quick minute, which the enemy responded to relatively weakly. Waving my handkerchief I again demanded their surrender. Too early! The enemy used the pause in fire to spread out into the bushes on both sides of the street and break out against us. At the same time several machine guns opened fire upon us from the hill just west of the street.

Now things would demonstrate who were better shots. The fire we gave from our concealed and highly elevated position had an excellent effect on the enemy who continuously massed close together on the right. After a five-minute gun battle I demanded yet again that the enemy surrender. And even this time it was in vain! Again the enemy's foremost units rushed towards us during the pause in firing. They were now lying about 80 meters across from us. Finally, after 10 minutes of extremely heaving shooting, the enemy felt vanquished and made signals to surrender. At that we stopped firing. Fifty officers and 2,000 men of the 4th Bersaglieri Brigade laid down their weapons on the valley road and came over to us. I charged the stalwart *Offizierstellvertreter* Stahl with assembling the prisoners and transporting them away across La Slava and Hill 1077 to Ravna. He could only take a few riflemen with him as an escort. We received reinforcements from the 3rd Company, which had jumped into the last part of the fight against the Bersaglieri from the hills east of the valley. Loud sounds of battle had been ringing out for what seemed like an endless time from the direction of Luico.

In order to clarify the situation, I armed one of

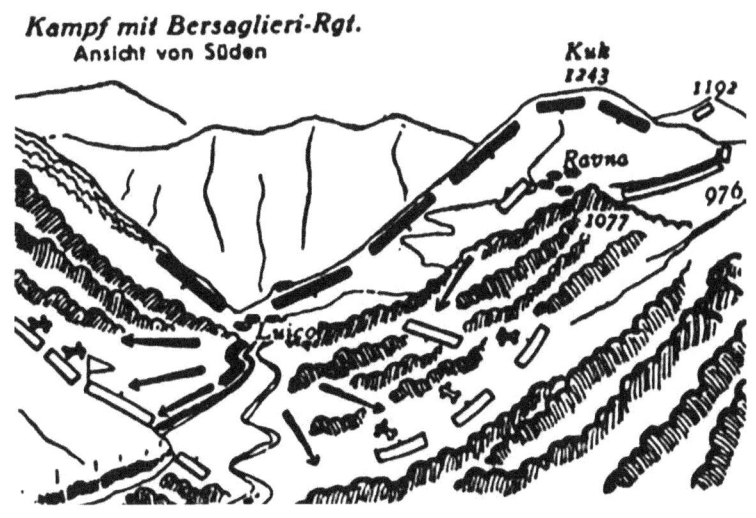

The fight with the Bersaglieri Regiment, shown from the south.

the captured vehicles with a heavy machine gun and drove forward in the direction of Luico. We only got forward very slowly over the street which was strewn for kilometers with abandoned Italian firearms and kit. Just south of Luico I came up against the Italian battery we had observed coming up from Ravna. The team of horses pulling it lay shot dead in the road. By the time I arrived in Luico at about 1530 hours, the remaining units of the Württemberg Mountain Battalion under *Major* Sproesser and the 2nd Life Guards Battalion had already reached Luico and the valley in the south in an attack from Ravna. I met *Major* Sproesser at the southern entrance to the town. At the same time, Life Guards units pushed the enemy on the Matajur road in the direction of Avsa.

I suggested to *Major* Sproesser to take all available units of the Württemberg Mountain Battalion from Polava on the shortest path cross-country via Jevscek to ascend toward Mount Cragonza and take it into our possession. If we had Mount Cragonza, the enemy located on Mrzli peak would lose access to the south and we could attack him in the rear while he was pinned down from the north and northeast by units of the 12th Division and the *Alpenkorps*. Furthermore, in taking Cragonza, we would seize the only mountain road leading to the Matajur massif and would thus cut off all Italians driving on the road or positioned in stationary batteries on the sides of it.

It looked far less advisable to me to advance against Mount Cragonza in parallel to the Matajur road via Avsa and Perati. How were things looking over here from the enemy's point of view? After surrendering the Luico pass, strong Italian formations were moving along the Matajur road toward the eastern slopes of the mountain range between the Mrzli peak and Cragonza, headed more or less to the rear. They probably wanted to occupy rear area positions that had already been prepared there. On the Matajur road, the enemy was in position with a relatively weak rearguard to keep pursuers from going for their throat. The enemy thus gained time to organize his units and to organize detailed plans for occupying the prepared positions. It could also be assumed that the positions in the first line on both sides of the Matajur road were occupied. These considerations compelled me to suggest the ascent to Mount Cragonza on the shortest route.

Major Sproesser agreed and placed under my command the units of the Württemberg Mountain Battalion located in and south of Luico (the 2nd, 3rd, and 4th Companies, the 1st , 2nd, and 3rd Machine Gun Companies and the Communications Company). At the same time, Detachment Gößler (consisting of the 1st, 5th, and 6th Companies of the Mountain Machine Gun Detachments 204 and 265) received orders to place themselves at the disposal of *Major* Sproesser and pull back towards Luico. *Major* Sproesser himself presented himself to the brigade in the Italian car captured near Polava, to report about the course of the fighting up until this point and the anticipated development of the attack, and also to secure artillery support for the anticipated battles.

Considerations:

- **The decision of the Italian leader** on Kuk to bring the German breakthrough into the Kolovrat position to a standstill, in which he **deployed numerous reserves in multiple lines to defend the eastern slope of Kuk, was wrong**. He created urgently needed breathing room for Detachment Rommel (by setting up his defense, organizing his units, bringing the reinforcements over). It would have been more practical to use these forces, once they were already on the move, to recapture Hill 1192 straight away. The necessary covering fire could have been given from the numerously positions on the northern slope of Kuk. If the enemy leadership had also managed to put an attack in motion against Detachment Rommel from the east at the same time, my detachment would have found itself in an extremely difficult situation. Furthermore it was impractical to put the three positions on the steep, barren, stony eastern slope of Kuk (a forward slope). Working for many hours, the Italian soldiers, although they were undisturbed by fire, **did not manage to dig deep enough into the ground**. A reverse-slope position on the western slope of 1192 would have been much more advantageous to the enemy. We would not have been able to get at them with either machine guns or artillery. And further still, the enemy neglected to block the high road on the southern slope of Kuk and to **monitor the barren hills below the high road with gunfire sweeps**.
- At the beginning of the attack against Kuk, between two and three Italian battalions with numerous machine guns stood across from Detachment Rommel in positions—some

well-constructed, others sloppily put together—at higher elevations. The detachment attacked them with the fire support of one machine gun company, six light machine guns and two heavy batteries, acting immediately and **with only two assault troops of about 16 men, probed possible avenues of approaching the enemy** and ultimately, with its main body, encircled the entire Kuk garrison. Ultimately the garrison was cut down in the early noon hours by assault troops of the Württemberg Mountain Battalion and the 1st Company of the Life Guards Regiment. During the attack, **the effect of machine gun fire and heavy artillery fire** proved **especially strong against the hastily dug-in enemy forces**. In many places they could not withstand this heavy test of nerves. However this fire did not do much to the Italian troops that were deeply dug in. Our own machine gun fire from Hill 1192 **drew far superior enemy fire and the whole attention of the entire Italian garrison like a magnet**. Thus the assault troop, and later the entirety of Detachment Rommel, managed to travel the camouflaged high road towards the eastern slope of Kuk under the enemy's nose and without suffering any casualties.

- **In Ravna, the communications within Detachment Rommel were severed because the leader of a machine gun company went off to capture pack animals.** Thus it came about that I was leading just over a third of my total strength in forces in the valley near Polava, and was only able to block the Luico-Savogna valley and then was forced to give up the area I had blocked on the Matajur road in the Mount Cragonza area. Of course the units who had gotten separated at Ravna later took part in the attack against the enemy at Luico—it's just that the results would certainly have been even greater, if Mount Cragonza had already come into our possession on October 25. Lesson: When a break-in, or even a total breakthrough, into a defensive position is achieved, the reserves must stick close to the front units and must not be allowed to hold back to take booty, etc. Also, the **utmost speed must be demanded from rearward elements in such moments in battle**. A regiment of the 4th Bersaglieri Brigade in a marching column unexpectedly stumbled across Detachment Rommel's roadblock in the narrow valley. Even if the front units had been pinned down by the mountain troopers' fire, the [Italian] **units further to the rear still could have mastered the situation by attacking the hills west or east**. A goal-oriented, firm leadership was lacking here.

On the afternoon of Oct. 25, 1917, the status of the battle was as follows: **Group Kraus:** Imperial Rifle Regiment 1 [*Kaiser-Schützenregiment* 1] was attacking Stol from Saga. The 2nd Battalion had taken Hum and the 1st Battalion had taken Pvrihum, both through storming attacks. The 43rd Brigade was ascending towards Hill 1450. The storming company of the Imperial *Jäger* Regiment 3* took Mount Caal, and their 13th battalion took Passio di Tanamea. **Group Stein:** The 12th Division advanced into the Natisone valley with Infantry Regiment 63 as far as the national border three kilometers south of Robic and there beat the Italian reinforcements back. The Italian positions on the northern slopes of Matajur were not attacked. Group Eichholz was still battling against Italian forces one kilometer north of Golobi. Only at a slow pace did they gain ground. At about 1700 hours, they attacked Golobi and at about 1800 hours reached Luico, where the 1st Life Guards Regiment and the rear units of the Württemberg Mountain Battalion had arrived in the meantime. With regard to the *Alpenkorps,* the Kuk garrison was mopped up by units of the Württemberg Mountain Battalion and one company of the Life Guards. At the same time, the 6th Company of the Württemberg Mountain Battalion had rolled up the Kolovrat ridge from Hills 1110 to 1114. Detachment Rommel had taken an enormous amount of units of the 4th Italian Bersaglieri Brigade prisoner after encircling Kuk, cutting off the Luico—Savogna valley and fighting near Polava. The bulk of the Württemberg Mountain Battalion and the 2nd Life Guards Battalion, attacking from Ravna, had taken Luico. The *Jäger* Battalion 1 and *Jäger* Battalion 10 battled with the enemy on the southern slopes of Hill 1114 and, during the course of the afternoon, took Hill 1044 and the entirety of Hill 1114 into their possession. With regard to the 200th Division, *Jäger* Regiment 3 fought with the enemy south of Hill 1114 in the Crai area and *Jäger* Regiment 4 took La Cima, located 800 meters south of Hill 1114, at about 1800 hours. **Group Scotti:** Grenadier Regiment 8 crossed over Trt. Judrio while attacking Mount Hum from Pusno. The 2nd Mountain Brigade took Mount Cicer and the 22nd Mountain Brigade took St. Paul. From all of this, the following is evident: The powerful 3rd Italian position on the Kolovrat hills south of the Isonzo was, on Oct. 25th, broken through—above all else—by the combat operations of the Württemberg Mountain Battalion, including in the west up until the Luico Pass, and torn open in the east as far as Hill 1114. Thus the attack of the *Alpenkorps* and the units of the 12th Division stalled north of Luico once again came into motion.—*Rommel*

* *Kaiserjägerregiment 3.*

"Nothing Ventured, Nothing Gained"—*The Storming of Mount Cragonza*

> ◊ One of the most striking descriptions in the book occurs when Rommel witnesses his men walking into the kill zone of a hidden enemy force. In short but powerful sentences, he describes a series of flashing scenes before his eyes, his sense of impending doom and desperation to save them.
> ◊ The reader gets a deeper sense of the strong bond between Rommel and Aldinger as they fight alongside one another. Aldinger is shot three times. Here we see the two of them, friends until the end of Rommel's life in 1944, both as young men relying on each other in a battle for their lives.
> ◊ Again Rommel has an opportunity to shoot unsuspecting enemies, but surprisingly calls out and offers them a chance to surrender, when it was arguably disadvantageous for him to do so.
> ◊ In his post-battle observations, Rommel draws attention to achievements of his own group by contrasting them to the rest of the *Alpenkorps*: after alluding to his troops' gains with some prominence, he calls out other units for sleeping during the night, standing still or just beginning to march. He makes the contrast clear. Ending his "lesson" with an extremely cheeky stroke of his pen, he draws a single conclusion from his description of the entire battle situation: the accomplishment of his own troops.

With the Württemberg Mountain Battalion units located in Luico, I moved as fast as I possibly could to the roadblock [set up earlier] north of Polava, organized my force, which then consisted of seven companies, and divided captured pack animals among the individual companies. Without taking any rest, we climbed in the direction of Jevscek—Cragonza; if we got going fast, we would probably not run into any enemies prepared to fight.

Despite the indescribably exhausting strains and deprivations of the past days, we gained ground in the steep, trackless terrain—partway over long stretches of meadow alongside impenetrable thorn hedges, partway climbing up stony gullies, and soon over hills. Once again this time, I had to demand a superhuman amount of effort from the fatigued troops, because the offensive could not be allowed to stall.

The higher we got, the more tiresome the ascent became. Deep furrows and thorn bushes forced us to make turns, which for the most part brought about considerable loss of elevation and expense of our strength. We climbed for hour after hour—evening came, all soon became dark. The troops by now were utterly and completely exhausted.

Should I stand down from my goal? No, Jevscek had to be reached and then additional intrepid men would be found to join us in taking Mount Cragonza.

The enormous disc of the moon shone brightly on the sharp slope of the hill, silvered the grassy turf and bushes and cast long, black shadows across the surface of the grass from behind the clusters of trees. The point men climbed onward, slowly and watchfully, and at last found a narrow footpath. The detachment followed at 50 meters distance. From time to time I had them all stop and we listened to the sounds of the night.

Once again we halted in the shadows of a hut just below the narrow path. Ahead of us lay a ravine, shadowed with a spooky darkness from high, thick bushes. Our path led right through it. We strained to listen. Clearly, we heard from the opposite edge of the ravine the buzz of voices, commands, and the noises of marching troops. It seemed like the enemy was not coming any closer, but was rather moving sideways alongside the opposite edge of the ravine. Only the narrow footpath could serve as an approach. This situation was hardly appealing to me. Furthermore, Jevscke and Mount Cragonza were most likely above us and to our right.

Under these circumstances I preferred to turn off the path again, going right. The point [men] stalked up a steep hill in the shadows of long rows of bushes. Ahead of us lay a large, grassy glade lit brightly by the moon, bordered by a half-circle of tall trees. Was it an illusion? Weren't those obstacles standing over there ahead of the tree line? Weren't there some brightly lit positions beyond the tree line? With exceptionally great caution, we crept closer and determined that we had indeed seen correctly. And then—we also heard Italian voices from the woods in front of us. Unfortunately it was not possible to reconnoiter further to find out whether the enemy was already in his positions.

In order to establish clarity about this, various different officer patrol units were sent out. During this time, the detachment dispersed and rested. Soon reports came that the enemy was, at that very moment, preparing to occupy the positions ahead of us and that the obstacles ahead of the position were very high.

To attack such a heavily fortified position uphill across the brightly lit meadow would have been an enormous gamble even with completely fresh troops. With the totally exhausted mountain rangers—who had been pulling off truly tremendous feats since the beginning of the offensive!—an attack was impossible for the next few hours. With this in mind it was indeed really questionable as to what purpose a break-in to this position in the early part of the night would have achieved and whether the results would prove useful enough. I declined to do it, deciding on several hours' rest. I also wanted to conduct completely thorough reconnaissance of the terrain and the enemy during this time.

Silently I drew the detachment into a wide basin offering cover from above against fire, located 250 meters in front of the enemy position, and ordered a rest period until midnight. The 4th and 2nd Companies provided security for our rest by posting sentries in a half-circle formation. Because the captured pack animals really made themselves conspicuous to a frightening degree through their whinnying, they were positioned at a significantly lower elevation. As we moved into our resting place, an extremely violent firefight unfolded in the valley below near Polava; the 1st Battalion of the Imperial Life Guards had stumbled upon an Italian position. There were even enemy forces down there again!

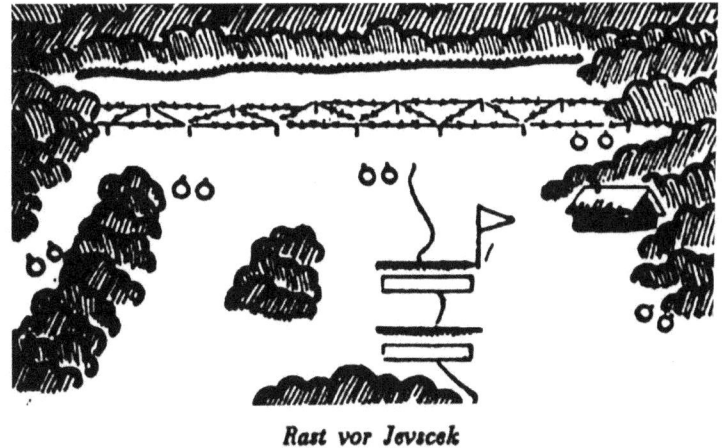

Rast vor Jevscek

The location of the rest before Jevscek.

Additional officer patrols were ordered to reconnoiter convenient means of approach to the enemy position, as well as the strength and depth of the obstacles, any gaps they happened to find, the type of garrison and the situation of the village Jevscek. They were to return with their reports no later than midnight.

I turned in for the night on a captured Italian sleeping bag, which my thoughtful buddy Reiher had loaded up on a mule near Polava. I lay in a spot just above the detachment. However, despite how exhausted I was, sleep was unthinkable; my nerves were just too worked up. The following exemplary report from *Leutnant* Aldinger which reached me as soon as 2230 hours brought me quickly to my feet: "Jevscek is sited 800 meters northwest of our campsite. The locality is heavily fortified and blocked with wire on every side. However it seems not to be occupied by the enemy. Italian troops are marching downhill in a southeasterly direction on the slope just west of Jevscek, as well as through the southern part of Jevscek."

I came to my decision quickly: "Up! To Jevscek!" Maybe we could still reach the place before the Italian troops headed there did. The rest was broken up, sentries were pulled back, and the companies were standing ready to march within just a few minutes. In the meantime the moon had sunk. The night was dark. Only the starry heavens gave off faint light.

The detachment climbed noiselessly to Jevscek along the path scouted by *Leutnant* Aldinger. The [company] leaders were briefed quickly about the situation. The 4th Company and the 3rd Machine Gun Company were our advance guard. The remaining companies followed with not much distance between them. First we crossed through a narrow strip of forest, then climbed up a steep hill in a forest meadow. Very soon, the point of our advance guard reached the obstacle, which was as tall as a man. *Leutnant* Aldinger reported that we were now only 300 meters away from Jevscek.

We halted and strained to listen to the darkness ahead of us for several minutes. No movement stirred in our immediate area, but we did hear the footsteps of climbing Italian infantrymen several hundred meters up on the slope above us. *Leutnant* Aldinger slithered through a narrow space in the obstacle into the position behind it and found the place empty. The point men followed. Afterwards I pulled the entire advance guard back and positioned them in a half-circle formation within the enemy position. Patrols were send out to reconnoiter the terrain in our immediate area and to provide clarify on the enemy uphill and the locale of Jevscek.

At the same time, the mass of the detachment (the 2nd and 3rd Companies and the 1st and 2nd Machine Gun Companies) also moved through the obstacle into the position. I had the Communications Company and the staff with the pack animals remain back on the hill beyond the obstacle.

I stalked towards the enemy on the hill above with a patrol troop. In the darkness we could only see a few meters ahead. The slope lay ahead of us like a ghostly black mass. Italian infantry were moving barely 100 meters ahead of us—probably in columns from right to left, climbing towards Jevscek. We crept closer.

Suddenly an enemy sentry called out to us. So we knew: the enemy was in a position ahead, and behind this were the marching columns. We crept back and turned left towards Jevscek. As we approached the first houses, a patrol troop returned with the report: "North part of Jevscek is clear of the enemy. Italian infantry marching through the southern part." Now I decided to move into the town with the goal of intercepting the enemy infantry in the southern part.

A few minutes later, the detachment moved slowly towards the town. Dogs all started barking from multiple farm buildings as soon as we got near the first of the houses. Shortly afterwards the enemy opened fire from a hill position above us and barely 100 meters away. Fortunately the sheaf of fire mostly struck in the forest to our left. For as long as we couldn't find cover, we lay flat on the ground and got our machine guns and carbines ready to fire—but remained completely still. We would only consider opening fire if the enemy was attacking. If he was not attacking—which is what I believed was most probable in that instance—he would soon cease firing and would think he had been deceived. During the enemy "fire magic,"[20] units of the mass [of the detachment] moved under cover through the unoccupied positions east of Jevscek towards the town. After a few minutes, the enemy fire fell silent. Soon the whole Detachment Rommel was in the town. Fortunately there were no casualties sustained due to the enemy's surprise fire.

I had the northern part of the town occupied in a circular formation, while avoiding

---

20. Another ironic reference to Wagner's *Feuerzauber* opera scene.

any possible collision with the enemy on the hill just to the northwest. Midnight had long gone by. The men not on sentry duty or in position rested with rifles in their arms in the houses, which were still occupied by Slovenian families. We all knew that for the most part we were only a hand grenade's throwing distance from a strongly occupied Italian position and that, in the event that the enemy probed towards Jevscek, a fight at close quarters could come at any second.

Since the surprise fire, the march of the enemy's forces on the hill to the northwest and through the southern part of the town had stopped. At this point only the enemy on the northwest hill had fired at us. By contrast, no shots were fired at us from the southern part of the town. Was there possibly a gap there in the enemy position that seemingly continued unbroken until Polava?

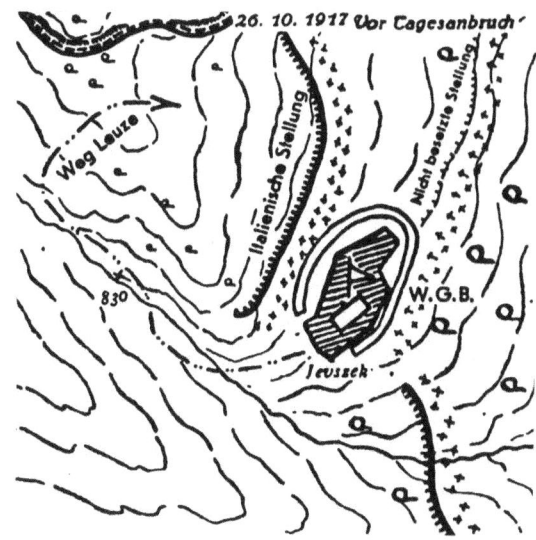

Jevscek and the Württemberg Mountain Battalion shown on Oct. 26, 1917 before daybreak. Leuze's path is shown on the upper left.

I meticulously studied the map in a house by the flickering light of an open fireplace. We were located one kilometer north of Polava in the northern part of Jevscek at about 830 meters elevation. Mount Cragonza lay about 500 meters to the west and 266 meters higher up. Because Jevscek was fortified on its eastern edge and enemies were in positions on the slopes northwest and southeast of the town as far as Polava, we were probably dealing with a rear Italian position planned out well in advance that was intended to block a breakthrough to the Luico pass. The enemy movements we had noticed during the night suggested that the Italians were extremely anxious to occupy this position all the way through. Judging from the manner of fortification, there was no doubt that Jevscek was part of this position. For some reason, it seemed like the garrison intended to occupy the town had not arrived yet. They could still come during the course of the night or in the early morning. Should we wait? Was not the "war luck"[21] of the courageous mountain rangers yet again offering its hand to us? Had we not, by taking Jevscek, already conquered a part of the enemy position that was supposed to block the way to us and the *Alpenkorps* to Mount Cragonza, Mrzli peak and Mount Matajur?

After these considerations, *Leutnant* Leuze received orders [from me] to find out whether the southwest part of the town was free from the enemy, and if so, to scout a <u>path to the rig</u>ht half a kilometer northwest of Jevscek in the rear of the Italian forces

---

21. *Kriegsglück*. Rommel seems to believe his men have a lucky star.

positioned just northwest of the town. He was to return in no later than two hours. *Leutnant* Leuze refused men to accompany him and went alone.

The exhausted detachment indulged themselves in an extension of their rest break. The main body sat—just a few meters from the enemy—in sturdily built houses beside open fireplaces with coffee and dried fruit that the very friendly Slovenians offered us. Now and then isolated shots rang out and occasionally an Italian hand grenade cracked. It seemed like the enemy lacked the desire to make forward push towards Jevscek to reconnoiter it. No shots were fired on our side. Total darkness enveloped the Italian and German forces standing so, so close across from each other.

At about 0430 hours *Leutnant* Leuze returned from his reconnaissance, bringing an Italian as a prisoner and reporting: "Southwest exit from Jevscek is free of the enemy. The path to the Hill 500 meters northwest of town has been scouted. I took this Italian prisoner on that hill but otherwise I did not run into the enemy anywhere." Thus he fulfilled his mission outstandingly.

Leuze's report brought me to the following decision: to immediately take the Hill 500 meters northwest of Jevscek with four companies of the detachment, leave behind the rest in Jevscek as backup, and attack the enemy just northwest of Jevscek at daybreak.

This decision did not come easy to me. If, after we had reached the hill in the darkness, the enemy positioned at higher elevations on Mount Cragonza overpowered us with fire at daybreak, we would have to fight on two fronts. Then things could easily go wrong. But nothing ventured, nothing gained.

It was still pitch dark when at 0500 hours the 2nd and 4th Companies, along with the 1st and 2nd Machine Gun Companies left Jevscek on the path scouted by Leuze, moving with soft footsteps. *Leutnant* Leuze led the point of the long column of twos. I left the 3rd Company and the 3rd Machine Gun Company as rear support in Jevscek under the command of the tried and true *Leutnant* Grau, with orders to pin down the enemy garrison northwest with fire as soon as Detachment Rommel started fighting on the hill above, and stop them from attacking the detachment from the east.

I gave these instructions as the detachment moved out of the village. As I joined the column of the 2nd Machine Gun Company, dawn broke on the summit of Mount Cragonza. The transition between night and day happened very fast in the mountains.

I had the ominous feeling that we would get there half an hour too late. I saw my companies ahead of me, moving in the usual column of twos in the open basin, going between boulders and climbing over Hill 830. Already, the uppermost cliffs of Mount Cragonza were laying bare in bright sunlight. I scanned those cliffs with my binoculars and—horror! On the slopes just a few hundred meters to the left above my detachment were enemy positions. Within them I perceived movement—and one steel helmet after another. If the enemy started shooting, the basin that my detachment was climbing through at that very moment offered hardly any cover at all, and heavy casualties would be unavoidable. Heavily—oppressively heavily—did the responsibility for the lives of my officers and men weigh upon me in this moment. I had to tear them away from the danger that they hardly noticed.

I got what men I was able to reach from the 2nd Machine Gun Company and deployed them as fast as possible to the right and above on the hill as fire

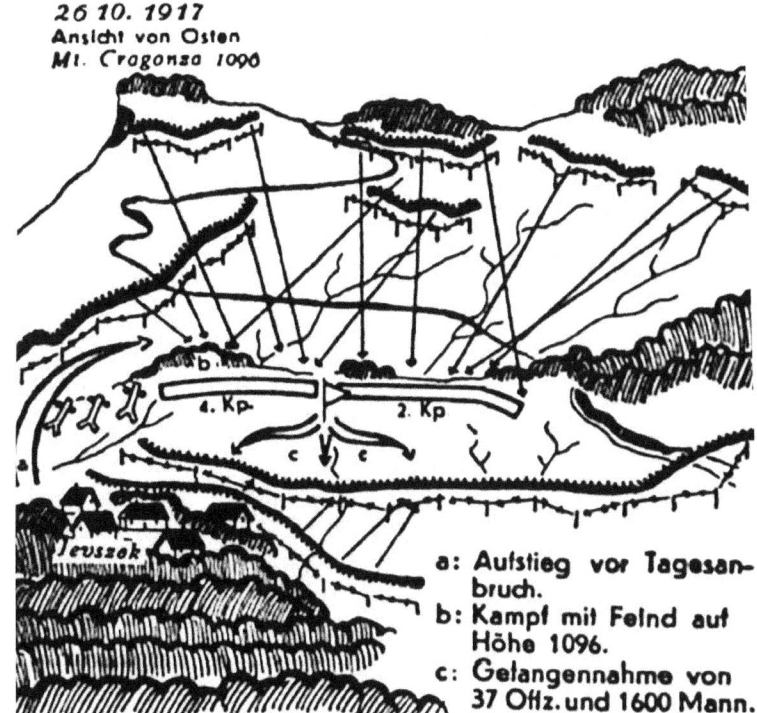

A view from the east of Mount Cragonza on Oct. 26. The village of Jevscek is shown on the lower left. Starting from the village and moving northwest: a) Rommel's ascent before daybreak; b) battle with the enemy on Hill 1096; c) capture of 37 officers and 1,600 men.

support. I gave them instructions to immediately put the enemy on the slopes above to the left under fire and pin him down just as soon as he opened fire. Then I rushed ahead with couriers and turned the points of individual companies to the right toward a hill overgrown with isolated clumps of bushes 500 meters northwest of Jevscek. Now was the time. Dawn was melting away into daylight.

As the last units of the companies left the basin, the enemy on Cragona showered the detachment with a rage of rapid fire. Because we were still on the slope facing the enemy and since the fire was given from a considerably higher elevation, there was nowhere to find protective cover. Only low-lying thorny scrub in isolated places offered the opportunity to at least get out of the enemy's sight. While being covered by the rapidly answering fire of the 2nd Company, the individual platoons dispersed, took the hill 500 meters northwest of Jevscek and there resumed the firefight. But against the massively overpowering enemy fire that came at us from a half-circle from the mountain positions in the northwest, west and southwest, we couldn't hold our own. By creeping sideways and making short hops, the infantrymen fighting in the front lines of the 2nd and 4th Companies tried to disperse and reduce the effect of the enemy's fire. Casualties mounted. Among others, the exemplary leader of the 2nd Company, *Leutnant* Ludwig, was very severely wounded.

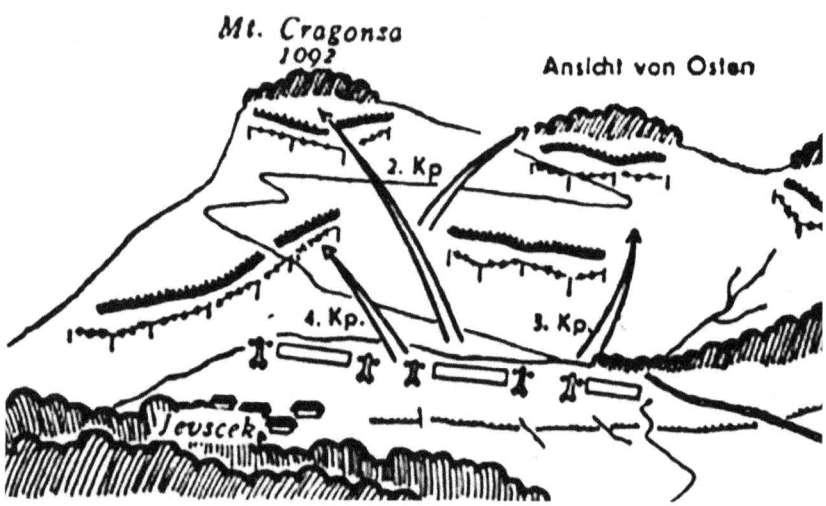

The attack on Mount Cragonza, view from the east.

In the meantime, to our rear, battle at Jevscek was also burning to its most violent point. In accord with their orders, the 3rd Company and the 3rd Machine Gun Company under *Leutnant* Grau had opened fire on the enemy northwest of Jevscek, thus pinning him down fast in his positions and stopping him from joining in the fight against the remaining companies, locked in fiercest battle, from the rear. I reached the hill 500 meters northwest of Jevscek with a few orderlies and found cover in a small group of bushes from fire aimed straight at us. Machine gun rounds clattered left and right away over the hill.

I had no groups left at my disposal. All of them were stuck in heavy firefights, shooting whatever they could get their hands on. And yet I still had to reach a decision, and quickly, otherwise the mountain rangers would totally bleed away.

I commanded, through my orderlies, that three light machine gun groups should pull out of the front lines of the 2nd and 4th Companies and get themselves over to the covered slope 50 meters east of my command post. Afterwards I created multiple storming troops out of them at top speed, and led them downhill to the rear of the enemy position northwest of Jevscek, whose front was taking fire from us from the east.

Climbing down through networks of bushes on the edge of a narrow scree—with light machine guns and carbines ready to shoot—we soon saw the enemy position below us. It was heavily occupied. Steel helmets crammed against each other. From above we could even see the bottom of the trench. If we shot now, the enemy would find no cover from us in his position. The fire from the mountain rangers on Hill 500 ripped over and away above us. Below, near Jevscek, we saw the 3rd Company and the 3rd Machine Gun Company firing at the Italians barely 100 meters below us.

The enemy had no idea what was menacing him. My assault troop got ready to fire. Then we called out to the enemy garrison to give themselves up. Horrified, the Italian soldiers turned around to their rear and stared up at us. Their firearms sank and dropped out of their hands. They knew their cause was lost and made signals to surrender. Not a single shot needed to be fired by my assault troop. But not only did this garrison—some three companies strong—between us and Jevscek stop fighting, but also, to our utter astonishment, the enemy forces in trenches further north as far as the Matajur road threw down their weapons. They had totally lost their heads at the rumpus of battle noises in their rear and the appearance of my weak assault troop on the northeast slope of the hill 500 meters northwest of Jevscek. The firefight between the Italian garrison on Mount Cragonza and the bulk of Detachment Rommel, as well as our appearance in their rear, probably made it seem to these enemy forces that the Germans were attacking from Mount Cragonza and already possessed the commanding heights.

An Italian regiment with 37 officers and 1600 men surrendered in the basin 650 meters north of Jevscek. They assembled with all their weapons and gear and it was hard for me to get enough people over to carry out the process of disarming them. In the meantime the battle a few hundred meters above was still churning on with the same violence as before. The Italian garrison on Mount Cragonza had seen absolutely nothing of all these happenings just northwest and north of Jevscek. They pressed my front line as hard as they had been before. But now we had the rear freed up.

The companies freed up by Jevscek were already moving and soon the rifle companies attacked Mount Cragonza head-on. A mighty clash of arms! Tenaciously the enemy clung to their strong, dominating positions, which our fire from below could barely touch. Over and away across steep cliffs went the mountain rangers, through a hail of lead, going straight for the enemy's hide.

As I had no more troops to send in, I went forward with 2nd Company, in the middle. The company was now led by *Leutnant* Aldinger in place of the severely wounded *Leutnant* Ludwig. We reached the lowest turn in the Matajur road. At it stood 14 Italian field guns and 25 ammunition wagons without horse teams. Were these the artillery groups from Avsa and Perati from the battles of the previous day? We had no opportunity to linger here. Flanking machine gun fire from the north hit between us. We stormed onwards. Shortly afterwards the 2nd Company lost its new leader; *Leutnant* Aldinger was severely wounded by three shots. I myself was targeted as a bull's-eye mark for

Italian machine gunners on the Matajur road for some time. There was no cover to be found from their sheaves of fire in any direction. By quickly jumping downhill to get around the bend in the road some 60 meters away, I still managed to escape from the well-positioned cone of fire.

As casualties mounted, so did the battle rage of the mountain rangers. They took trench section after trench section, machine gun nest after machine gun nest. At about 0715 hours, our difficult task was accomplished. The intrepid 2nd Company, now led by *Vizefeldwebel* Hügel, had stormed and taken the peak of Mount Cragonza. Thereby the fate of the enemy forces on the northeast slope and eastern slope of Mrzli Peak only became a question of time.

I could only guess how things were looking for our neighbors. Judging from the noises of battle, which had been rising continuously on our right since daybreak, I estimated that units of the 12th Division and the *Alpenkorps* were attacking Mrzli Peak from the northeast and east, maybe also climbing alongside the Matajur road from Avsa here towards Cragonza. Should I wait for them to arrive? Should I pull together my own forces east of Cragonza and organize the mixed-up units? Hadn't they earned a rest on the peak after this extraordinarily difficult attack? On the other hand, shouldn't I worry that strong Italian forces on my right flank would counterattack or plan and mount an attack to retake Mount Cragonza? I considered it effective for preventing all enemy countermeasures to lead an attack on the slopes leading toward Mrzli Peak with all available forces I had (half a company) without delay.

Considerations:

- **During the night ascent toward Jevscek, the Italian troops betrayed themselves by their loud calling and noisy marching**. Thus they directed us to the right path [to approach] and we managed to avoid an unwanted collision.
- **As the exhausted troops rested, the officers were tirelessly active in order to make exact determinations about the enemy and the terrain**. The reconnaissance activity even continued from Jevscek after midnight. Thereby they created the basis for the success of the attack northwest of Jevscek and for the storming of Mount Cragonza.
- During the night of Oct. 25–26, I knew very little **about our neighbors**—didn't know where they were, what they were doing or wanted to do. Also there was **zero connection** to my superiors available. But it was clear to me that everything needed to be arranged to get the attack back in motion again on the Oct. 26.
- Also, the **truly desperate situation** the mountain troopers faced at daybreak between the enemy positions without cover turned ultimately to our advantage. A few groups

of brave-hearted men made this turnaround come to pass. The attacking power of the Württemberg mountain rangers proved itself, above all, during the frontal attack against the Italians in dominating and exceptionally good positions on Mount Cragonza. The 2nd Company did not falter even as their officers fell out of action.

On Oct. 26, 1917, at 0715 hours—the time that Mount Cragonza was stormed—the overall battle situation was as follows:

- **Group Kraus**: The Stol (1,668 meters) fell during the night of Oct. 25–26 at 0300 hours to the 2nd Battalion of Imperial Rifle Regiment 1. At 0600 hours the 2nd Battalion reached Bergogna. The 1st and 3rd Battalions of the regiment and the 43rd Brigade followed. Reached Bergogna at 0800 hours.
- **Group Stein:** The 12th Division stood with Infantry Regiment 63 as on previous says at the national border in the Natison valley. The 2nd Battalion of Infantry Regiment 62 and the 23rd Infantry Regiment closed on the outposts of the 2nd Battalion of the Life Guards near Avsa and got ready to march.
- *Alpenkorps*: The Württemberg Mountain Battalion broke into the enemy position from Mrzli peak—Jevscek—Polava at Jevscek at daybreak, ripped this position open by one kilometer to the northwest and stormed Mount Cragonza at 0715 hours. The remainder of the Württemberg Mountain Battalion marched from Luico via Avsa to Mount Cragonza. The 2nd and 3rd Battalions of the Life Guards prepared to march and afterwards fell in behind the 1st and 3rd Companies of the Württemberg Mountain Battalion during the advance to Mount Cragonza. The 1st Battalion of the Life Guards was at an outpost near Polava.
- The *Jäger* Regiment 2 (without *Jäger* Battalion 10) moved from Ravna toward Luico. The *Jäger* Regiment 1 and *Jäger* Battalion 10 spent the night at Hill 1114 and there made themselves ready to march. The 200th Division: *Jäger* Regiment 3 advanced across Drenchia to Trusgna, and reached there at 0800 hours. *Jäger* Regiments 4 and 5 moved from Hill 1114, where they had spent the night, to Ravna at 0430 hours. There they rested until 0800 hours.
- **Group Scotti:** Grenadier Regiment 8 took La Klava at 0500 hours with its 1st Battalion and just then began attacking Mount Hum with all three battalions.

From this, we conclude: The Italian position running from on the northern slope of Matajur—Mrzli vrh—Jevscek—Polava—St. Martino was, just like the Kolovrat position on the previous day, broken open by the forward units of the Württemberg Mountain Battalion in the earliest morning hours near Jevscek, and ultimately Mount Cragonza, the key to all Italian positions on the Mrzli Peak and Matajur massif, was stormed.—*Rommel*

"We Risked The Attack"—*The Capture of Hill 1192 and the Mrzli Peak (Hill 1356), and the Storming of Mount Matajur*

> ◊ Rommel persists in denying his men rest and loses contact with his troops. He disobeys orders to keep attacking. Previously Rommel had been more circumspect in making decisions. Now he speaks openly of risks and gambles.
>
> ◊ It appears his disobedience proves a sticking point with *Major* Sproesser, who previously supported him. Rommel had been more complimentary of Sproesser earlier, but at this significant point, all praise disappears, with a somewhat terse reference to Sproesser "demanding" a report from him after he reaches the Matajur peak. It is easy to imagine why Sproesser might have been angry. Rommel focused on the actions of his own unit while Sproesser acted with a more birds-eye view of the situation directing larger forces. Another sticking point for Sproesser may have been that his other troops were gaining on the same area from a relatively short distance behind Rommel. The matter of who deserved credit for taking Mount Matajur was likely contentious—especially judging from Rommel's writings. Rommel makes the case that he and his men were there first. Rivalries and debates about who deserves credit for battle achievements are common in the military.
>
> Some historians view Rommel as a schemer, accusing him of "stealing" credit from Sproesser and getting an undeserved *Pour le Mérite*. Others perceive Sproesser as a dull superior who tried to steal credit. The truth is certainly more nuanced. Both men deserve credit—and both ultimately received medals for it.
>
> ◊ Rommel's post-battle lessons become a paean for his troops. Rommel has formed a close bond with his men and believes they are the best the German military has to offer. Rommel developed a similar esprit de corps with his Afrika Korps in World War II.

Even though the mountain rangers were totally fatigued after storming Mount Cragonza, I still couldn't grant them that well-deserved rest on the peak. The exceptional *Vizefeldwebel* Hügel rose to his new task with that eager fighting spirit so unique to him. Without waiting for support, he attacked alongside the rising slopes between Hill 1192 and the Mrzli Peak so as to take ground as long as it was possible with our weak forces.

I first sent orders to the detachment via courier to move as fast as possible over Mount Cragonza in the direction of Mrzli Peak to the Matajur road, which led to the peak's west side. Then I went along with the 2nd Company. We ran into the enemy after only a few hundred meters; he had nested within a forested knoll crowning the ridge. The sounds of battle on the eastern slope to our right grew significantly in ferocity. It seemed like rearward units of Detachment Rommel, climbing from Jevscek to Mount Cragonza, were shooting or attacking. Or maybe it was units of the *Alpenkorps* from Luico trying to ascend Mount Cragonza from the Matajur road.

Vizefeldwebel Hügel did a masterful job of tying down the enemy, superior in numbers and weapons, with frontal fire and at the same time hitting him in the flanks and rear with assault troops. These movements played out in only a few minutes and inevitably resulted in the enemy being thrown back and retreating northeast, downhill in the direction of Luico.

Because we came to grips with the enemy quickly wherever we found him, our connection with our rear was soon severed. A report reached me that the detachment was severely hindered by strong machine gun fire from Italian positions northeast of Cragonza and had fallen behind by about a kilometer. Should I stop the 2nd Company? No, we would maintain our attack towards Mrzli Peak until we encountered strong enemy forces.

At about 0830 hours, the 2nd Company, gradually reduced to a single platoon lumped together with two light machine guns, tore itself away from the enemy on Hill 1192 two kilometers west of Avsa. A strong enemy in position 800 meters southwest of Mrzli Peak's summit prevented further penetration. He took the knoll we had reached under fire with numerous machine guns. It was fierce fighting to our right and below on the slope, as well as right to the rear in the direction of Jevscek. It seemed like units of the *Alpenkorps* were now attacking. If I wanted to approach the enemy on the southeast slope of Mrzli Peak, I would need at least two rifle companies and one machine gun company. In order to get these forces together, I hurried back along the Matajur road. Hügel had orders to hold Hill 1192. For far and wide I came across no man to connect me with the bogged-down Detachment Rommel.

Turning a curve 600 meters south of Hill 1192, I found myself suddenly standing close in front of an Italian detachment coming from the direction of Avsa and just now crossing the street. The Bersaglieri snatched the rifles from their shoulders and shot...A quick jump into the bushes just below the street drew me out of the fire aimed at me. Some enemies followed me downhill through the bushes. But as they hurried down towards the valley, I was already climbing up towards Hill 1192. Arriving there, I tasked a stronger patrol with establishing contact with the remaining units of Detachment Rommel and transmitting orders to the individual company commanders to close up the gap to Hill 1192 as fast as possible. (In the meantime, units of the *Alpenkorps* and the 12th Division located in the area from Perati—Avsa—Luico started advancing on the Matajur road in the direction of Mount Cragonza. Marching at the point of the column, the 2nd Battalion of Infantry Regiment 62 came into contact with the enemy in a strong position

1.5 kilometers south of Avsa and attacked him. The units following (the staff and Detachment Gößler of the Württemberg Mountain Battalion, Infantry Regiment 23, and the 2nd and 3rd Battalions of the Life Guards) managed to come forward along the Matajur road in the direction of Mount Cragonza.

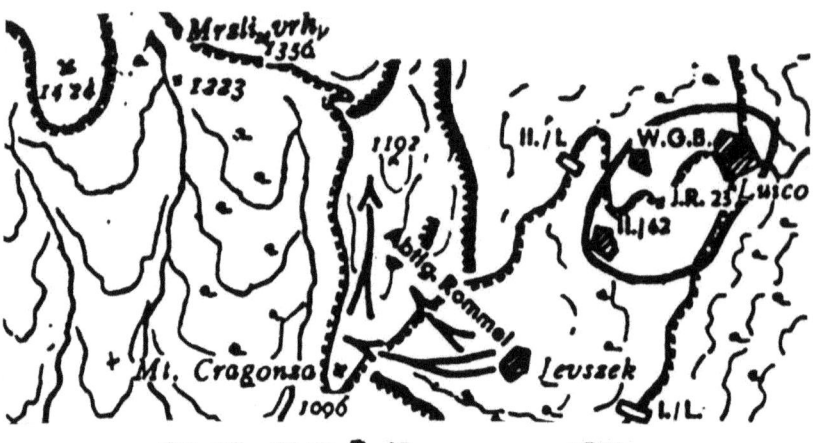

The storming of Hill 1096 on Oct. 26, 1917.

The 1st Battalion of the Life Guards was still at a standstill before an Italian blocking position near Polava. (This was part of the position from Jevscek—Polava—Mount San Martino.) It was already about 1000 hours by the time I had assembled a fighting force of a strength of two rifle companies and one machine gun company. The units had been pulled together from all companies of Detachment Rommel. Their movement over to Hill 1192 had been very much delayed because the individual units were constantly getting entangled in fights with the enemy, who was trying to retreat in a southwesterly direction across the Mount Cragonza—Hill 1192 line.

I now felt myself strong enough to take up the fight with the Italian garrison on Mrzli Peak. Using light signals, we requested artillery fire on the enemy positions on the southern slope of Mrzli peak, with the astonishing result that German shells began falling there very soon. Then the machine gun company nailed down the enemy garrison firmly in their positions with fire from Hill 1192 as both rifle companies under my leadership ran straight for the enemy's hide in the patch of forest just below the road. We managed to envelop the enemy's western flank. Afterwards we wheeled around against the flank and rear of the enemy position. However the enemy cleared out at top speed just as he saw us coming from that direction and pulled back to the eastern slope of Mrzli Peak. We took several dozen prisoners. Because I had no intention of following the retreating enemy across the eastern or northern slopes, I broke off the fight, advanced along the road further to the southern slope of the Mrzli Peak and brought the machine gun company up.

During our attack we had already observed hundreds upon hundreds of Italian soldiers next to a vast camp of tents within the saddle formed between the two highest knolls of

Mrzli Peak. They were standing there apparently indecisively and inactive and looked as if they had been turned to stone as they watched our advance. They had not expected Germans from a southern direction—thus in their rear. A distance of only about 1,500 meters separated us from this assembly of troops. The Matajur road wended aloft over the partially forested

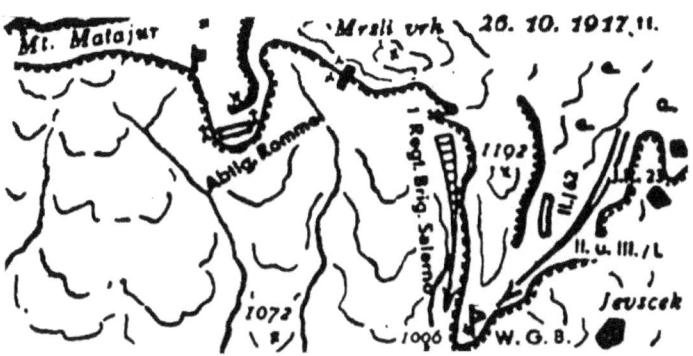

Detachment Rommel is depicted in the center of the map near the Mrzli peak on Oct. 26.

southern slope of the Mrzli Peak in multiple loops. It led westward just below the vast enemy camp to Matajur.

The enemy massed in the saddle loomed ever larger. The group of Italians there had to be between two and three battalions strong. Since they didn't enter into battle against us, I drew closer to them, going along the street waving a handkerchief with my detachment widely dispersed in depth.

The three days of the offensive had already taught [me] how one had to come to grips quickly with new enemies. We came within about 1,000 meters distance. Over on the enemy's side nothing changed. Did they really have no intention to fight? Their situation was most certainly not hopeless! If they threw all their forces to attack, they would crush Detachment Rommel and could take back Mount Cragonza. Or at least they could pull back to the Matajur massif with the fire support of a few machine guns. None of those things happened. Every man of the enemy group stood over there as if turned to stone and didn't stir from their places. Our waving with handkerchiefs was not reciprocated.

We came ever closer. At 600 meters distance from the enemy, a tall forest enveloped us and removed us from sight of the enemy, who was still at about 100 meters higher elevation than us. The street now turned sharply east. What did the enemy up there really want to do? Was he perhaps still deciding whether he wanted to fight us? If he suddenly rushed downhill at us, a fight would develop in the forest at the shortest imaginable distance—a hand-to-hand fight. The enemy was fresh, had monstrous superiority and also had the advantage of being able to fight from higher ground. Under these circumstances I considered it urgently necessary to quickly take the tree line below the enemy encampment. But my mountain rangers with the heavy machine gun toward the rear were so dead tired, that just could no longer expect them to make the steep climb

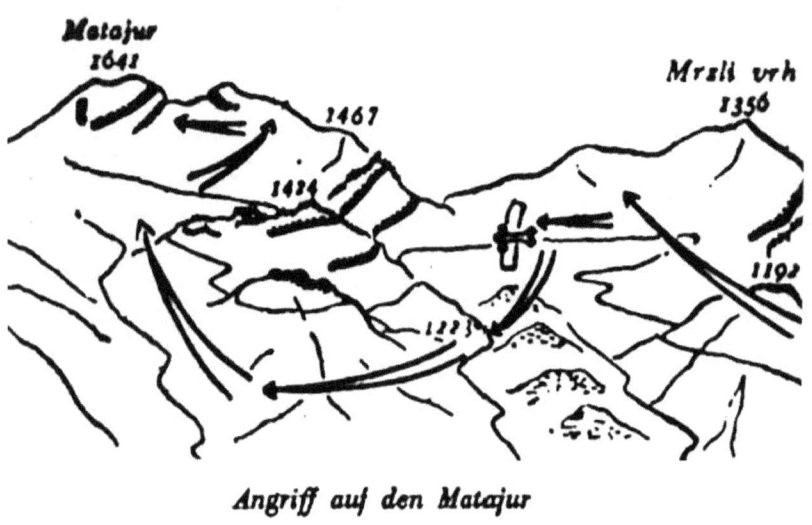

Angriff auf den Matajur.

The attack on Mount Matajur.

through dense thickets. Therefore I had the detachment march down the road. With *Leutnant* Streicher, *Oberarzt* Dr. Lenz and a few mountain rangers I ascended in a broad front—about 100 meters space between each man—through the forest on the shortest path to the enemy. During this *Leutnant* Streicher surprised an enemy machine gun crew and took them prisoners. We reached the tree line unhindered. A distance of about 300 meters still separated us from the enemy troops standing above the Matajur road—a mighty mass of men. There was a lot of shouting and gesticulating among them. All of them still had weapons in their hands. It looked like a large group of officers were standing in front of them. The tip of the enemy detachment couldn't be arriving here at this point. It had probably just reached the sharp turn in the road 600 meters east.

Feeling that I needed to handle the situation before the enemy decided to handle it, I emerged from the tree line, walking unhesitatingly and continuously towards them, calling out and waving my handkerchief as I demanded they surrender and cast down their weapons. The mass of men gawked at me and didn't move. Now only 50 to 100 meters separated me from the tree line. It wasn't possible for me to retreat if the enemy fired. I had the impression that I couldn't allow myself to stand still, otherwise our cause would be lost.

Now I was only 150 meters from the enemy! Suddenly there was movement among the masses above me. They came rushing downhill, and their protesting officers were dragged right along with them. The majority of men threw their weapons away. Hundreds hurried towards me. In the wink of an eye I was surrounded and found myself sitting on Italian shoulders. *"Evviva Germania!"* cried a thousand throats. An Italian officer who hesitated to give himself up was shot dead by his own men. For the Italians on Mrzli Peak, the war was over. They exulted with gladness.

Now the point of my column emerged from the woods on the right out onto the street. With customarily quiet, but long mountaineer strides they went forward despite the heat of the sun and their severe exertions. Using a German-speaking Italian, I had the prisoners positioned below the Matajur road with their front facing east. They were 1,500 men of the 1st Regiment of the Salerno Brigade. I did not allow my own detachment to stop for anything; I only called one officer and three men out of the column. Two mountain rangers were charged with the transportation of the Italian regiment away over Mount Cragonza to Luico; *Unteroffizier* Göppinger was in charge of the disarming and transport of the 43 Italian officers, who had been separated from the men. At the sight of the weak Detachment Rommel, the Italian officers had become eager to fight and tried to get their men under their control again. But now it was too late. Göppinger discharged his duties with astuteness and strictness.

As the disarmed regiment marched toward the valley, Detachment Rommel moved right past and below the Italian tent encampment. A few captured Italians had informed me shortly before that the 2nd Regiment of the Salerno Brigade was located on the slopes of Matajur—a very famous Italian regiment, that was supposed to have been repeatedly boasted about by Cadorna[22] repeatedly in his order of the day due to outstanding performance in the face of the enemy. This regiment would most certainly shoot at us. We should be wary.

Their estimations proved correct. The point men of Detachment Rommel had barely arrived on the western slope of Mrzli Peak when mighty machine gun fire struck from Hills 1467 and 1424. The sheaves of enemy machine gun fire were spread exceptionally well across the street and soon emptied it. Dense bushes below the road removed us from fire aimed at us. I got my men quickly under my control. I now marched with them just below the Matajur road in the direction of Hill 1467, otherwise turning sharply towards the southwest. I wanted to move across Hill 1223 to the sharp bend in the road sound of Hill 1424 at the highest possible speed. Once there, the 2nd Regiment of the Salerno Brigade wouldn't be able to escape from us anymore and would be in a similar situation to the 1st Regiment a half hour ago, only with the exception that they would be stopped from marching away south over the barren slopes of Matajur by our fire in contrast to the Italians on Mrzli Peak who still would have been able to make a concealed march away through the woods.

To deceive the enemy, I had a couple machine guns start playing from the western slopes of Mrzli Peak. I reached the bend in the road 600 meters south of Hill 1424 with the rest of my detachment undisturbed by enemy fire, since the enemy could not see movement through the dense thickets. Once there I prepared a surprise attack on the garrison of Hill 1424, which kept shooting at the rear units of Detachment Rommel and our machine guns on Mrzli Peak. The success on Mrzli Peak made one forget all the strains and tiredness, the wounds on our feet battered from walking and the heavy loads on our

---
22. Italian general Luigi Cadorna.

shoulders, which were torn apart as if we had been flogged.

As I organized the preparations for the attack with utmost energy, put machine gun platoons into position and prepped my assault troops, the command came through from the rear. "Württemberg Mountain Battalion, about face, march!"

Major Sproesser had now reached Mount Cragonza. The great number of prisoners taken by Detachment Rommel (estimated at up to 3,200 men by now) were coming past him from the opposite direction and gave him the impression that enemy resistance on the Matajur massif had already been broken.

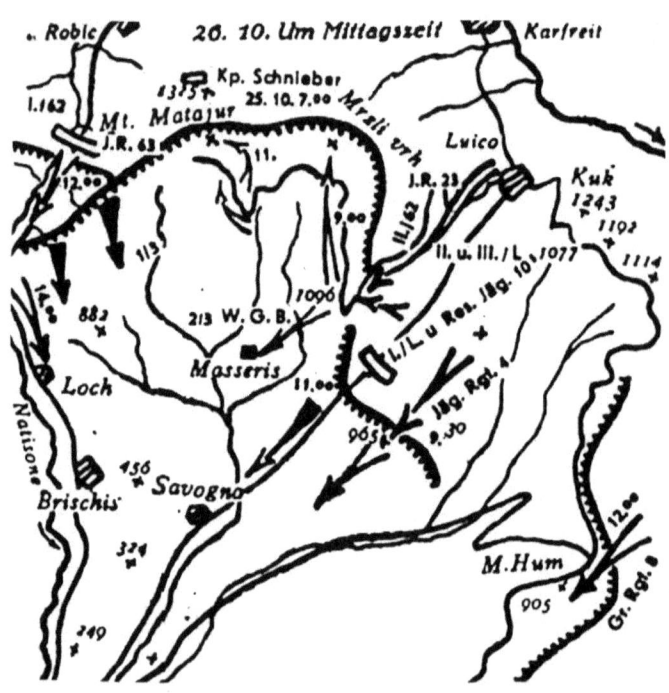

The overall situation in the region on Oct. 26 around noon.

The battalion's order to turn "about face" and "march" would have the result of withdrawing all elements of Detachment Rommel apart from 100 riflemen and six heavy machine gun crews with me back in a rearward march to Mount Cragonza. So, should I break off the fight anyway and pull back to Cragonza? No! The battalion command had been given in ignorance of the battle situation on the southern slopes of Matajur. There was still a lot of work to be done here. Indeed I could not count on additional reinforcements in the foreseeable future. But the terrain was extremely favorable to my planned attack and—every Württemberg mountain ranger was worth more than 20 Italians. We risked the attack despite our laughably few numbers.

The defenders over on Hill 1427 and 1467 were positioned between immense granite boulders with their front facing east. As the fire of our machine guns struck amid the enemy columns from the south by surprise, the enemy quickly took full cover. The huge splinters in the rocks over there must have considerably multiplied the effects of individual bullets. The enemy countermeasures were meager. Our machine guns were so well-nested in thick, tall bushes that the enemy could only find them with extreme difficulty.

I observed the exceptional effect of our fire with my binoculars. As the first Italians tried to retreat on the north slope of Hill 1424, I broke out with the riflemen astride the

Matajur road to the west slope of Hill 1424. We went forward quickly under the strong supportive fire of the heavy machine guns. The enemy was completely clearing out his positions over on the right on the eastern slope. His fire died.

We remained in attack. I had the heavy machine guns brought forward in multiple stages. An enemy battalion wanted to move from Hill 1467 to the southwest via Fta. Scrilo. But the fire of our machine guns, given from 50 meters ahead of the column, forced the enemy to a halt. They had understood us. A few minutes later we approached the rocky summit 500 meters south of Hill 1467 waving handkerchiefs. The enemy had stopped firing completely. Two machine guns guarded and observed our advance from our rear.

An eerie silence reigned. Now and then you could see an Italian scuttling in a flash through the rocks. Now the road itself twisted through the boulders. We could now only see the terrain in front of us by a few meters. As we turned a sharp corner, our view extended further left. Ahead of us stood—barely 300 meters away from us—the 2nd Regiment of the Salerno Brigade, who had all assembled and set down their weapons.

On the roadside sat their regimental commander shaken with deep distress, surrounded by his officers and weeping with rage and shame at the disobedience of his soldiers of his once-so-proud regiment. Before the Italians could become aware of the small numbers of my group, I quickly separated the 35 officers from the 1,200 assembled men and had the men marched off at top speed down the Matajur road to Luico. The captured colonel was so enraged when he saw that we were only a handful of German soldiers that he practically foamed at the mouth.

Without pausing I launched the attack on the peak of Matajur. This was still about 1500 meters away from us and 200 meters higher in elevation. Using binoculars one could see the enemy garrison in position on the rocky crest of the summit. It seemed like the enemy here did not want to follow the example of their comrades on Matajur's southern slope who had given themselves up and were already marching. With a few machine guns, *Leutnant* Leuze provided fire support for this attack, which we attempted on the shortest path from the south. But the enemy defensive fire was too nasty here and made potential avenues of approach unfavorable, so I instead chose to turn east onto a domed hill unseen by the enemy garrison and tackle the summit position from Hill 1467. During our movement, small troops of Italians with and without weapons continuously strove to the position where the Salerno Brigade's 2nd Regiment had laid down their weapons a quarter of an hour earlier.

On the sharp eastern spine of the Matajur, 500 meters east of the peak, we surprised an entire Italian company. Without knowing about what had taken place in their rear, they stood with their front facing north on the northern slope below the ridge from Hills 1467 to 1643 in a firefight with patrols from the 12th Division who were climbing toward the Matajur from Mount della Colonna. Our sudden bobbing up in their rear with weapons ready to shoot on the hill above forced this enemy to rapidly surrender without resistance.

As *Leutnant* Leuze shot at the garrison on the peak with numerous machine guns from the southeast, I climbed westward with the remaining units of my little flock to the ridge leading to the peak. Additional machine guns on a rocky knoll 400 meters east of the peak provided additional fire support for the assault troop deployed in position on the southern slope. Yet, before we had opened fire, the enemy garrison on the peak motioned for surrender. An additional

An aerial photo shows Mount Matajur during World War I not long after Rommel captured it. Rommel's writing on this image and the arrow he has drawn from the bottom toward the central ridge show the path he took to ascend to the summit on Oct. 26, 1917 up the eastern spine of the ridge.

120 men patiently waited to be taken into custody next to the collapsed house on the Matajur peak (a border guard hut) at 1,641 meters elevation. A patrol of Infantry Regiment 23, made up of one *Unteroffizier* and six men, crossed paths with us as they climbed up from the north.

At 1140 hours on Oct. 26 1917, three green flare and one white one announced that the Matajur massif had fallen. I arranged a one-hour rest period for my detachment on the peak. It was well-deserved.

We beheld the violent world of the mountains all around us in the radiant light of the sun. Our view swept far and away. In the northwest, nine kilometers away, lay the Stol mountain at 27 meters higher elevation, assailed by Group Flitscher. In the west we could see Mount Mia (1,228 meters) far, far below us. We could not see down into the Natison valley; it was only three kilometers away, but 1,400 meters lower. In the southwest lay the fruitful fields[23] around Udina, the headquarters of Cadorna. In the south shimmered

---

23. *Gefilde.* A poetic term used to refer to the Elysian Fields.

the Adriatic Sea in a narrow sliver. In the southeast and east lay the mountains we knew so well—Cragonza, Mount San Martino, Mount Hum, Kuk and Hill 1114.

We were reminded that war was all around us by the presence of the prisoners sitting amongst us, the distant artillery fire and an aerial battle, in which an Italian plane plunged burning into the deep. There was nothing to be seen of our neighbors.

I dictated the combat report to *Leutnant* Streicher that *Major* Sproesser demanded daily.

Considerations:

- **The storming of Mount Matajur came 52 hours after the start of the Tolmein Offensive. My mountain rangers fought in the front lines nearly ceaselessly during these hours** and formed the attacking spearhead of the *Alpenkorps*. In doing so they overcame an elevation difference of 2,400 meters uphill and 800 meters downhill and traveled a distance of 18 kilometers as the crow flies through unique types of enemy mountain positions, all while carrying heavy machine guns on their shoulders.

- **One after another, 5 fresh Italian regiments were overpowered by the weak Detachment Rommel in battle within 28 hours.** The number of prisoners and war booty taken by Detachment Rommel is as follows: 150 officers, 9,000 men and 81 artillery guns. This number does not include the enemy units which, after they were cut off on Kuk, around Luico, in positions on the east and northern slopes of Mrzli Peak and on the north slopes of Matajur who cast aside their weapons and willingly joined the columns of prisoners streaming to Tolmein.

- Above all, the behavior of the 1st Regiment of the Salerno Brigade on Mrzli Peak was incomprehensible. Irrationality and inactivity had led to catastrophe here. **The masses' council of war undermined the authority of the leader.** A single machine gun manned by officers could surely have rescued the situation, or at least guaranteed that the regiment went down honorably. If only the officers of this regiment had so much as led their 1,500 men to attack Detachment Rommel, Mount Matajur would hardly have fallen at all on Oct. 26.

- During the battles from Oct. 24–26, 1917, various different Italian regiments, whenever they saw themselves being attacked in the flank or even in the rear, regarded their situation as hopeless and gave up the fight early. This was due to a lack of decision-making ability on behalf of Italian commanders. They were unaccustomed to rapidly changing attack tactics and also did not have firm control of their men. Added to all this was the fact that the war with us Germans was not popular among the common people. Many Italian soldiers had earned their daily bread in Germany before the war and had found a second home there. Clearly, that sentiment towards Germany found expression among the ordinary soldiers in their cry of *"Evviva Germania!"* on the Mrzli Peak.

• A few weeks later the mountain rangers in the Grappa area faced Italian troops who hit hard and exceptionally well, and behaved like real men in every respect. No successes like at Tolmein could be wrought here.

• Today the Italian army is one of the best in the world.* They have been animated by a new spirit and have passed the test of their high ability in the exceptionally difficult campaign against Abyssinia.

• The appraisal of the Württemberg mountain rangers' success in the first days of this great battle is reflected in the order of the day of the German *Alpenkorps* (General von Tutschek) on Nov. 3, 1917, which states the following: "The storming of the Kolovrat ridge broke the entire structure of the enemy resistance into a collapse. The Württemberg Mountain Battalion under its goal-oriented leader, *Major* Sproesser, and his courageous officers contributed to this in the front lines. The taking of Kuk, the occupation of Luico, and breakthrough into the Matajur position by Detachment Rommel led, in a large part, the inexorable pursuit." The casualties of Detachment Rommel in the three days of attack were thankfully few: six dead, among them one officer, and 30 wounded, among them one officer.

At noontime on Oct. 26, 1917, the battle situation between Flitsch and Tolmein was as follows: **Group Kraus:** The foremost units were resting in Bergogna. Enemy attacks at Passo di Tanamea were repelled. **Group Stein:** Among the 12th Division, an attack on Loch by Infantry Regiments 62 and 63 was underway in the Natison valley from the national border across Stupizze. They reached Loch at 1400 hours. No forces were deployed to attack the Italian positions on the Matajur—Mrzli line from the north. Infantry Regiment 23, marching toward Matajur across Cragonza, had reached Cragonza around noon. Among the *Alpenkorps,* Detachment Rommel of the Württemberg Mountain Battalion had taken the Mrzli Peak and Matajur. The majority of the Württemberg Mountain Battalion under the command of *Major* Sproesser was ascending from Mount Cragonza toward Masseris. Following him were the 2nd and 3rd Battalions of the Life Guards. The 1st Battalion of the Life Guards and the Reserve *Jäger* Battalion 10 started advancing to Polava at about 1100 hours, after the enemy had cleared out his positions there. With the 200th Division, *Jäger* Regiment 4 had already taken Mount St. Martino at 0930 hours and then went further forward in the direction of Azzida. **Group Scotti:** During the course of the morning, Grenadier Regiment 8 took Mount Hum. The Imperial and Royal 1st Division began the attack on St. Jakob via Cambresco.

From this we gather: the forces of the 12th Division and the *Alpenkorps* around Luico first came forward in a southwesterly direction, then took the Italian positions on Mount Cragonza and the Salerno Brigade on Mrzli Peak and Matajur, which the front elements of the Württemberg Mountain Battalion had cut through. Also the attack by the 12th Division in the Natison valley northwest of the Matajur massif, which they reached during the night of Oct. 24–25, was only able to gain ground after the enemy had been ejected from the positions on Matajur [by Detachment Rommel].—*Rommel*

* This unusual sentence stands out for being out of touch with Rommel's pattern of thinking about the military skills of Italian soldiers and for referring to "current events" in the 1930s, which Rommel does not do unless he is discussing tactics. I speculate that this sentence and the one following it may have been added by Rommel's publisher to avoid embarrassing Germany's then-Italian ally. Since Rommel controlled details included in the book and was sensitive to making political statements, he probably would have had to approve such a change.

Rommel (center, towards left) poses for a candid photo among his Italian troops in North Africa during World War II.

6

Pursuit across Tagliamento and Piave 1917–1918

"Seething With Fury"—*Masseris, Campeglio, Torrente Torre, Tagliamento, Klautana Pass*

◊ Rommel invites captured officers to dine with him—something he also did in World War II, notably after capturing St. Valery in 1940. Some German officers might have found it beneath their dignity to share a table with captured enemies. The gesture is unique to Rommel, who seems to have wanted to demonstrate a lack of hostile feelings.

◊ Sproesser and Rommel meet for the first time after Mount Matajur. Rommel is ambivalent about their exchange, only stating his forces were no longer being considered to help Sproesser during a battle. He gives no reason why—unusually. In the past, Sproesser was willing to hear Rommel's ideas and allow him flexibility. Now when Rommel rushes over on horseback in the face of machine gun fire to offer assistance, Sproesser shuts him down. Rommel's forces are not needed—or not wanted. Why no combat orders for Rommel? Was Sproesser upset over Rommel's disobedience of his orders leading up to the assault on Mount Matajur? It seems likely. Rommel does not say.

◊ Rommel is taken out of action and chafes at it. His troops are removed from the advance guard and not even used for patrols. This is a very sharp contrast to duties he has been allowed—and indeed even asked—to perform previously. Detachment Gößler takes the lead instead. It appears that Sproesser was punishing Rommel for his previous disobedience.

◊ Upset, Rommel learns credit for storming Matajur is taken by another unit. He mentions the matter "was put right" with "higher powers"—essentially admitting he complained about it. While some have accused Rommel of being underhanded about claiming the victory, Rommel clearly believed others were taking away hard-earned honor from himself and his men, and that it was an injustice. (Recall the emphasis in German culture about personal honor.) He draws attention to this, acknowledging there was a dispute and making it clear he sought redress.

◊ Rommel learns a hard lesson about the limits of human endurance when he fails an attack for the first time—because his troops are too exhausted to function. He admits his men need food and rest to fight well...but it seems he fails to take this lesson to heart. While the failure of troops to perform due to exhaustion would have been a good point to bring up in his post-battle notes for students, Rommel overlooks this important point in his teaching. It would come back to bite him later in World War II.

◊ Rommel risks his life to rescue an Italian POW from drowning. Anyone else probably would have let the man get swept away, but Rommel rides his horse into a raging river to pull the enemy soldier out because, he says, he felt sorry for him. It is a unique moment—almost ridiculously compassionate, very impulsive, and very much Rommel.

On the peak of Mount Matajur, *Leutnant* Autenrieth brought over the command from *Major* Sproesser to move toward Masseries. It was 800 meters below us. The descent there demanded the last bit of strength from my dead-tired infantrymen. We took the captured officers of the 2nd Regiment of the Salerno Brigade along with us, since they gave the impression that they could not accept their new situation.[1] Thus I chose not to send them off to Luico through largely coverless terrain in which thousands of weapons were still lying scattered around.

Climbing down on a narrow footpath, we reached the charmingly arranged little village of Masseries in the early afternoon hours without encountering the enemy. Quickly the companies had been split up and arranged in the few farm buildings, the most immediately necessary measures for security and communication with the other units of the Württemberg Mountain Battalion already marching toward Pechinie were arranged, and then the tired troops rested.

I invited the captured staff officers to a simple dinner. No conversation was forthcoming during it. Also they barely touched the modest meal. These gentlemen were just too shaken about their fate and that of their proud regiment. I understood them completely, and cleared the table early.

Before daybreak on Oct. 27, Detachment Rommel was on its way to the Natisone valley. The remaining units of the Battalion were sent forward to Cividale and had a considerable lead. As the hills west of the Natisone were hotly contested in battle, Detachment Rommel moved to Cividale without resting from the march or stopping to eat. Riding ahead of the detachment, I came across the staff and detachment troops of Detachment Gößler of the Württemberg Mountain Battalion near San Quarzo in a fight with the enemy, who still held the Purgessimo.

I rode across the battlefield with *Leutnant* Streicher. Now and again, a sheaf of fire from an Italian machine gun would cause us to hasten our pace. I met *Major* Sproesser just east of San Quarzo. There was no longer any consideration of deploying my detachment here.

At about 1400 hours the battle for Purgessimo came to an end. After resting for several hours on the northern edge of the burning Cividale, Detachment Rommel moved into Campeglio around midnight, where the remaining units of the Württemberg Mountain Battalion were already providing security in the directions of Faedis and Ronchis. In the wee morning hours of Oct. 28, the pursuit to the west was continued. Cloud-shattering rain crashed down, soaking us to the skin. Sometimes the men shielded themselves with umbrellas, which the resourceful lads had managed to dig up from somewhere. Soon however, higher powers forbid us from using this novel type of kit. We marched on ahead in streaming rain without any enemy before us.

---

1. i.e., These Italian officers seemed rebellious and potentially dangerous.

Rommel is pictured on horseback with an aide nearby during World War I.

In the afternoon, Italian rearguards blocked the way near Primalucco at the swollen waters of the river near Torre. The river usually did not have much water but had become a roaring, 500-meter-wide flood due to the continuous heavy rain. The enemy on the other side shot at everything that appeared on the eastern banks.

We billeted in Primalucco, fixed ourselves with some dry kit in an Italian laundry depot and lay down early to sleep. The strains of the last few days and nights had hit us hard. However, one hour before midnight came an order from *Major* Sproesser: "Detachment Rommel, strengthened by a platoon of mountain artillery, must cross over the Torre river while it is still night, at the very latest by daybreak."

Alarm! Feverishly the detachment worked during the remaining portion of the night. As the artillery platoon shot multiple shells at the Italian force on the western bank, we made a bridge across the numerous branches of the river using various vehicles of all types in the area. The enemy hardly disturbed this work. He seemed to have moved off during the first bombardment of our grenades onto the west bank. As day dawned, the tip of our makeshift crossing was just 100 meters from the west bank and the enemy had moved out.

Leutnant Grau rode as the first one through the last branch of the river, which was an extremely strong torrent. Because the vehicles shoved in were not enough to finish the crossing to the west bank, a strong rope was pulled over the last stretch. The infantrymen held onto this rope while crossing through the rushing waters of the mountain stream, which would certainly have swept away individual men.

During the crossing, one of the Italian prisoners carrying a large medical bag near the rear was torn away by the strong water and dragged, screaming, lying on his back, away downstream. It appeared that the man was unable to swim. Furthermore his heavy bag was dragging him down to the depths. The poor devil moved me to pity.[2] I gave my black steed the spurs, galloped after the Italian and I managed to come up next to him in the rushing waters. In terror of death, he seized onto the stirrup of my horse, and the brave animal brought us both safely back on land.

A quarter of an hour later the entire detachment had completed the crossing. We advanced through Rizzollo, where the population gave us a very friendly welcome, and through Tavagnaco to Faletto. There we ran into the remaining units of our battalion, who had crossed over the bridge at Salt. Without making contact with the enemy, the battalion moved west toward the Tagliamento and reached Fagagua in the late evening. I was accommodated in good quarters with my staff. The owners of the place had already taken off and only the domestic servants were still there. We ate and slept.

On Oct. 30 the battalion reached the Tagliamento near Dignano by way of an advance through Cisterna. The bridge there was destroyed. Strong enemies occupied the west bank of the broad, swollen river. Attempts to cross failed. To the north we found the street leading to the bridge of Pietro via St. Daniel utterly clogged with Italian columns and vehicles of all kinds. Here, columns of military trucks and horse-drawn heavy artillery wagons, pack animal trains and refugee vehicles were all wedged together. On the street, as well as on both sides of it, the vehicles had driven up against each other so that none could move either forward or backward along a stretch of several kilometers. Italian soldiers were nowhere to be seen. They had probably sought safety in the open field. The horses and pack animals had been standing there crammed together for days and, starving, ate anything they could reach—saddle blankets, mantles, things made of leather!

A nightly thrust of Detachment Rommel cross-country to the Pietro bridge, already planned, was unfortunately cancelled again on higher orders before it took place. We grieved not to be getting any exciting experiences and that night hunkered down in Dignano.

In the coming days it became known that a troop unit of the 12th Division was credited in an official army communiqué for the storming of Mount Matajur. This matter was put right with higher authorities.

Over the next several days, all attempts to pull off a crossing of the Tagliamento River failed. Only on the night from Nov. 2–3, 1917, did Battalion Redl of the Bosnian Infantry Regiment 4 have the luck of setting foot on the west bank in the Cornino area.

---

2. *"Der arme Teufel dauert mich."* A very colloquial and South German turn of phrase.

The Württemberg Mountain Battalion split off on Nov. 3 from the German *Alpenkorps* group and received orders to break through as the advance guard of the Imperial and Royal 22nd Rifle Division to the Carnic Alps via Meduno—Klaut and reach the Piave valley by Longarone as quickly as possible in order to block the southern retreat of Italian forces on the Dolomite front.

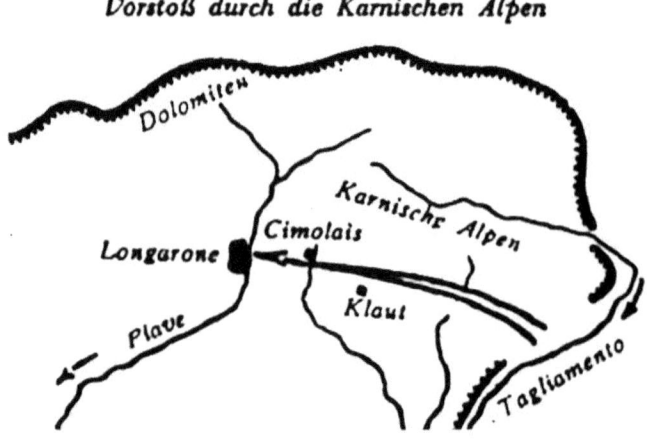

The forward thrust through the Carinthian Alps.

The Württemberg Mountain Battalion was one of the first battalions to cross the Tagliamento by Cornino. Strong patrols were sent forward to Meduno on captured Italian bicycles. Moving beyond this place, the advance guard of the Württemberg Mountain Battalion managed to take 20 officers and 300 men prisoner at Redona. Afterwards the weak Italian rearguard took to their heels through the wildly jagged Klautana Alps, using narrow paths to get to the Klautana Pass.

My detachment marched in the main body, while Detachment Gößler took the advance guard. They reached Pecolat on the evening of Nov. 6. Early on Nov. 7, the Württemberg Mountain Battalion ascended in this order towards the Klautana Pass. As the lead units of the advance guard neared the 1,439-meter-high crag of the pass, a strong enemy showered them with fire from the heights on both sides of the pass. Machine guns and artillery fire made it difficult to make the distance along the narrow road between Pecolat and the advance guard at the pass. The road had many winding curves and the advance units were 900 meters above us. Soon every movement on the road, as well as in the rocky terrain on both sides of it, was made impossible by Italian fire. The enemy was sitting in exceptionally well-sited nests high up on the vertical rock wall of Mount la Gialina (1,634 meters high) and on the northeast spine of Mount Rosselan (2,067 meters high) with a breadth of about 2,000 meters astride the pass.

Major Sproesser deployed Detachment Rommel, located in the main body and consisting of the 1st, 2nd and 3rd Company as well as the 1st Machine Gun Company) to bypass the enemy on the pass south of Mount Rosselan. The advance into Silisia was already very hindered by enemy machine guns and shells. One by one we sprang from rock to rock. At last we managed to find cover from enemy fire in the side valley leading to Hill 942. Soon however the vertical rock walls, several hundred meters high, of Mount Rosselan stood ahead of us and prevented further climbing. Bypassing the enemy from the south proved to be impossible. There was nothing else left to do but approach the enemy in the pass frontally.

For hours we climbed among the rocks in order to approach the enemy from south of the main pass road. The brave infantrymen carried heavy machine guns on their shoulders over spots that I myself without my rucksack had a hard time getting through. Shortly before the fall of darkness Detachment Rommel, totally exhausted, reached the snow-covered knoll 600 meters southeast of the pass and established communication with Detachment Gößler, at the same elevation as us but stuck several hundred meters north on the pass road. Mountain bushes concealed my men from view of the enemy, who occupied the rocky hills ahead of us in a half-circle formation.

Night attack on the Klautana Pass. A view from the east.

I granted the troops, flattened with exhaustion, some rest and in the meantime scouted the area with *Leutnant* Streicher and numerous patrols for possible means of surprise night attack on the pass.

The night was dark. The skies were covered with clouds. It was a good thing that the snow between the low bushes gave some light! However the sound of the snow crackling under our footsteps soon drew the fire of the defenders at us. But, due to this I was soon able to clarify the organization of the enemy position.

I was able to discern individual machine gun positions, barely 100 meters away from the pass ridge and many meters high above it. Throughout many hours of work, we set up our fire support for the night attack with utmost vigilance and precision. The entire Machine Gun Company would be deployed for this. At the same time, the 1st and 3rd Companies, some 300 meters distance from the pass and covered against enemy fire from above, were prepared for the attack.

At about 2400 hours all machine guns of the Machine Gun Company were to pin down the enemy at the pass ridge for a span of two minutes and then swivel to fire at the enemy on both sides of the pass. The 1st and 3rd Companies were to launch a storm attack on the left and right of the ravine leading to the pass just as soon as the heavy machine guns fired, and take the pass using hand grenades and bayonets.

Unfortunately I stayed too long with the platoons I set up for fire support. As they let loose continuous fire from their machine guns, I was still on the rocky slope several hundred meters away from both storm troops companies—which of course were supposed to attack whether I was with them or not. I hurried forward as fast as I could but, to my great astonishment, found both companies still in their launching positions. Had their commanders bungled it, or had it even been the troops? Two minutes of harassing machine gun fire had already gone by. Now the forward rush of the storm troops was no longer in sync with the machine gun fire. The enemy in the pass could no longer be pinned down. No wonder, then, that the mountain rangers' attack was repelled with casualties after a tough hand grenade fight. After the failed attack I sent both companies back to their launch position.

I was seething with fury about being driven off like this. This was the first attack since the beginning of the war that I had failed. Hours of toilsome work had been for nothing. To repeat the attack during the night seemed pointless and also nothing more could be demanded from the exhausted troops. After the strains of the day and night they needed rest and nourishment in order to get fit to fight again. Of course, 1,400 meters high up in ice and snow, right in front of the enemy, both rest and food were fairly impossible. It also appeared probable to me that strong forces would amass at the pass during the day.

After these considerations I broke off the fight. As they had before Detachment Rommel had attacked, the 5th Company took over providing security along the pass. I moved back into the valley near Pecolat with four companies. On the way I reported to *Major* Sproesser, whose command post was located halfway along in a rock crevice, about the miscarriage of the night storming attack.

The detachment arrived in Pecolat shortly before daybreak. The few shoddy huts there were jam-packed with troops. We camped in an open field. We called over the staff minding the pack animals and then we received coffee from heated, boxed pots. The hot drink tasted delicious. Two hours went by. Day came. The sun threw its first beams in the narrow valley. Then I was summoned to the telephone. Through it I heard from the battalion: "The Klautana pass has been cleared of the enemy. Detachment Rommel is to get ready to march at once and join up with Detachment Gößler. The Battalion is following through Klaut."

Shortly before daybreak the reconnaissance patrols of the 5th Company had determined that the enemy had cleared out from the pass. The joy about the enemy abandoning this exceptionally good position without a fight gave me new strength. Soon

Detachment Rommel was marching again. After hours and hours, this time ascending along a street, we reached the top of the pass and could there see for ourselves the good effect of the 1st Machine Gun Company's fire on the enemy pass position. One of the machine guns had seemingly swept the street just northwest of the pass across several hundred meters and had thereby caused casualties. The numerous bloody bandage wraps on both sides of the street testified to this.

> **Considerations:**
> - **The night attack** of Detachment Rommel on the Klautana Pass **failed, because the coordinated fire from the Machine Gun Company and the forward thrust of the storm troop companies were not cohesively timed.**—*Rommel*

"Utterly Fatigued"—Pursuit to Cimolais

> ◊ *Sproesser again shows reluctance to cede any authority to Rommel. He is no longer allowing Rommel to march to his own tune.*

It was incredible with what self-reliance the mountain rangers carried their heavy loads. Without long rest periods they had been moving for more than 28 hours as well as in battle. They had climbed the Klautana Pass twice in this time—a total of 1,800 meters altitude! Now we went downhill with wide steps. Detachment Gößler, also the advance guard on Nov. 8, had a considerable lead on us.

At midday we joined up with the advance guard near Klaut. Soon we went onward. Detachment Gößler encountered the enemy near Il Porto and attacked. However it didn't develop into a serious fight because the enemy withdrew north. While Detachment Gößler (consisting of 5th Company and the 3rd Machine Gun Company) moved toward Il Porto, Detachment Rommel (the 1st, 2nd, and 3rd Companies and with 1st Machine Gun Company) was sent out from S. Gottardo to Cimolais as part of the vanguard of the 1st Battalion of Imperial and Royal Rifle Regiment 26, which was being reinforced by the Württemberg Mountain Battalion.

In developed formation, Detachment Rommel followed the retreating enemy through the broad valley which became narrower towards Cimolais and had stone cliffs almost 2,000 meters high on both sides. The enemy was retreating to Cimolais on the western edge of the valley. Bushy terrain on both sides of the path concealed our movement from the sight of the enemy. Some bicyclists under *Leutnant* Schöffel and as many of the detachment staff as had ridden up found themselves forming a type of security line ahead of the developed companies.

It was already getting dark when we reached the eastern bank of the river Celina just east of Cimolais. The riverbed, more than 100 meters wide, was nearly dry. The enemy seemed to have moved off in the direction of Longarone and the town of Cimolais didn't seem to be occupied.

With the bicyclists I crossed the river in a broad front. No shots fell. Afterwards *Leutnant* Streicher rode with me into Cimolais. The local community representative greeted us with exceptional courtesy. He said everything had already been prepared for the German troops. He wanted to place the key for the town hall directly into my hand. But could we trust this? Could the enemy not perhaps have laid an ambush?

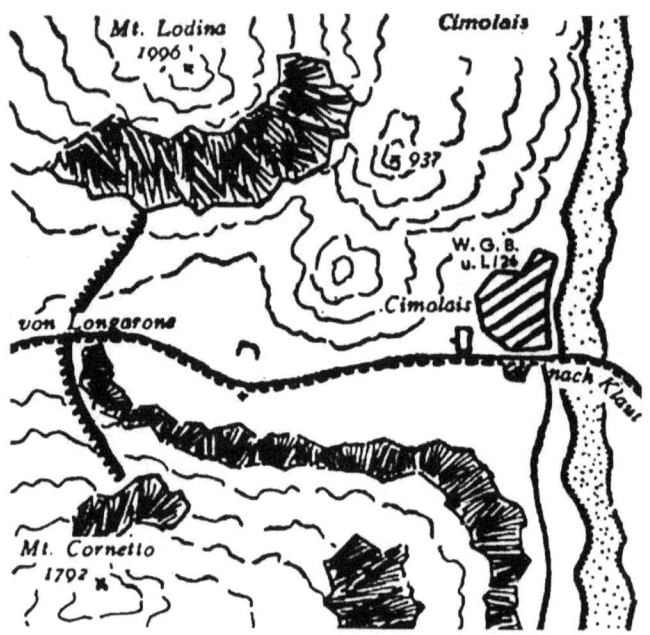

An overview of the situation around Mount Cornetto and Cimolais.

For security I sent the bicyclists along another stretch of road going far to the west toward Longarone. Then the dead-tired troops of Detachment Rommel moved into emergency response quarters in the southern part of town. They secured the road to Longarone and the way to Fornace stadium. The quarters were good and the food was plentiful. After the monstrous exertions that lay behind Detachment Rommel—32 hours spent in battle and marching without long rest periods—a few hours of sleep had to made the men totally fighting fit again. Who knew what lay ahead of us in the Piave valley only 10 kilometers ahead?

The staff of the Württemberg Mountain Battalion, the communications company, Detachment Schiellein (the 4th and 6th Companies and the 2nd Machine Gun Company) and the 1st Battalion of the Imperial and Royal Rifle Regiment 26 took shelter in the northern part of Cimolais. This latter battalion took over security towards the north. Meanwhile nightfall had long passed. The bicyclists of Detachment Rommel under *Leutnant* Schöffel reported that the enemy was positioned on the slopes of Mount Lodina (1,996 meters high) and Mount Cornetto (1,793 meters high) and was working hard at digging earthworks. The report was passed on to the battalion. Arriving around midnight, the battalion command sounded as follows: "The 3rd Company on the morning of Nov. 9 will attack the enemy west of Cimolais from the western edge of the town. As this is happening, the detachment will outflank the enemy positions there as follows: Detachment Rommel (consisting of the 1st and 2nd Companies with the 1st Machine Gun Company) will go over Mount Lodina, ascending before daybreak; Detachment Schiellein (the 4th and 6th Companies with the 2nd Machine Gun Company) will cross over Mount Cornetto (Hill 1793), Mount Certen (Hill 1882) and Erto; Detachment Gößler (the 5th Company and 3rd Machine Gun Company) will cross over Hill 1995, Hill 1483 and Erto.

An ascent at night over trackless, jagged mountains at 2,000 meters elevation (a 1,400-meter-high difference from our current position) with my detachment's utterly fatigued troops was something I regarded as undoable. Thus I presented myself to *Major* Sproesser shortly after midnight and requested that he modify his orders. I suggested that my whole detachment launch a frontal attack on the enemy west of Cimolais on the morning of Nov. 9. Only grudgingly did *Major* Sproesser change the orders—which then stated that one company of Detachment Rommel would carry out the outflanking movement over Mount Lodina, while the remaining companies would remain in reserve for me to carry out the frontal attack.

"Could I Keep My Promise?"—Attack Against the Italian Positions West of Cimolais

> ◊ Rommel teaches his men to scatter and hide, making themselves difficult to kill or capture as they melt away into the landscape. He used this technique retreating in North Africa.
> ◊ Horses that Rommel and his men are riding are killed by surprise fire. Rommel is usually straightforward but declines to describe the exact fate of the animals, making a vague reference that the "something" which happened to the horses would have happened to him and his men, too. Notably, he becomes unusually vengeful towards enemies on Mount Cornetto after this incident and does not make his normal extra efforts to give the Cornetto garrison opportunities to surrender. I suspect that Rommel was embittered by the fate of the horses, including his own.
> ◊ Anxious to get back into the action, Rommel gives *Major* Sproesser a guarantee he can get results on the battlefield—without actually having seen the lay of the land. It seems Rommel was desperate to get back in the fray and would seize any opportunity.

Three hours before daybreak the hardy 2nd Company led by *Leutnant* Payer moved to outflank the enemy position to the north via Mount Lodina, guided by a local resident.

At about 0500 hours *Leutnant* Schöffel established that the enemy west of Cimolais was utterly silent and inactive. He estimated that the positions had been cleared out, as had happened the previous day.

At this I prepared the companies for battle and ordered the company leaders to ride to the southern exit of Cimolais. I rode out, with security provided by bicyclists, in order to determine whether the enemy had actually pulled back, specifically to scout the attack zone in front of the enemy positions on both sides of the pass. Day began to dawn as we were trotting out from the southern exit of the town. The street rose gently uphill. The bicyclists had a lead of between 50 to 100 meters.

As we reached the chapel of La Crosett 150 meters west of Cimolais, the hills ahead of us flashed in a half-circle motion through the dawn light. Fire from machine guns and rifles clattered onto the street and whistled right past our ears. In a few seconds the bicyclists were off their bicycles and the riders off their horses; the animals went galloping without their masters back toward Cimolais.

Soon the whole reconnaissance staff found itself inside the La Crosett chapel. Nobody had been hit. The walls of the small chapel protected us from the fierce fire which now concentrated itself on our place of refuge. Soon the roof tiles began splintering under sheaves of Italian machine gun fire and the debris fell down on us below. With every passing minute the enemy's view of us improved. His closest position was only 200 meters away. A single enemy shell would be enough to send us all into the afterlife. If we waited that was what would surely happen.

As the rifle and machine gun fire let up somewhat, I decided we would spring back to Cimolais one by one and arranged the order of who would go in line after who. *Unteroffizier* Brückner was supposed to be the first. I would be the second. Although the enemy would surely fire vehemently at each one of us, if we all went scattering in different directions and never left cover wherever we could reach it, we would all manage to get back to Cimolais in one piece. Only a few horses had gotten hit during the scouting ride. If the Italians had let us get 100 meters closer to their positions, the same surely would have happened to us all.

Meanwhile daytime had completely come. The observation troop of the detachment staff led by *Vizefeldwebel* Dobelmann had already used the detachment's binoculars (40% magnification, taken as booty from Tagliamento) during the firing to determine the exact layout of the enemy positions west of town. The flashing of the shots in the morning light had lit up the binocular glass considerably. Dobelmann pointed out the enemy to me from the church tower.

The enemy was at about battalion strength and sitting well-nested on both sides of the road from Cimolais to Erto in well-built and fortified positions. The position rested against

Feuerüberfall auf den Erkundungsstab

The enemy surprise fire on the reconnaissance staff, who ended up taking refuge in a church.

a vertical rock wall several hundred meters high about 800 meters north of Cimolais at the southern slope of Mount Lodina. The position stretched across a steep slope of boulders, crossed the large street 500 meters west of Cimolais and then lay south of the street across a ridge declining vertically toward the east. The continuous and fortified position ended 150 meters south of the street. From here outward, the northeast slope of Mount Cornetto was occupied by a line of enemy riflemen of about one company's strength plus some machine guns. The individual riflemen had nested themselves in cleverly with their front facing Cimolais. However, due to the rocky ground surface, they had not dug into earth all too deeply. Their positions mainly consisted of stones and boulders piled around them. The enemy positions on the slope of Mount Lodina and on both sides of the street were protected by barbed wire. The positions on the slope of Mount Cornetto had no defense worthy of notice, because vertical rock faces and rocky gullies steep as rooftops made this section of the position nearly impossible to approach.

Despite all this I had pledged last night to *Major* Sproesser that I would take these positions with a frontal assault. Could I keep my promise?

I had imagined that the task would be considerably easier. Now I had to get ready in the face of these difficult circumstances. A frontal attack across a broad front could only be considered against the barbed-wire positions on Mount Lodina and on both sides of the street. The attack would be exposed to flanking fire from the garrison on Mount Cornetto. However, there was still the prospect of preventing this through an attack by our own machine guns at a very high altitude from the top knoll of Mount Lodina 700 meters north of Cimolais; the enemy had not included that knoll in his position. However the opportunities to provide fire support for the attack against the position fortified with barbed wire were very meager. To advance against the positions on Mount Cornetto seemed totally pointless. A single rockslide caused by the defenders there would be enough to wipe out every attack column, not to mention the flanking fire we would get from the Mount Lodina position. To bypass the enemy position on its flanks over Mount Lodina would be, aside from revealed in the daylight, totally and completely grueling and

A map showing the position west of Cimolais from the east.

time-consuming. Undertaking a similar enterprise across Mount Cornetto offered absolutely zero prospect of success either. The eastern slope of the mountain consisted of vertical rock walls which no human being had probably ever climbed.

There was nothing more to be seen of the 2nd Company which had already started climbing over Mount Lodina during the night. From all outward appearances they were already far to the north. It could be evening before they attacked. I couldn't count on the outflanking columns of Schiellein or Gößler to attack either.

The only suitable hill to support a frontal attack against the enemy positions west of Cimolais lay 700 meters to the north of the town. It was a forward knoll of Mount Lodina, 937 meters high and overgrown with low-lying bushed on its upper half. After I had thoroughly examined the attack zone from the church tower at Cimolais using the binoculars, I arrived decisively at the following conclusion: "We will surprise the garrison on Mount Cornetto by launching concentrated fire from multiple light machine guns from high positions on the hill 700 meters north of Cimolais, deal with them and then move into attack on both sides of the road in the valley."

During the course of the next few hours I brought the light machine guns of *Leutnant* Triebig's 1st Company, unseen by the enemy, into the bushes on the knoll 700 meters north of the town, put it into position and briefed the company about the attack plan and their tasks. Afterwards the remaining units of my detachment (the rest of the 1st Company, the 2nd Company and the 1st Machine Gun Company) were prepared on the covered slopes just northwest of Cimolais for attack and briefed the individual units of their projected tasks. No men were immediately sent into action. The detachment command post was located with the 1st Machine Gun Company. The communications troop set up a telephone connection to the light machine gun crews as well as to the 1st and 3rd Companies.

As we were making these preparations, mountain rangers, in addition to multiple machine guns of the 1st Battalion of the Imperial and Royal Rifle Regiment 26, opened fire from the area near the church as well as from the church tower of Cimolais at the Italian positions in the pass. They did so without establishing any contact with Detachment Rommel beforehand or letting us know anything. Because this spur-of-the-moment battle did not fit into my attack plan, I went personally to appeal to *Major* Sproesser's command post in Cimolais and obtained

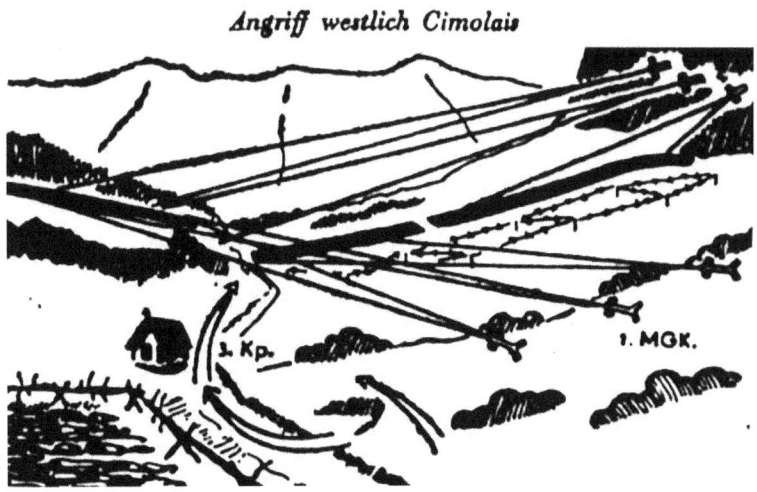

The attack west of Cimolais. The 1st Machine Gun Company is shown firing from the right.

that this firing was brought to a stop. At 0900 hours, I ordered the gun crews of the 1st Company to begin firing. In accord with their instructions, four light machine gun crews laid immediate fire on the enemy guns furthest left on the slopes of Cornetto, while two light machine guns pinned down the remainder of the Cornetto garrison. Although the distance from our light machine guns was considerable (over 1,400 meters), the effect was, as we expected, excellent. We observed the results from different points using binoculars.

The Italian riflemen on the southeast flank were without cover and exposed to fire from the higher elevation. They were not hit, but even so they were so rattled by the fire that they speedily abandoned their gun nests and sought cover near their neighbors to their left in a zone which had not been threatened up to then. The sheaves of fire from our light machine guns followed them. Soon things were too hot for the Italian riflemen even in their new foxholes. They cleared out as fast as possible to the fortified positions south of the pass road, where they hoped to find protection from the effects of our fire.

First there were only a few Italians running. Soon however it was a whole platoon tumbling away. This was exactly what I had been lying in wait for. The 1st Machine Gun Company got orders to join in the battle as quickly as possible from the hill just west of Cimolais. Until this point nobody had been able to set foot on this little hill because the Italian garrison on Cornetto could hit them with fire from a higher elevation. Now the Cornetto garrison was as good as done for.

As the first light machine gun pitched into the fight, a herd of Italians of at least company strength about 600 meters away on Cornetto came scrambling down in a panic toward the southern end of the fortified position above the rock 150 meters south of the pass road. The effect of our weapons noticeably increased. Afterwards a heavy machine gun joined in the fight. Then came fire from six light machine guns from much higher elevation. Men dashed down below, scurrying to the narrowest trenches, all squeezed together with no room between them. Even these positions offered insufficient cover from the light machine guns, which struck into the steepest crannies.

Now the 3rd Company was tasked with immediately leading the storm attack on both sides of the pass road. They had nothing more to fear from the Cornetto garrison and the Machine Gun Company was now taking care of pinning down the other Italian positions. As the 3rd Company crept forward in depth, concealed from the fire of the Italian garrison on the slopes of Lodina, the Machine Guns did a lot of work. They hit the enemy position south of the road, which was completely stuffed with men, from the front and from above. They held the enemy down north of the road and diverted him.

Now the Italian position south of the street started to empty itself towards the rear. Except, the enemy would find it difficult to pass through the tight loops of German machine gun fire at 500 meters distance. Most of those who fled were mowed down after just a few minutes.

I had leadership of the gun crews firmly in hand because I was lying down between the machine gun companies and had a telephone connection to the light machine gun crews on the hill to the left behind me. The 3rd Company had already reached the enemy obstacles and was now breaking into the pass position, supported in an outstanding manner by light and heavy machine guns. We had won!

I had one part of the gun crews keep firing. I followed as fast as possible into the seized position with all remaining troops and was soon traveling the same path as the 3rd Company. The enemy garrison on the slopes of Lodina was still hanging on. A report of the successful storm attack was sent to the battalion and at the same time bicyclists, dispatch riders and horses to ride were called forward. As I arrived in the newly taken position, the enemy garrison of Lodina consisting of two officers and 200 men cast aside their weapons.

A special source of happiness was that we had sustained very few casualties; we only had many light wounds to show. I hadn't expected that we could take the enemy position so cheaply. Units of the enemy garrison withdrew west. To follow them, to overtake them and to seize the Piave valley as quickly as possible was what I saw as my new task.

> **Considerations:**
> - If the **battle reconnaissance against the enemy** west of Cimolais during the night of Nov. 8–9 been done more thoroughly, the enemy's surprise fire on our reconnaissance staff would have been avoided.
> - **On the other hand, the enemy's surprise fire worked to provide instant clarify about their disposition**. The analysis of the enemy fire by *Vizefeldwebel* Dobelman, the detachment's self-motivated observer, was done in an especially clever manner.
> - The **technique of battle** caused a big headache for the attack at Cimolais until the key was found. What worked here was the considered **moral effect of light machine gun fire even from across a great distance**. The first Italian riflemen to flee Mount Cornetto caused panic among their units.
> - The **simultaneous use of weapons during the attack** against the enemy west of Cimolais was implemented superbly. **Strongest fire was focused on the break-in point** shortly before the 3rd Company pressed through. The **well-prepared telephone network enabled firmly controlled leadership of the assault**. —*Rommel*

"Blown Up into the Air"—Pursuit through Erto and the Vajont Canyon

> ◊ Rommel leads a haphazard pursuit through a canyon where the Italians start demolition operations. One German officer, Fischer, is killed in a horrific manner trying to pull the fuze from an enemy explosive set up to destroy a bridge. Could a more methodical approach from Rommel have prevented Fischer's demise? Rommel stresses that speed was a priority. He reasons that enemy demolition operations could have held the troops up for days and resulted in a net total of more casualties.

There was no time to organize my units. If we let the fleeing enemies off even for a few minutes the Italian commanders could get their soldiers back in order again. I got whatever of my troops that I could reach to start the pursuit. The rearward units and gun crews received orders to pull back to the street at top speed.

Machine gun fire from the slopes of Mount Lodina halted our advance 300 meters west of the position we had taken. It was coming from units of the 2nd Company who could no longer tell the difference between friend or foe due to their considerably high position. We could find no cover from their fire, which was an extremely awful feeling. It was good that after several minutes they realized their error. Because of this pause, the distance between us and retreating enemy had become somewhat longer.

So we hurried after the Italians at an even faster speed than before. We could simply not allow them to stop us before Longarone again. At 1010 hours, *Leutnant* Streicher and I, along with the front units of the 3rd Company, reached St. Martino. At the same time bicyclists and dispatch riders arrived from Cimolais with the horses of the staff. The street took a very wide turn north and ended 800 meters west of St. Martino in the village of Erto e Casso. Right and left the mountains now began to recede greatly. Small units of Italian troops were retreating at a fast pace on the street 500 meters ahead of us in closed formation.

Quickly I had a light machine gun positioned for fire support but told him not to fire except if things came to a fight. Then we ran down the street after the enemy. With our horses and bicycles we soon caught up to the fleeing Italians. It did not come to a fight. Calling out to them to give themselves up, signaling them to disarm and giving them a direction to march back as prisoners was enough. We galloped to Erto and galloped right through it. Here and there were Italian pack animals standing tethered up. No shots fell here. All the prisoners we took gave themselves up without resistance.

Ahead, at the point of the column, the whole pursuit looked like a racing derby between horses and bicycles. From further back it looked like the ultimate battle of an army baggage train. Wheezing, the infantrymen hauled their burdens plus the light and heavy machine guns. Detachment Rommel had dispersed itself across a length of several kilometers. Every trooper knew that this was now a question of completely overrunning the enemy and that the success of the pursuit depended on speed.

After Erto the valley got narrower. The road sank toward the Vajont gorge. We were still four kilometers away from the Piave valley—our goal. The most difficult part of the terrain lay ahead of us: the Vajont canyon. It was 3.5 kilometers long and extraordinarily narrow and deep. The street, jumbled throughout with vertical rock walls from 200 to 300 meters high, now led to the north side. The middle of the canyon straddled a 40-meter-long bridge 150 meters above the roaring mountain stream. After the bridge the street ran to the south side of the gorge. Various side canyons had bridges. For the most part the street led through long tunnels. A single explosion at a suitable place would be enough to block the road to Longarone for several days. But yes, a machine gun placed at the entrance of a tunnel could hold us up here for a long time. All this could have been learned from studying the map, but until this point I'd had no time for a detailed examination of it.

After passing through Erto there were more bicyclists than riders on horseback on the steep road. They overtook more Italian troops at a bend in the road. Then they disappeared from our view. Shortly afterwards, the sound of gunshots fell.

Further ahead in the canyon, an Italian truck was visible, driving west. We made the horses go as fast as they could on the precipitous road and chased through the first of the pitch-dark tunnels. Once were got there—a massive explosion just a few hundred meters ahead of us nearly ripped us off the horses. We probed the dark tunnel—which, as we learned later, was full of Italians—trying to find an exit. We saw the results of the explosion 50 meters ahead to the west. A deep abyss yawned in front of us. The enemy had managed to blow up a bridge that spanned a side ravine of the Vajont canyon.

But where were my bicyclists? A firefight spinning up further west answered that question. Dispatch

The destroyed bridge blocking Rommel's path through a ravine in the Vajont canyon.

rider Wörn received orders to lead all units of my detachment that he found to the front just as fast as possible. Then we climbed into side ravine, heading right, passing over the ruins of the destroyed bridge and getting to the street over on the other side. We ran forward to where the shots kept falling.

On the north end of the bridge spanning the Vajont canyon we met the bicyclists behind a building beside the bridge. The bicyclists were in a firefight with the occupants of an Italian truck, which had just driven over through a tunnel from the other side. It looked like this was an Italian explosive team who had the task of detonating various structures in the Vajont canyon which were already rigged to blow up. Soon the bicyclists reported that they had ridden over the last bridge just a few seconds before it blew up and that *Unteroffizier* Fischer, seeing the burning fuze and trying to detach it from the explosives, had been blown up into the air along with the bridge.

Another bridge lay ahead of us. It spanned 40 meters and was 150 meters above the wild waters of the river. Thus it was the most important bridge for the Italians. We could clearly see the piles of explosives in deep, four-cornered cases in the middle of the road. Had the fuze already been lit?

The enemy on the opposite side of the bridge stopped firing and was no longer visible at the tunnel entrance. Had they retreated?

If the bridge ahead of us went flying into the air, then it could take days before we managed to get near the Piave valley. The bridge needed to be snatched quickly.

I knew *Unteroffizier* Brückner of the 2nd Company was an especially daring and valiant-hearted soldier, so I gave him the following task: "Take a shovel, run across the bridge and hack apart any and every wire you can see on the bridge. As soon as you do this, we will all come over in closed formation and rip the fuze out as we go on our way."

The bridge that the enemy had rigged to blow up.

Because of the many cables hanging off the bridge into the deep, I feared an explosion of the bridge caused by electricity. The exemplary *Unteroffizier* Brückner carried out his orders immediately. As the last cable fell, I ran forward with the bicyclists and tore the fuze out of the explosive box as we ran past. The bridge thus fell into our hands unscathed.

Onward we went in utmost hurry to the Piave valley. We had to stop the enemy explosive commandos from making any other explosion anywhere. *Unteroffizier* Brückner was sent forward with some bicyclists. The detachment units to the rear received orders to step up their marching speed. Again we went through various tunnels. The street sank as we got to the exit of the canyon. The vertical rock wall that the street had been blasted through now reached a height of 450 meters. No shots fell ahead of us near Brückner's patrol. He must have reached the exit of the canyon a long time ago.

At about 1100 hours I reached, accompanied by numerous bicyclists, troopers of the 3rd Company and the detachment staff—10 carbines altogether—the exit from the Vajont canyon one kilometer east of Longarone.

Ahead of us was an overpoweringly beautiful sight: in the glowing light of the midday sun lay the Piave valley. The bright green mountain river with its many branches rushed 150 meters below across its broad, stony bed. On the opposite side lay Longarone—a small, widely dispersed little town, with rocky cliffs of about 2,000 meters high towering behind it. Right at that very moment the Italian explosive commandos were driving their truck over the Piave bridge.

An endlessly long column of enemy soldiers was marching on the wide valley street on the west bank of the Piave—carrying all types of weapons, coming from the Dolomites in the north, moving south through Longarone. The town of Longarone and its train station, as well as Rivalta, were densely occupied with troops and stationary columns.

"Not to My Taste"—**Battle At Longarone**

> ◊ As Sproesser receives orders to wait in position, Rommel gets embroiled in fighting. Sproesser, in an awkward predicament, faces a tough decision. Again Rommel's detachment has run away with itself and is on the brink of catastrophe. Sproesser pursues, sends in backup and bluffs to the Italians, securing a surrender. Rommel proudly interprets Sproesser's decisions as acts of faith. Perhaps Sproesser, faced with a situation spiraling out of control, was trying to cut his own losses. At one point in the general confusion, Sproesser hears Rommel and his whole force are captured. Rommel is insulted at the idea, indignantly calling out the "rumors" as "most unbelievable" and citing the din of battle and fires as testaments to his undiminished fighting. This vision of burning fire and yelling might have appealed to his fierce personality, but may not have been exactly comforting to Sproesser at the time.
> ◊ Rommel states his mountain troopers insist on a torchlit procession for "their commander." Because this sentence appears in Rommel's description of his own detachment, Rommel was the likely honoree. It would be characteristic of Rommel to refer to himself indirectly here.

Very few soldiers who fought in the World War faced demands similar to we did in the Piave valley: thousands of enemy in an orderly retreat in a not-too-wide valley, blocked right and left by 2,000-meter high mountains which were mostly unclimbable, not knowing what dangers lay on the flanks.

We mountain rangers felt our hearts nearly leap for joy. The enemies over there were no longer going to get past us, that was for sure. Soon I nested myself in with 20 riflemen in dense bushes 100 meters south of the street. Then our fire struck from across 1,200 meters among the columns of enemies marching on the road from Rivalta to Pirago. We maintained firing at a spot that it was impossible for the enemy to retreat from: on their right were rock walls, on the left the Piave river! The front troops of the 3rd Company were already arriving breathlessly at the exit from the canyon pass and strengthening our firing line.

Our rapid fire ripped the caterpillar of enemy forces in two after only a few minutes. The north half marched back toward Longarone and the southern half sped up their march. Minutes later the enemy deployed a large number of machine guns against us. Because we were very well-nested inside clumps of bushes on the front slope and away from the street leading out from the canyon, the enemy guns did not find us. All of their fire hit the street and into the canyon. But the advance of Detachment Rommel was considerably delayed, even so.

Now the enemy in Longarone tried to move south in small groups. A platoon of the 3rd Company with two light machine guns, now in position south of the canyon, made this departure of the enemy considerably difficult.

The battle at Longarone shown from the east.

Suddenly one of my couriers observed Italian infantry in company strength climbing up the rock walls in our rear (from the direction of Hill 854). As fast as possible, I brought some riflemen and a light machine gun from the firing line facing west to turn and face the new enemy. Presently the enemy climbed up the steel rock wall, one man behind another, towards us. They were about 300 meters away. As soon as we shot, the first enemy we hit would fall back over the rock face and take his comrades down with him as he fell. Such a result was certain in my mind. However I did not shoot at once, but instead demanded that the enemy surrender by calling out to him. The enemy quickly realized that he would certainly lose here and surrendered. If we had discovered the enemy five minutes later, he would already have gotten over the steep cliff and thus could have truly endangered us.

In the Piave valley the enemy detonated another bridge east of Longarone. An enemy attempt to move off in the direction of Mudu with closed units was foiled by our fire. It was only possible for the enemy to filter out in tiny groups on the road to Mudu and Belluno, as well as on the train tracks going south. Multiple enemy gun batteries from Longarone joined the fight. That changed nothing, because they could not find our positions either.

Dozens of shells fell ahead of the canyon road and into the canyon, as well as on the rocks above the canyon road. Despite the extremely uncomfortable effect of the enemy's machine guns and artillery fire, which was increased by the falling rocks and pieces of boulders, the remaining units of the 3rd Company got to the Hill 1000 south of the canyon exit road, as did the 1st Company and a platoon of the 1st Machine Gun

Rommel (center with arms folded behind his back) is pictured among his troops circa 1917, possibly in the Italian Alps.

Company. They arrived up until 1145 hours.

To block the street and rail line to Belluno on the west bank of the Piave and captured all enemy units coming from the north, I sent out the 1st Company, strengthened by the heavy machine gun platoon, to the west bank of the Piave, in the area of Pirago, via Dogna. The whole 3rd Company took on the duty of providing supporting fire for this advance and stopped the enemy from marching away in closed groups.

In rows and at very close intervals, the 1st Company hurried over the steep grassy hill, on which only a few clumps of bushes offered cover from sight, in the direction of Dogna. Soon Italian machine guns and batteries turned their fire on the company. Nevertheless they reached the cover of the houses at Dogna with no casualties. Watching this, the enemy machine gun fire and artillery fire now became even stronger. Most of it aimed at the Vajont canyon.

Now one could see the 1st Company advance into the Piave riverbed west of Dogna. The riverbed however offered zero concealment and even less protection from fire. Very soon the Italians in Longarone showered the 1st Company with a hail of shots. The company was only saved by pulling back to Dogna with heavy casualties.

During these proceedings I personally rushed forward to Dogna with the detachment staff. A telephone connection to the 3rd Company which we left in its position was set up as we went along. Shells and machine gun fire sped up our pace. The enemy shot at each individual man.

In Dogna I met the 1st Company, which had just returned from the banks of the Piave. The misfortune could not be allowed to discourage us. If a whole company could not manage to get across the Piave, perhaps it could be done by a few courageous troopers. They could make better use of the terrain and perhaps get further south.

I immediately brought the heavy machine gun platoon forward into the upper floor of a house and arranged it in such

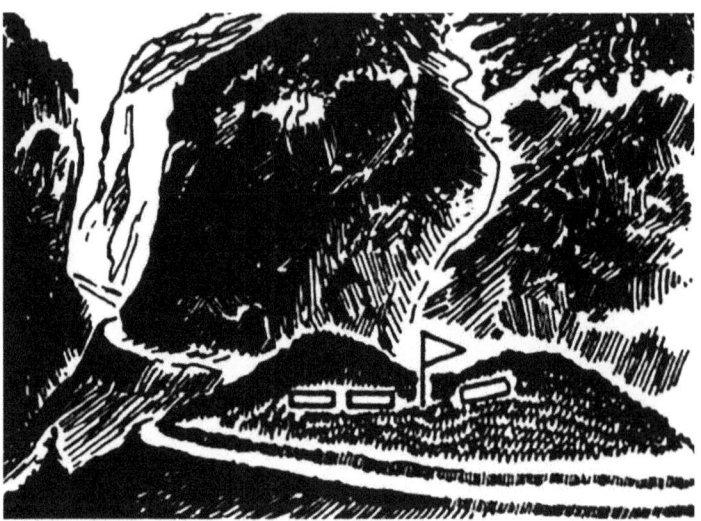

A view of the position oustide of the Vajont canyon.

a position so that, lined up to fire through certain rooms, it could hit the rail line and bridge near Pirago 900 meters away where small groups of Italians had been dispersing. Their task would be to prevent large numbers of enemy detachments from marching away south. With our low supply of ammunition—only 1,000 rounds per gun—we had to be sparing.

Then I deployed multiple patrols under especially plucky commanders across the Piave. They were supposed to cross the Piave in very light formation and, upon arriving on the west side, reach the Pirago area and there catch whatever enemies they could who were fleeing over the bridges. As soon as they had gathered a large number of prisoners, they were to move off to the east bank of the river in the direction of Dogna. This task was very difficult and would demand a totally extraordinary amount of spunk and ingenuity from officers and men.

Under the effect of strong enemy fire the individual patrols—a total of five—advanced extremely slowly. Under these circumstances I doubted whether even one of them would reach the west bank of the river.

Meanwhile *Major* Sproesser had come through the mountain pass and was at the exit with the communications company and the 1st Battalion of the Imperial and Royal Rifle Regiment 26. At my request the communications company took over the 3rd Company's duties at the pass exit. The 3rd Company rushed in very small groups to Dogna.

Nothing more was to be seen of the patrols on the riverbed. However enemy machine gun fire kept sweeping the bare, 800-meter wide gravel bank. At about 1400 hours I attacked from Dogna with the 1st and 3rd Companies in a broad front in the direction of Pirago. I hoped to bring at least a few units over the river in this manner and to block the west side of the valley road with the fire of the whole detachment. The attack broke out from its launch point on both sides of Dogna with the element of surprise. However strong enemy machine gun fire and artillery fire from a few hundred meters away forced us to the ground and to use our spades when the troops were not coverless in the face of the enemy's fire. We had managed to create a broad front within 500 meters distance of the street the enemy was using to march away on. Furthermore we had caused fire against our patrols further south to be diverted [towards us].

Because I very much doubted that even one of the five patrols had reached the west bank, I sent additional patrols out under *Leutnant* Streicher and *Leutnant* Triebig. However the first leader was very soon put out of the fight by the blast pressure of an Italian shell at the main branch of the river and the second was wounded by machine gun fire. It seemed impossible to even get a single man over the river. From two side, the Italian artillery chopped the land in which were lying to pieces: from positions just south of Longarone and from the direction of Mount Degnon in the southwest. The enemy seemed to have no shortage of ammunition.

The detachment staff had nested itself in behind a small stone wall in the Piave riverbed. The place became a special target for one Italian battery. Various gaps in the little stone wall proved that the enemy was bracketing his fire very well. For this reason we had already been putting our spades to good use for quite a long while. *Vizefeldwebel* Dobelmann scanned the terrain south of Longarone very sharply with the binoculars. My adjutant was out on reconnaissance. I was dictating to *Unteroffizier* Blattman, who was being trained as our detachment's clerk, the after action report from Cimolais. The enemy's fire maintained its uninterrupted fury. Above all the 3rd Company was suffering from it. Across among the enemy one could constantly see individual men and vehicles rushing amid the positions targeted by our fire.

At about 1430 hours the 3rd Company and the 1st Machine Gun Company of the Imperial and Royal Rifle Regiment 26 arrived in Dogna at my disposal. The leading officers reported to my command post. Because I did not want to expose any more troops to the enemy fire at the Piave riverbed, I had them reinforce my troops in Dogna and remain at my disposal, and only sent out one heavy machine gun platoon to strengthen the box barrage that Württemberg Mountain Battalion units had already been laying on the street and rail line of Longarone. I hoped to reach the opposite riverbank with all units by the onset of darkness at the latest.

Seven patrols had been trying to get across to the other side for hours. I had no reports from any of them. Had a single one of them made it over the river? Just as before we could see small enemy troops making their way south. Unfortunately we could not stop them. Our ammunition was getting very scarce, above all for the machine guns. We had to be sparing. Minutes stretched on and on. The enemy fire kept raging, claiming its victims in one place and then another.

At about 1500 hours *Vizefeldwebel* Dobelmann reported that he believed he had recognized some of our mountain rangers on the other side of the river, headed in a southwesterly direction. At the same time it seemed that an Italian coming from the hill west of Fae at been taken prisoner by one of our troopers behind a house near the railway line. I had to see it to believe it. Now everything was in order. Now I knew that the little groups of Italians slipping away south through our detachment's fire would not get out from Fae.

In vain we waited for the prisoner to be sent packing towards us across the eastern bank of the Piave. I hoped here for a considerable slackening of the tough situation in the riverbed. Maybe some of my units could make it across the river while the prisoner was being sent over from the other side.

At last at about 1530 hours we saw a dense mass of captured Italians two kilometers south of us in a wide bed of the Piave river. Most of them were already on the eastern side and marching toward Dogna. However I soon had cause to be angry that this exchange would have no results for us, because Italian artillery around Longarone opened fierce fire on the mass of prisoners. Obviously the Italian artillery had mistaken them for Germans. The fire resulting in the prisoners retreating back across the west bank of the river. So this incident did not reduce the tough situation for us even one bit. The enemy kept us pinned down as before with shells and machine gun fire.

Shortly before the onset of darkness a large number of Italian prisoners materialized near an old dam on the western branch of the Piave in the area of Hill 431 a kilometer north of Fae. They began crossing the river. What I had been hoping for the whole day was now coming to fulfillment. I sent all remaining units of Detachment Rommel capable of reaching our goal straight over to that dam. We no longer were concerned about the enemy fire aimed at our positions and at the western edge of Dogna.

The hundreds of prisoners crossing over the main branch of the Piave would shield us from further enemy fire. Quickly the detachment's river crossing was underway. The prisoners showed us how to best cross this wild river with its many offshoots, which in some place was strong torrent and it its deepest areas was breast-high. Even those men who were excellent swimmer could only reach the opposite side with extreme difficulty. The strong currents easily dragged them. Soon the Italian prisoners on the opposite side formed a human chain holding hands and walked diagonally into the river facing the current; they slanted their bodies forward more or less according to the strength of the current. Soon we were going over the various branches of the river in the same manner and then hurrying as fast as possible to Fae. The ice-cold dip in the Piave—whose higher waters lay in the snow-covered mountains—sped up our footsteps considerably.

We had tremendous joy at being reunited with our patrols at Fae. They soon reported all that had happened to them during the last several hours. *Offizierstellvertreter* Huber and *Vizefeldwebel* Hohnecker had, with 16 men of the 1st Company, managed to swim across the Piave 1.5 kilometers south of Pirago despite violent enemy machine gun fire from Longarone, keeping great distances between each man. They were now in possession of Fae Castle. During this action, the trooper Hildebrandt fell in battle.

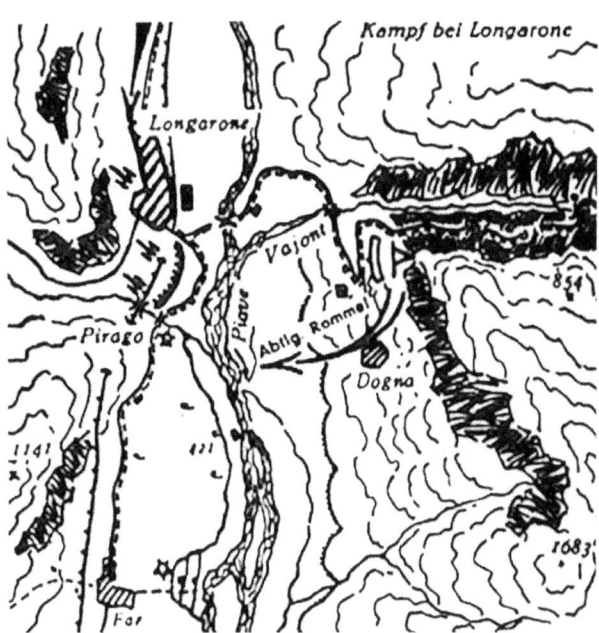

A bird's-eye view of the battle at Longarone.

Once in Fae, the little flock [of men] had blocked the road and rail line toward Belluno and captured small groups of Italians coming from Longarone who believed themselves to have reached safety. *Leutnant* Schöffel arrived later. During the course of the afternoon, 50 Italian officers, 780 men and a rich assortment of vehicles of all types at Fae were captured by units of the 1st Company.

Now they were very happy that reinforcements had arrived. Meanwhile the mood among the few men guarding such a large number of prisoners was pretty uneasy. The Italian officers, especially, had required that a very close eye be kept on them. Until this point it had not been possible to transport them away. The officers were on the ground floor of the castle under the watch of two mountain troopers. I had more important things to do than to worry about them now.

The [enemy] telephone connections between Longarone and Belluno had by this time been destroyed by our patrols with utmost attention to detail. Despite this I did not believe we could exclude the possibility that an enemy relief force intended for the troops in Longarone was on its way already. At the very least the enemy battery on Mount Degnon had a very precise awareness of what was going on around Longarone. Therefore I ordered the 3rd Company of the Imperial and Royal Rifle Regiment 26, reinforced by a heavy machine gun platoon of the Württemberg Mountain Battalion, to provide security and reconnoiter southward. Our foremost combat outposts were to be 800 meters south of Fae with the strengthened company itself located in the vicinity of Fae.

I couldn't count on receiving any additional troops to command. It was unlikely that the outflanking detachments of the Württemberg Mountain Battalion (Detachment Gößler, Detachment Schiellein and the 2nd Company) could reach the entrance of the Vajont canyon, 1,000 meters east of Longarone, even if they did not encounter the enemy at all. At the moment *Major* Sproesser was there with the remainder of the 1st Battalion of the Imperial and Royal Rifle Regiment 26, our battalion's communications company and the Mountain Howitzer Detachment 377—which in any case had no more ammunition.

Should I just content myself with blocking off the western banks of Piave valley the north and south near Fae? Should I wait until the enemy attacked me? No—that was not to my taste.

To bring about the decisive moment around Longarone, I decided to make a night thrust towards the town with all remaining units of the forces at my disposal (the 1st and 3rd Companies of the Württemberg Mountain Battalion and the 1st Machine Gun Company of the Imperial and Royal Rifle Regiment 26.)

Meanwhile it had gotten pitch dark. After we had crossed the river, the enemy's march from Longarone to Fae had ceased. Italian artillery shot rapid fire into the area of the Piave riverbed where we had made the crossing. The enemy was probably clearly aware that the path to Belluno was blocked. In the evening light he had probably most certainly seen the 800 captured Italians and Detachment Rommel switching sides across the river. What would he be getting up to now? Would he try to make a breakthrough during the night? I had to predict him!

I gave orders via telephone to the heavy machine gun platoons near Dogna, which as previously were giving harassing fire on the streets and railway bridge near Pirago as well as in the rock-cut part of the road several hundred meters north of that town, to cease firing now. I told them the detachment itself was now planning to advance on Longarone.

Then we left Fae, moving north. I led the point men myself. They were echeloned as follows: light machine gun crews marched on the right side of the street with one machine gun loaded and ready to unleash continuous fire, while on the left were a group of infantrymen marching in single file along the roadside with 10 meters distance between each of them. After that the companies followed in single file at 50 meters distance. The detachment staff were echeloned among the lead troops of the companies. We slinked forward as soundlessly as we could because we knew the enemy sentries could hear from especially far distances in the quiet of the night.

Despite all our caution, the point was shot at by an Italian sentry 300 meters south of Pirago. Only the flash of a single shot was visible in the black night—then my light machine gun on the right started hammering. Its sheaf of fire burst against the surface of the road, against the wall of a house on the right and against the steep rock left of the street, sparking fire from the stone. Not a single shot more came from the enemy after this—he had been blown away.

Vor der Sperre

The scene ahead of the barricade that Rommel ran into.

We continued our march forward, reached Pirago without running into another enemy and crossed the bridge that we had been blocking with our fire all day. Our machine guns at Dogna were silent now, most likely because of the orders I had transmitted to them via telephone.

With the most exceeding caution we stalked forward along the street. On the rocks above the left side of the street—barely a few hundred meters away as the crow flies—Italian artillery were firing shot after shot in the direction of the place where we had crossed the Piave river.

The shell fuzes drew a one-of-a-kind luminous streak behind them as they flew into the night sky. It looked like a pretty firework.

Only a few hundred meters now separated us from the first houses in Longarone. We crept forward very slowly. There—in the light of the firework, we saw a black wall crossing the bright street barely 100 meters ahead of us. Did the street turn left or was this a roadblock? We approached within 70 meters of it before it was clear to me that this was indeed a roadblock. Apparently we were expected to arrive there.

I ordered everyone to halt and brought the machine gun company forward. The company commander, an *Oberleutnant*, received orders to bring multiple heavy machine guns over on both sides of the street without making a sound and to prepare for surprise fire on the barricade. After a short fire for effect I wanted to launch a storm attack with the 1st and 3rd Companies and seize the southern access to Longarone.

The preparations for this undertaking went underway. Just as the crews of four heavy machine guns were bringing their guns forward to be positioned 70 meters ahead of the barricade, machine gun fire suddenly hit us from the flank. Our own heavy machine gun platoon in Dogna was shooting at us! Apparently my order to cease firing had not been transmitted to them.

Bullets hit to our right against the wall, against the street, and against the rock face on our left, sending sparks spraying everywhere. We tried to take cover as fast as possible. The machine gun chattered on and bullet casings smacked violently against the ground! The fire lit up the whole barricade right in front of us. Now multiple enemy machine guns in the area we were lying in started to reply with fierce counterfire. To be within 70 meters of machine gun fire with no opportunity to take cover is enough to make you go bonkers! Death stands very close to a person in such moments.

We ourselves could not fire at all. The heavy machine gun equipment hadn't been assembled. For minutes we lay in the most stomach-turning crossfire. An attempt to get rid of the enemy behind the barricade with a hand grenade failed. The distance was too far. To storm against the enemy machine guns on the narrow street was impossible. We sought cover under outcrops of the semicircular roadside wall and, when the fire from our own men hit this, we took cover in the gutter on the left. If we threw hand grenades the fire from the barricade would only get heavier.

Casualties piled up. Among others, the leader of the Machine Gun Company of the Imperial and Royal Rifle Regiment 26 was lying in the gutter on the left, severely wounded. The only good thing was that the darkness of night at least made the Italian gunners' aim worse.

Our undertaking was now completely pointless. Now it was only a question of breaking off the attack as fast as possible and without taking too many losses. I myself was pinned down by the fire. I relayed my order from man to man to retreat as far as the Pirago bridge. The units further back would find it relatively easy to dispel the enemy. This situation was very difficult for the main body of my detachment which was located right in front of the barricade. The moments that the enemy fire slackened even just a little bit were quite rare. These moments only allowed you to take a short hop backward. Often you would only get a few meters before sheaves of enemy machine gun fire forced you to the ground.

Countless hops brought me, uninjured, back around the last safe turn in the road—safe at least from the enemy fire. Unfortunately our heavy machine guns at Dogna made life sour for us even here from time to time. Now their fire was blocking the Pirago bridge. I had only a few of my mountain rangers with me. A portion of them had already gone off towards Pirago. A large number of them had to be left behind, lying near the barricade.

To our complete astonishment the enemy at the barricade stopped firing. Shortly afterwards we heard the murmur of voices coming from that direction, and soon getting even closer. Those were no German mountain troops. Strangely not a single man from my detachment was returning.

Rommel's men brace themselves to halt a rushing mass of Italians before being swept away.

Now I hurried back as fast as I could to Pirago. On the way I found some of my mountain rangers and brought them along, including a man with a flare pistol. I met nobody else from my detachment at the Pirago bridge. It seemed like my orders to halt here had not been carried out.

A herd of rapidly approaching Italians were hooting and hollering behind me, getting ever closer. Was it the enemy making a breakthrough or were they people who had given themselves up? What had happened to the front units of Detachment Rommel (the 3rd Company and Machine Gun Company of the Imperial and Royal Regiment 26)? I hoped to clarify the situation using a couple of flares.

I shot the flares just to the right of the bridge from a low-lying wall leading toward a mill. In the light of the flares I saw a dense mass of men waving handkerchiefs rushing toward Pirago. The first among them were barely a hundred meters away. I must have been clearly recognizable to them in the flare light. No shots were fired from the Italian side. They continued to approach rapidly, hollering. I still was not clear what I was looking at just then—an enemy who wanted to break through or Italians who had surrendered.

It was impossible for me to bring this crowd of people to a halt with the four or five troopers I had with me. The rest of my detachment seemed to have retreated all the way toward Fae. Now I hurried back down the street as fast as I possibly could. I wanted to catch up with the majority of my people and use them to bring this enemy to a halt.

A few minutes later I had assembled about 50 men near a group of houses 300 to 500 meters south of Pirago. A house on the right side of the street was occupied quickly under the leadership of *Leutnant* Streicher. Half of the troopers were deployed to create a roadblock by standing in the street themselves. The men stood shoulder-to-shoulder with carbines ready to fire. *Leutnant* Schöffel was to the left near the rock wall. *Vizefeldwebel* Dobelmann and I stood on the right near the house. We had no more flare pistols or flares available. It was not possible for the enemy mass to make a left turn. It was impossible to see how things looked on the right due to the darkness and the shortage of time. The Piave river had to be rushing somewhere over there. We only had a few seconds left to prepare and anticipate. Soon the shrieks of the Italians got nearer.

In the darkness of the night you could barely see 50 meters ahead. The terrain to the right and left was as black as evil. As the enemy came within 50 meters distance, I commanded, in as loud a voice as I could yell: "Halt!" and demanded their surrender. Did the bellowing of the crowd mean a yes or a no?

No shots fell. The mob, screaming, swept nearer. I repeated my demands with the same result. At 10-meter range, the Italians started shooting. At the same time a salvo burst from our side. We had no machine guns, either light or heavy, available. But before we could load our weapons a second time, we were flattened by the crowd—overrun.

Whatever men were standing on the street, some of them wounded, all fell into the enemy's hands. The garrison in the house on the right of the street had not been able to get into action in our defense at any speed because the windows on the ground floor had all been painted black. Most of my men inside saved themselves by crossing the Piave in the darkness. The Italians ran down the street, heading south.

At the very last moment I whisked myself away from Italian captivity by jumping over the wall near the street. Now I ran in the darkness with the Italians charging down the street racing against me. I went cross-country over plowed fields, tiny streams, over hedges and fences. The 3rd Company of the Imperial and Royal Rifle Regiment 26 and a heavy machine gun platoon of my battalion were at Fae, still 1,400 meters away, facing south without any idea of the looming danger. The thought of losing this last remaining bit of my force gave me superhuman strength. Presently I found a country road beneath my feet and tore towards Fae.

I managed to get to Fae before the enemy did. With all forces I could reach, I created a new front to the north as quickly as possible—a firmly closed front, which was to be held even to our last breath. The 3rd Company of the Imperial and Royal Rifle Regiment 26 had barely occupied the northern edge of Fae when we heard the screaming Italians approaching. I let them get between 200 and 300 meters of us before I had my men open fire. The enemy onrush faltered.

Soon Italian machine guns clattered. Their sheaves of fire sprayed the walls behind which the Styrians[3] were positioned. The enemy seemed to be attacking right and left of the street. Their cry sounded from a thousand throats: *"Avanti, avanti!"*

---

3. The Austrian troops of the Imperial and Royal Rifle Regiment 26.

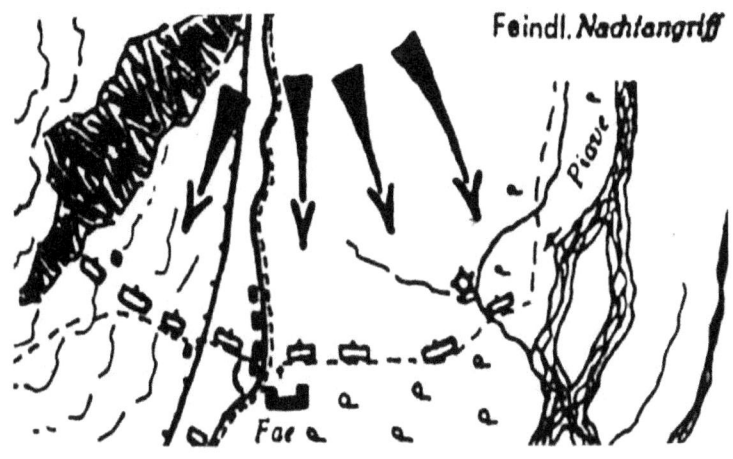

The enemy night attack.

If I wanted to thwart an enemy breakthrough to the south, I had to use my reinforced company to hold a line that stretched from the lumber mill on the Piave 350 meters east of Fae Castle across the northern edge of Fae up until the rock walls of Mount Degnon, 250 meters west of Fae—in total, a 600 meter-wide front. In the middle of this sector the reinforced 3rd Company of the Imperial and Royal Rifle Regiment 26 were already fighting on both sides of the street. On the flanks reaching toward the river and Mount Degnon, vast expanses were still unoccupied. My last reserves consisted of two or three groups of the 1st and 3rd Companies, and the rest had been sent forward during the failed night attack on Longarone.

In order to identify enemy outflanking attempts and create better visibility conditions for the battle, I had a group of mountain rangers set fires across the whole front from the Piave to Mount Degnon. The troopers knew it was an all-or-nothing situation here. Soon the mill on the Piave was burning. The flames spread from a huge haystack 50 meters to the right of the street and from various houses and sheds above the street on the left.

Units of the Austrian rifle regiment were pulled out of the frontline and redeployed to occupy the entire sector, even if they could only do so sparsely. Despite the manic enemy fire we managed to close all the gaps.

My brave buddy Unger volunteered to bring over reinforcements from the east banks of the Piave. He was a good swimmer and trusted his knack for daring deeds.

In the meantime dozens of enemy machine guns hammered against the walls of the castle. The enemy infantry lay densely massed and ready to storm from about 100 meters ahead of us in trenches and ploughed furrows. Again and again their rallying cry rang out above the clattering of the rifles and machine guns: "*Avanti, avanti!*"

During this battle the severely wounded *Vizefeldwebel* Dobelman dragged himself into our line from across the ploughed farmland in the area of the mill. This exemplary man had taken a shot through the breast during the night battle on the street 1,400 meters north of Face. Yet despite being taken prisoner, he had gotten away from the enemy in the darkness and managed to press through to us.

I held a few mountain rangers in reserve in case this overpowering enemy managed to break through our thin line at any point. Two infantrymen were still keeping the 50 Italian officers in check on the ground floor of the castle. Because their own troops were so near, the Italian officers were getting quite combative but still hesitated to attack our two men.

Another view of the Italian night attack.

Shots that struck the north face of the castle pattered like hailstones. The majority of the Styrians [Austrians] were in position at a wall on the north edge of Fae and fired shot after shot—even if they weren't aiming—over the wall at the enemy. Whenever we heard the Italians' battle cry, the fire got even fiercer. This type of fighting naturally cost us a terrible lot of ammunition. The companies would have shot all their ammunition long ago, if we had not been able to fall back on the rich stock of weapons and ammunition in the castle courtyard—taken as booty by Huber's and Hohnecker's patrols in the afternoon.

With the help of the few mountain troops I had available, I managed to arm our whole front against Italian guns and ammunition during the course of the battle. It was pretty bad, however, that the heavy machine gun platoon positioned on both sides of the street only had 50 bullets per rifle.

For officers, I only had the Austrian regimental commander and *Offizierstellvertreter* Huber. All the others seemed to have fallen into enemy hands. I was even missing *Leutnant* Streicher.

The battle raged around Fae for hours with undiminished fury. The enemy constantly tried to storm us head-on in a dense mass. Our ceaseless rapid fire stopped enemy breakthroughs at all points. To the south, only six men of the Austrian regiment were providing security. No more men were available. Soon it was getting towards midnight. New fires were lit ahead of the front because the old ones were threatening to go out. In vain I waited for the reinforcements I had sent for. Strong units of the 22nd Imperial and Royal Rifle Division certainly were on the east bank of the Piave and also the remaining detachments of the Württemberg Mountain Battalion could have arrived there by now. No telephone connection to *Major* Sproesser's command post existed.

Just after midnight the enemy fire diminished noticeably. We exhaled in relief. Thanks to the clever use of the scant cover we had available, our casualties were bearable. Now we worked feverishly to strengthen our positions. We observed rearward movements among the enemy.

As the fire just about totally stopped, the Austrian regiment sent patrols forward. One of these patrols lost their intrepid commander from shots at point-blank range. Another patrol brought about 600 Italians back at about 0100 hours; these men had cast down their weapons just ahead of our front in trenches and furrows in the farmland. The main body of the enemy seemed to have pulled back toward Longarone.

At about 0200 hours reinforcements finally came: the entire 2nd Company under *Leutnant* Payer, which had completed its outflanking maneuver over Mount Lodina, plus units of the 3rd and 1st Companies, who had withdrawn over the Piave after the failed night attack south of Pirago. Furthermore the remainder of the 1st Machine Gun Company came, with a rich supply of ammunition, and also the 1st and 2nd Companies of the Imperial and Royal Rifle Regiment 26 led by *Hauptmann* Kremling.

Our entire defense was now reorganized and the castle itself was prepared for defense. Ammunition was prepared en masse. One company of the Austrian Rifle Regiment 26 took over reconnaissance and security to the south. Furthermore, the 50 captured Italian officers, who had been silent witnesses to the battle at Fae, were sent off over the Piave to its east bank. They went over the roaring ice-cold waters only very reluctantly that November night (from the 9 to 10).

We were neither surprised nor unprepared when a new, extremely strong enemy attack hit us around 0300 hours. The Italians brought more artillery to support them from very close distances. Dozens of shells burst among our lines. Walls split apart. Roofs buckled. Shortly afterward enemy storming columns ran headlong towards various points. Hand-to-hand combat ensued. But the front held. It had been considerably stabilized due to our increase in forces. I did not need to send in the reserves at all. After one quarter of an hour the attack was finally beaten off. But would the enemy try it again?

This time however the Italian leadership contented itself with this one attack only. They broke off as it failed with heavy casualties on their side and withdrew their troops to Longarone. Unfortunately the Italian artillery fire had cost us some casualties.

Shivering in the frosty cold, we awaited the morning in our wet clothes. To warm ourselves up we swigged numerous bottles of Chianti with the Styrians, drinking to our comradeship-in-arms. Before daybreak the 1st Company reconnoitered a path above the rail line leading up to the iron railway bridge of Pirago. Patrols of the 2nd and 3rd Companies reported that the terrain between the Piave and the street to Longarone was free of the enemy. They brought two dozen prisoners back.

At about 0630 hours an additional battalion of the Imperial and Royal Rifle Regiment 26 arrived at Fae Castle. It strengthened our security to the south. At the same time, Detachment Rommel with the 2nd and 3rd Companies and the 1st Machine Gun Company began another advance against Longarone on the road with the 1st Company traversing the path on the slope above the rail line. We wanted to fasten the shackle around the enemy at Longarone even tighter now.

On the way *Leutnant* Streicher ran into us. He had managed to escape from Italian captivity at the battle south of Pirago, but while attempting to swim across the Piave he had been swept away by the river current and carried for kilometers before washing up on the banks unconscious.

As we got closer to the bridge at Pirago, the enemy blew it up clear into the air. Secured by the 1st Company on the slope above to our left, we soon reached the bridge's position. We found a badly wounded mountain ranger there, covered with the rubble from the explosion. However we could discern no enemy on the other side.

With cover from heavy machine guns positioned on the cliff just south of the bridge, we climbed over the rubble of the detonated railway bridge. As we reached the spot on the other side where the barricade had stood in the street at night [during our failed attack], *Leutnant* Schöffel came riding towards us from Longarone on a mule. A hundred handkerchief-waving Italians were following after him. Schöffel, who had been captured during the night battle south of Pirago, brought with him the happy tidings of the capitulation of all Italian forces around Longarone, put down on paper by the Italian commander:

"*Commando piazza Longarone.*
Al Comando delle Truppe Austriache e Tedesche.
Non essendo le truppe che si trovano in Longarone in condizioni di potere oltre resistere questo Comando si mette a disposizione di cotesto et attende disposizione da cotesti.
Maggiore Lay."

(Garrison Longarone. To the Commander of Austrian and German troops. Because the troops in Longarone are no longer in a condition to be capable of offering further resistance, this command places itself at your disposal and awaits your instructions.— *Major* Lay.)[4]

Our happiness about the outcome of the hard battle and, above all else, that our comrades who fell into enemy hands that night were free again was indescribably immense. Amid the sounds of jubilation among the dense mass of Italians, who lined up on either side of the road like an honor guard cheering *"Evviva Germania!"* as we passed, we moved off

---

4. Rommel gives the words of this order in the original Italian.

down the street to Longarone. The leader of the 1st Machine Gun Company of the Austrian Rifle Regiment 26, who had been severely wounded just ahead of Longarone and had fallen into enemy hands with the majority of his company, was driven past us in a motor ambulance.

We got forward only slowly through the mass of thousands of Italians who were filling the streets. With the ambulance driving ahead, I reached the market square of Longarone, where I met units of my detachment who had been captured who once again were fully equipped with their weapons and kit.

Rommel (left) and a comrade on horseback during the war. He saved many pictures of these two horses in particular, which were named Moritz and Max after two troublemakers in a German children's story; the horses seem to have been attached to one another and usually stand nudging their heads close together.

A few minutes later Detachment Rommel moved in as the first troop to enter Longarone and occupied a group of houses south of the church as its quarters. It began to rain. It took a very long time for the masses of Italians—there were thousands upon thousands of them—to be sent off from Longarone into the Piave valley to the east. The remaining units of the Württemberg Mountain Battalion began marching out from the Vajont canyon, with the 22nd Imperial and Royal Rifle Regiment behind them.

During the pursuit from Cimolais across Erto to Longarone, and during Detachment Rommel's battles in the Piave valley, the remaining units of Group Sproesser had done everything in their power to pitch into the fight. As soon as the Italian positions west of Cimolais were captured, *Major* Sproesser took up the pursuit with the communications company of the Württemberg Mountain Battalion and the 1st Battalion of the Imperial and Royal Rifle Regiment 26—against the orders of the 43rd Rifle Brigade. In view of the contact with the enemy at very close proximity and the manner of leadership in battle, it was not possible for the commander of the Württemberg Mountain Battalion [Sproesser] to allow himself to be relieved by another troop. Upon arriving in the St. Martino area, *Major* Sproesser again received orders from the 43rd Rifle Brigade: "The Württemberg

Mountain Battalion will halt, be rationed and spent the night in the Erto millworks. Rifle Regiment 26 will take over the advance guard." At this, Sproesser sent a report back to the Brigade: "The reinforced Württemberg Mountain Battalion is engaged in a battle in front of Longarone and requests infantry support on the pass road and the deployment of the Imperial and Royal Mountain Howitzer Detachment 377."

The tenacity with which *Major* Sproesser held fast to his task and his absolute refusal to be moved from it by the orders of the 43rd Brigade provoked *Hauptmann* Kremling, commander of the 1st Battalion of the Imperial and Royal Rifle Regiment 26, to make the following assertion which would be often repeated: "I don't know what I should marvel at more—your courage before the enemy, or your courage before your superiors."

At about midday *Major* Sproesser reached the exit from the Vajont canyon 1,000 meters east of Longarone. It took a very long time before the communications company and units of the 1st Battalion of the Austrian Rifle Regiment 26 could get to work outside of the canyon, which was under heavy enemy fire. Afterwards the communications company took over from the 3rd Company which was storming Dogna around this time and shot at the withdrawing enemy from the hills just south of where the road emerged from the Vajont canyon.

As the foremost companies of the 1st Battalion of the Rifle Regiment 26 had put the Vajont canyon behind them at about 1400 hours, *Major* Sproesser sent them on to Dogna to reinforce Detachment Rommel. *Major* Sproesser had nearly no other forces available at his disposal. Detachment Gößler (consisting of the 5th Company and 3rd Machine Gun Company) had scaled Cra Ferrona (Hill 995) and the Forcella Simon (Hill 1483) from Il Porto. During this action, their exemplary leader, the expert mountaineer *Hauptmann* Gößler, while hurrying ahead of his detachment across an icy slope, fell to his death.

Detachment Schiellein (consisting of the 4th and 6th Companies and the 2nd Machine Gun Company) had climbed Mount Gallinut (Hill 1303) from Fornace stadium and had reached the Vajont valley via Cra Ferrona (Hill 995). The 2nd Company under *Leutnant* Payer was then in the process of descending from Mount Lodina towards Erto.

After Detachment Rommel's thwarted night thrust on the west banks of the Piave, a most unbelievable report came to *Major* Sproesser, whose command post was near the canyon pass: The enemy was supposed to have broken through just south of Longarone and the majority of Detachment Rommel, including its leader [Rommel], had allegedly been captured. Soon, however, the sounds of battle near Fae and the sight of fires burning dispelled these rumors.

Upon hearing the report of infantryman Unger sent from Fae, *Major* Sproesser sent additional units of the Rifle Regiment 26 to Fae via Dogna. He also later sent the 2nd Company which had just arrived from their outflanking movement across Mount

Lodina. Units of the 1st Battalion of Rifle Regiment 26 established a crossing over the Piave west of Dogna. Before morning on Nov. 10 *Major* Sproesser was standing on the heights 900 meters east of Rivalta, prepared for battle, with the newly arrived Detachment Schiellein (the 4th and 6th Companies with the 2nd Machine Gun Company), the Württemberg Mountain Battalion's Communications Company, four infantry gun batteries of the 1st Battalion of the 26th Rifle Regiment and the Imperial and Royal Mountain Howitzer Detachment 377. Detachment Grau (consisting of the 5th Company and 3rd Machine Gun Company) was approaching in a march from Erto. During the night, *Major* Sproesser had sent an Italian POW back to Longarone with the following written message translated into Italian by *Oberarzt* Dr. Stemmer: *"Longarone is surrounded by troops of a German-Austrian division. All resistance is futile."* When, at daybreak, *Major* Sproesser recognized that Detachment Rommel was advancing against Longarone again and that the enemy there were casting away their weapons, he started marching towards Longarone with the units of Württemberg Mountain Battalion located 900 meters east of Rivalta. During the course of the day the 43rd Brigade of Imperial and Royal Rifle Division 22 followed him.

It rained before noon on Nov. 10. Slowly the streets of Longarone were emptied of Italian soldiers. Mountains of weapons lay in the marketplace—even Italian artillery guns were dropped off here. The Piave valley east of Longarone was now full of prisoners. In total more than 10,000 men—a whole Italian division—had cast away their weapons. Our booty consisted of 200 machine guns, 18 mountain artillery guns, two revolver cannons, more than 600 pack animals, 250 fully loaded vehicles, 10 transport trucks, and two motor ambulances.

Detachment Rommel's casualties in the fights near Cimolais, in the Vajont canyon, near Dogna, Pirago and Fae all amounted to six dead, two severely wounded, 19 lightly wounded, and one missing. The casualties sustained by the 1st Battalion of the Imperial and Royal Rifle Regiment 26 are unknown to me.

During the attempt to stop the Italians south of Rivalta, *Leutnant* Schöffel was taken captive among others. At first the Italians beat him up. Upon complaining, and at his request, he was brought before a company commander, who did not apologize even once for the bad treatment but instead wanted a "souvenir" for himself from the German officer. Afterwards Schöffel had to march with them in the front line towards Fae. As a fight developed here, Schöffel lay on the side of the street right next to an Italian officer. He attempted to escape but this officer thwarted him. Schöffel was especially upset to be under fire from our own. As the Italians then broke off the fight at Fae around midnight, Schöffel was also taken back to Longarone, where he met with the other mountain rangers and Styrians taken prisoner. Around morning the prisoners had to march under heavy cover toward the south again. However, there was soon a new hang-up, because the Italians again failed to break through at Fae. So the prisoners were taken back—again—to Longarone. Throughout the morning hours, the Italian officers became very friendly to Schöffel, who made exaggerated estimates to them about our strength. Finally, he was sent back to us with the written piece of paper stating the capitulation of Italian troops around Longarone.

Around noon on Nov. 10, Longarone was packed full of German and Austrian troops. Detachment Rommel was forced to defend its chosen place of shelter by standing guard with fixed bayonets. The majority of troopers had taken off their thoroughly wet clothes and tended to themselves in a well-deserved rest in good and nourishing quarters. In the evening the mountain troops insisted on holding a torchlight procession for their commander.

Considerations:

- After the breakthrough into the enemy position west of Cimolais had been achieved, **mobile units** of Detachment Rommel (riders and bicyclists) took over **front line pursuit of the retreating enemy**. They managed **to overtake** the enemy and **stop the detonation** of constructions **by Italian engineering commandos** in the Vajont canyon – except for one bridge. Without these mobile forces, the pursuit would have quickly come to a standstill.
- The Italians deployed numerous machine gun units and batteries against the few mountain rangers at the western exit from the Vajont canyon, who had brought the rearward march of the enemy division to a standstill. However, because the **troopers** had **cleverly nested** themselves in, **the mass of fire amounted to nothing but a puff of smoke**. Here, as previously on Kuk, the **defensive conduct of the enemy** was **wrong. An attack from the enemy** with a unit of their forces against the western exit of the canyon **could still have redeemed the situation** for them.
- Detachment Rommel's attack across the coverless Piave valley west of Dogna remained stuck under very strong enemy fire. The troops had to make speedy use of their spades. During this, light patrols on the west riverbank captured the enemy fleeing southward into the valley from Detachment Rommel's fire.
- **During the defensive night battle** at Fae, **light that was needed for shooting** was created **by large fires** and the **rising shortage of ammunition** was remedied by **the requisitioning of Italian firearms and ammunition. Both** of these things were done a**mid the heaviest enemy fire**, an exceptional achievement for the mountain rangers.—
Rommel

"To My Great Pain"—*Battles in the Mount Grappa Area*

◊ Rommel's narrative winds down rapidly without providing much context. Rommel has less operational control and is more introspective. There are no in-depth descriptions of his decision-making. Rommel is asked to regroup with Sproesser's main force but is not on the frontlines. Eventually Rommel reveals—almost in passing—that he has been serving as the battalion's rearguard. This is the opposite of what he has been tasked with throughout the war. It's hard to escape the notion that Rommel has irked higher powers and is being slowly removed from front line duty.

◊ Rommel is faced with a situation he hates—static warfare. He tries to make a deal with Sproesser to allow him to maneuver, promising to complete a task assigned by Sproesser. However Rommel steps on the proverbial toes of some nearby Austrians, which causes a stir.

◊ A key moment in Rommel's life is the arrival of his Pour le Mérite medal. He receives the medal at a low point in his career while trapped on a snowy mountain. The tiny package containing the medal meant a great deal to Rommel—probably more than he could express in writing. The medal became arguably one of Rommel's most precious possessions. He wore it constantly throughout his life until the paint chipped off, the blue faded and the surface collected innumerable scratches—which still can be seen on it today. Rommel does not describe any sense of personal accomplishment when mentioning the medal. First he gives credit to Sproesser, then focuses on the honor the awards reflect on fellow Württembergers in the battalion.

◊ Rommel goes home on leave, but makes zero mention of happy reunions—no gladness about seeing his mother, siblings, or wife. He expresses only bitterness—anger even—at being parted from his military family. At a time when he wants to stay fighting by their side, he is taken away. He makes a point of saying he was "not allowed" to return to the mountain ranger troops.

◊ Rommel uses the words *"Volk und Heimat."* The concept of *"Volk"* (indicating a community of German people) existed in history long before the Nazis hijacked it. What is noteworthy here is Rommel chose the word *"Heimat"* to refer to Germany instead of *"Vaterland."* The Nazis used "Fatherland" to emphasize Hitler's concept of a unified German nation. The word *"Heimat,"* by contrast, is a rather folksy term for "homeland" that had existed practically forever and can also be used colloquially to refer to one's local region. Therefore the Nazis didn't use the word *"Heimat"* much. Although Rommel could have scored points with the Nazi Party by saying "Fatherland," he uses the humble German notion of *"Heimat."*

Upon the orders of the Imperial and Royal Rifle Division 22, the Württemberg Mountain Battalion now moved into the 2nd line and had a rest day in Longarone on

Nov. 11, 1917. That day the battalion bore its dead to the cemetery at Longarone for their final rest.

The momentum of the attack began to ebb away. The speed of the pursuit slowed down although the enemy initially could not afford to remain in any place for long.

During the course of the next few days the mountain troops marched to Feltre via Belluno. Here they were merged into the formation of the German *Jäger* Division. On Nov. 17, they moved from Feltre downward along the Piave. The sounds of a violent battle could be heard from the direction of Quero and Mount Tomba. Soon it was difficult to advance forward through the narrow Piave valley which was jam-packed with troops. We came within range of Italian artillery which from time to time scattered heavy bombardments of shells across the valley road. It became known that the foremost units of Austrian troops had encountered strong enemy forces on Mount Tomba.

While at Ciladon, the Württemberg Mountain Battalion was ordered by the commander of the German *Jäger* Division to break through to Bassano via Mount Grappa.

In the afternoon the battalion moved in developed formation into the expanse just north of Quero, which lay under the most heavy type of bombardment from Italian artillery. The strong Italian artillery had very good observation posts on Pallone and Mount Tomba at its disposal. No wonder that they shot with good aim at the narrow spaces near Quero and all key points within their reach.

Major Sproesser deployed Detachment Rommel (consisting of the 2nd and 4th Companies, the 3rd Machine Gun Company, one third of the Communications Company, two mountain artillery batteries and one radio station) across Quero—Campo—Uson—Mount Spinucia—Hill 1208—Hill 1193—Hill 1306. The main body of the Württemberg Mountain Battalion moved via Schievenin—Rocca Cisa—Hill 1193 to Hill 1306.

As darkness fell we hurried at a fast march in light columns of two under heavy Italian artillery fire across the rubble of Quero, which had been utterly blown to pieces. Shell holes from five to 10 meters in depth were no rarity here. Dead and wounded *Jäger* troopers lay in great numbers along our path. Already multiple Italian searchlights were transforming night into day. At the same time, enemy guns of the heaviest calibers were striking in the areas around Quero, Campo, Uson and Alano. Searchlights probed the valley ceaselessly from the directions of Spinucia, Pallone and Tomba while heavy shells hurtled towards us from across great distances. Faced with the searchlights and shells we only had a few seconds of opportunity at a time to rush in the direction of the enemy.

During this time we lost contact with both mountain gun batteries. *Unteroffizier* Windbühler was tasked with reestablishing contact with them and bringing them up to Uson. The rest of Detachment Rommel managed to reach the tiny little village of Uson without any casualties.

Like Quero and Campo, this locale had been completely cleared of local inhabitants. Ghostly emptiness reigned among the houses. Searchlight beams focused nearly unceasingly on the spaces my troops reached. Widely dispersed in groups, the troops rested in the shadows of houses and trees. Heavy artillery fire burst at extremely close quarters. Splinters howled through the air. Clumps of earth and stone fell down on us. The shelling posed a hard test on our nerves.

Light patrols were sent out with telephone troops in radial directions, including *Leutnant* Walz who headed toward Spinucia. It was clear to me that there could no longer be any talk of an easy-breezy breakthrough to Bassano across Mount Grappa. The enemy front was strong and consolidated. We had come too late. (Six French and five English divisions had meanwhile rushed to the Italians' aid).

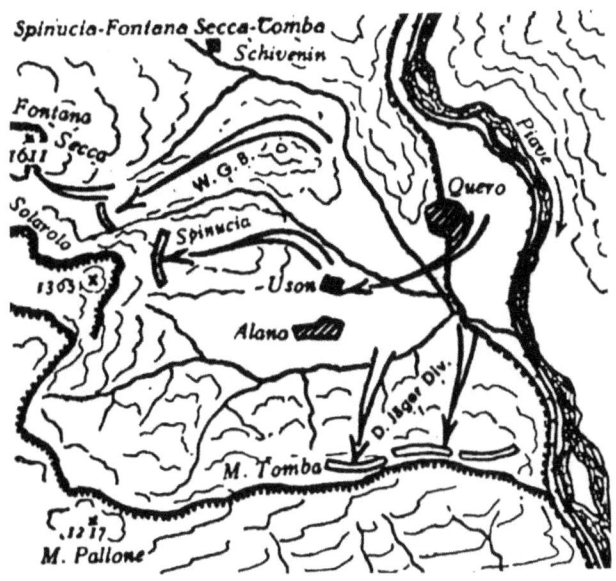

Rommel's map showing the overall situation west of the Piave.

Around midnight couriers arrived. Contact with our neighboring troops near Alano at been established. *Letunant* Walz had climbed all the way up the eastern foothills of Mount Spinucia without running into any enemies. *Unteroffizier* Windbühler was bringing both mountain gun batteries to Uson. He initially entered the Uson—Ponte della Tua valley with them and had come across a well-lit barracks there. Windbühler ordered the batteries to halt, went creeping over to the building alone and discovered sleeping Italians inside. As a man without fear, he drew his pistol, woke the enemy up and took 150 men prisoner. In addition to this, two machine guns fell into his possession.

During the second half of the night from Nov. 17–18, Detachment Rommel climbed the eastern foothills of Mount Spinucia. Here, on the early morning of Nov. 18, the front units of the detachment ran into an enemy force nested very tightly into the rocks of the sharp ridge running from the east across the highest summits of Mount Spinucia. They were located 700 meters east of the highest summit. A frontal attack without artillery support or *Minenwerfer* mortars proved to be pointless. The enemy dominated the ridge completely with deeply dispersed machine guns and mountain gun batteries from

Fontana Secca and Pallone. Opportunities to launch outflanking attacks did not exist here. We were stuck in a gridlock!

Strenuous efforts to further carry out the attack across the slopes of Mount Spinucia were undertaken until Nov. 23, 1917. Because no support from our own artillery was forthcoming and *Minenwerfer* were not available, the results remained the same. On Nov. 21, *Unteroffizier* Paul Martin of the 6th Company fell dead next to me, hit by a sharp-edged splinter from an Italian mountain shell, as we conducted forward reconnaissance. At the same time, a Hungarian artillery *Oberleutnant* was severely wounded.

On Nov. 23, 1917, Detachment Rommel moved to join the battalion at Rocca Cisa. In the meantime Detachment Füchtner had stormed Hill 1222 and, in unison with the Imperial Rifles and Bosnians, taken the Italian positions on Fontana Secca.

At daybreak on Nov. 24, 1917 the entire Württemberg Mountain Battalion lay under my leadership on the northeast slope of Fontana Secca just behind the troops deployed on the front lines, which consisted of the Imperial Rifle Regiment 1, which had been placed at the disposal of Group Sproesser. After the Imperial Rifles had carried out an attack against Mount Solarolo, the Württemberg Mountain Battalion was supposed to break through in the direction of Mount Grappa.

For hours the battalion waited in the snow and ice at Fontana Secca, in the most biting cold, and under severely trying fire from Italian mountain batteries, for the Imperial Rifles to get results. Their attack against Solarolo didn't get forward. Our own artillery support was too meager while the enemy's artillery was too strong. It was probably around noon when the report came to us from *Major* Sproesser that the Imperial and Royal Mountain Brigade 25 had taken Mount Solarolo in an attack from the west.

Meanwhile the situation on the southern slope of Fontana Secca had not changed one bit. The Imperial Rifle Regiment still hadn't gotten forward by any noticeable distance. There was zero prospect that this situation would change while it was still daytime. Because of all this I requested permission, from Group Sproesser, to move toward Solarolo toward the right of the Austrian Mountain Brigade 25, and then, afterwards, attack in the direction of Mount Grappa from there. *Major* Sproesser was in agreement.

Soon the whole Württemberg Mountain Battalion was on the march. It proved impossible to take the shortest path, namely because the almost vertical rock walls on the western slope of Fontana Secca couldn't be crossed. So there was nothing else we could do except to descent into the Skizzone valley. We stepped lively but were surprised to be overtaken by the onset of darkness near Dai Silvestri. I ordered the tired troops of the Württemberg Mountain Battalion to rest here and sent out *Leutnant* Amman (of the 6th Company) to reconnoiter the situation among our own troops at Mount Solarolo. My intention was to march onward as early as possible so that the battalion would be rested by daybreak on Nov. 25 and ready to continue to attack from Solarolo.

However, as *Leutnant* Amman returned from his reconnaissance—which was very detailed and successful—the situation had completely changed. The Württemberg Mountain Battalion were reviled a great deal for maneuvering into the attack zone of the successful brigade neighboring them. The outrage about this was so immense that no other choice was left to *Major* Sproesser but to request to be relieved from our attachment to the Imperial and Royal Rifle Division 22, which was done. The battalion spent a few days in rest quarters east of Feltre, then started moving on Dec. 10 towards the front at the Fontana Secca massif, again proceeding down the Piave valley.

During the night of Dec. 15–16, my detachment encamped in snow and ice at a height of 1,300 meters elevation. On Dec. 16 the positions on "Pyramid Knoll," Solarolo (Hill 1672), and "Star Knoll" were reconnoitered. The enemy continuously and tenaciously held the most important areas of these dominating heights.

We got snowed right into our tents during the night of Dec. 16–17. The next day Group Sproesser attacked. The attack succeeded in penetrating the enemy positions on Star Knoll, capturing 120 Bersaglieri of the Ravenna Brigade and repelling very strong enemy counterattacks.

The exemplary *Unteroffizier* of the 2nd Company, Quandte, never returned from a patrol. He probably had been wounded and fell to his death.

In icy cold, we held out on the ragged slopes of Star Knoll amid heavy Italian artillery fire right up until the evening of Dec. 18, 1917. Then the Württemberg Mountain Battalion moved off in a march into the valley toward Schievenin.

There the field post delivered two small packages. They contained the *Pour le Mérite* for *Major* Sproesser and me—an unheard-of distinction for a single battalion at that time.

We spent Christmas Eve in a small village northeast of Feltre. An the first day of Christmas the mountain rangers marched under the leadership of their Old Alpino—as the *Major* was nicknamed—again through the narrow Piave valley south of Feltre to the front. My detachment was deployed to relieve the Prussian *Jägers* in the Palone sector on the left flank of Mount Tomba. Positions were barely available. Individual machine gun nests and nests for riflemen lay in small holes on the steep barren slopes which hardly offered any cover. Snow was all around! But the cold was bearable. During the day the riflemen had to lay masterfully camouflaged beneath their tent squares because the enemy could see across the whole terrain. To light fires was forbidden. Rations were brought forward only at night. Tracks in the snow had to be whisked clear with great care at all times. Woe to us if Italian artillery or even *Minenwerfer* zeroed in on one of our nests! The companies had to some extent been lumped together in groups of 25 to 35 men. Despite it all they performed their hard, perilous duty with the greatest sense of personal responsibility.

Rommel (right) poses for a wartime photo with *Major* Sproesser after they both receive the Pour le Mérite, Germany's highest valor decoration. This image was taken somewhere in the mountains. Rommel looks confident in this photo. Little did he know that soon afterwards he would be taken to Germany on leave and then transferred away from his unit to serve on a large staff far away from the action.

On Dec. 28, 1917, an Italian attack was repelled ahead of the Württemberg Mountain Battalion's front. The next day brought heavy fire into the battalion sector. The heavy Italian mortars shot from three kilometers away were most especially unpleasant. That day the enemy artillery even heavily bombarded the rearward terrain near Alano were Sproesser's staff was located. Gas shells were fired repeatedly.

On Dec. 30, 1917, the enemy increased his fire at Mount Tomba to maximum force. Formations of enemy planes dove toward our positions and those of our neighbors and shot at our garrisons with machine guns. After hours of battle French Alpine troopers managed to seize the positions of the Imperial and Royal Mountain Brigade 3 next to us on the left. We were capable of holding our own, but we were now hanging with our left flank totally in the air. If the enemy thrust further in the direction of Alano from Tomba, we would be cut off and would have to break through to our own line at night. It snowed, and was getting colder!

On the early morning of Dec. 31, reserves moved into the yawning gap to the left of us. But they had to suffer under the fire of Italian artillery from the direction of Pallone.

Leadership decided to pull back the front some two kilometers to the north.

Until the late night hours of Jan. 1, 1918, the mountain rangers held the positions on Pallone and Tomba in the bitter cold. Two of the bravest fell even in the last minutes at forward machine gun posts: *Unteroffizier* Morlok, and *Schütze* Scheidel. While repelling an enemy assault troop of some 30 men, the heavy machine gun failed. It came to a close quarters fight. As part of the garrison in the position fended off the vastly superior enemy with pistols and hand grenades, Morlok and Scheidel tried feverishly to get the frozen heavy machine gun into working order again. An Italian egg hand grenade fell between them and mortally wounded them both. The enemy was driven back.

Shortly before midnight Detachment Rommel, the rearguard of the Württemberg Mountain Battalion, arrived with the bodies of both of the fallen at Alano, and then moved silently across the fields of the dead at Campo and Quero up the Piave.

Eight days later I drove with *Major* Sproesser via Trient on leave to go home, from where, to my great pain, I was never again allowed to return to the mountain troopers. Due to orders from the highest authorities I was transferred to *General Kommando* v.b.V. 64[5] and integrated as an assistant chief of staff within a very large staff organization. From there, with a heavy heart, I followed the destiny of the Württemberg Mountain Battalion: the great battle in France, the storming of Chemin des Dames, the attack on Fort Conde, the attack on Chazelle and the Paris position, battles in the forest of Villers-Cotterets, the crossing of the Marne, the retreat across the Marne, and the battles at Verdun. These battles tore enormous gaps in the ranks of the victors of Mount Cosna, Kolovrat, Matajur, Cimolais and Longarone. Only a small portion of them were permitted to see their homeland again.

In the west, east and south rest the German infantrymen who followed the path of most loyal fulfillment of their duty for their people and homeland[6] up to the bitter end. Forever they urge those of us who survive and the coming generations not to fall short of them when it comes to making sacrifices for Germany.

The End[7]

---

5 An army corps command post.
6. Lit. *Volk und Heimat*.
7. Rommel included this.

One of the last photos of Rommel taken during his lifetime in 1944.

Conclusion: Rommel at War's End

Rommel is shown alone in his car in an undated photo during World War II.

The last impression Rommel left us with at the conclusion of his war memoir was one of death and of separation. As he was no longer able to spread his wings, his writings turned to focus on grim conditions around him. He reflected with great sadness on the deaths of two of his comrades. He also described *Totenfelder*—"fields of the dead"—that he saw on the march. Rommel's tone at the conclusion of his book was downright gloomy.

Rommel was clearly unhappy about how the war turned out for him. He was obviously very upset, seemingly almost embarrassed, about his transfer to a staff position. Apparently these feelings didn't age with the passage of time, nor did age give him a different perspective when he penned his First World War memoir as an older man.

Rommel evidently had such a low opinion of staff work that he didn't even think what he did for the rest of the war was worth writing about. The book ends after he leaves the front. At this point Rommel effectively considered that his war was over. This says something about how Rommel viewed war. Frontline combat was everything to him. He showed no desire to learn to be good at staff work—which might have served him well, in the grand scheme of things. He desired only to be in action. If he wasn't actually fighting at the front, then it wasn't a real war to him nor worth recording for history.

It has been commonly assumed that Rommel was "promoted" to a staff position because of his abilities. Rommel's son Manfred wrote that the *Pour le Mérite* award made his father a hero and a desirable candidate to have as a staff officer, hence the transfer. While it is true that the *Pour le Mérite* was a recognition of Rommel's heroism in combat and earned him acclaim—as well as envy—I find it difficult to believe that Rommel's abrupt transfer to a staff position was actually a promotion. In view of his clashes with Sproesser towards the end of his frontline service, it's difficult to escape the impression that Rommel,

Rommel is pictured with members of his Old Comrades Association, a group he founded for veterans who served with him in World War I.

after a series of remarkable victories, and also of disobedience of orders, was "kicked upstairs." Prior to his abrupt transfer, he was placed on rearguard duty. Additionally, Rommel was not informed that he would be transferred until he was miles away on leave and not able to do anything about it. Sproesser of all people would have known that Rommel was unsuited to staff work. What "higher powers" had arranged for Sproesser's talented junior field commander, trained in mountain warfare, to be removed from doing what he excelled at? Was this transfer indeed a promotion or was Rommel simply being removed from the frontline after stepping on others' toes or making superiors jealous? Did Sproesser play some role in arranging for Rommel to be gotten out of his way?

After escorting Rommel back to Germany on leave—effectively seeing him off—Sproesser, himself a *Pour le Mérite* recipient, returned to action and was badly wounded in February 1918 at Chemin des Dames. He died in 1933. Rommel's book was published years later. Although respectful and at times complimentary of Sproesser, Rommel did not include any special statement to memorialize his deceased commanding officer nor mention that Sproesser was severely wounded later in the war. If there was some degree of rivalry between the two of them that would not be unusual; as anyone familiar with military history will know, it is not always the case that commanding officers are personal friends with their junior officers. (Also, as an assertive personality and talented military mind, the rising young leader Rommel may well have made Sproesser feel threatened or insecure enough to try to stop him from distinguishing himself further in battle.)

Whatever factors may have played a role in Rommel's undesirable transfer, it cannot be said that Rommel was considered a military failure by his peers. He demonstrated considerable talent and was viewed as a military hero due to his *Pour le Mérite* after the war ended. He also was selected to be part of Germany's severely reduced military force during the Weimar Republic. Rommel was also well-respected in veteran circles.

Rommel the Veteran

I don't believe that World War I ever really ended for Rommel. It dominated his mind for years afterward, even up until World War II. His actions demonstrate this. He visited his former battlefields, drawing the landscapes and taking photos. He even climbed Mount Matajur and sat on top of the peak again—there is a photo of him there with his wife Lucie, in which the couple is sitting some distance apart from each other and Rommel appears rather distressed. He visited the graves of his fallen friends and photographed them. Ostensibly the point of Rommel's travels was to gather material to write his book about tactics, but there seems to have been much more to it than that.

Rommel and Lucie are shown visiting Mount Matajur. Lucie looks calm and is sitting some distance away. Rommel (close-up on right) appears distraught, frowning and huddling with both arms together in closed body posture.

It was not enough for Rommel to revisit past battlegrounds and work on his book. He wanted to be among his former comrades constantly. Rommel helped to found a veterans' group called the Old Comrades Association. He became deeply involved in this. He went out of his way to reconnect and share in activities with other men who had shared his wartime experiences. He and the men did many group activities together and raised money to help wounded war veterans. They built a lodge in the countryside near Cannstadt in South Germany and used it as a type of clubhouse; the lodge is still standing today.

Surrounded by memories of the war, Rommel analyzed and discussed his wartime experiences in military lectures. He wore his *Pour le Mérite* medal constantly until it faded and chipped. There seems never to have been a moment in either his professional or private life when the memory of World War I was not present or did not loom over him in some way—a result of his deliberate choices. He kept revisiting that time in his life on purpose. He collected materials about the war and dwelled on it constantly. There was something about the war he wouldn't—or couldn't—let go of.

Having read many veterans' memoirs written in today's times, I am of the opinion that Rommel's preoccupation with the war indicates that he struggled to come to terms with his experiences in some way. One plausible theory is that Rommel felt some survivor's guilt after being transferred (against his wishes) to a staff position safe behind the lines while his friends had to continue serving in combat and lost their lives. In addition to this he may also have also struggled to come to terms with things he had personally seen or experienced in combat.

Rommel's Insomnia and PTSD

Rommel's lack of sleep was abnormal. It indicates an underlying problem. He frequently described hypervigilance and insomnia. Insomnia is a common symptom among combat veterans suffering from PTSD—a particular difficulty is sleep onset, or the process of falling asleep. This is exactly what happened to Rommel. He frequently mentioned this in *"Infantry Attacks!"*—he had a lot of trouble trying go to sleep and usually gave up. His hyperactivity enabled him to operate at a high energy level for days before crashing. He seemed to have gotten used to not sleeping and habitually stayed awake until his body started shutting itself down. He also alluded more than once to terrible nightmares that occurred throughout the war.

I do not believe these problems resolved themselves after the war. His wife Lucie claimed to biographers that he did not have nightmares nor problems related to war trauma; I do not believe this. Judging from how often he was unable to sleep during World War I, Rommel's case of PTSD-related insomnia seems to have been pretty severe. I highly doubt it vanished after the war. It might even have gotten worse.

An indicator that the problem stayed with him is the fact that Rommel brought up the issue of his sleep regularly in World War II letters. Sleep was in fact a constant theme in his writing to his wife. In Belgium and France in 1940, Rommel barely slept.

Rommel is shown wearing his Pour le Mérite medal, which he wore so much during his life that it faded and became very chipped. He suffered from chronic insomnia during his life.

This repeated itself in North Africa. Rommel described getting by on a mere two to three hours of sleep. In letters to Lucie, Rommel admits to lack of sleep in somewhat apologetic tone with phrases such as: "You will understand that I can't sleep for happiness," following some 1941 successes and later on, "It's going to be a hard fight, so you will understand that I can't sleep." He explained that being active all day and night was "very necessary" to excuse his lack of rest. He also complained regularly about being "very tired" because of lack of sleep. He indicated that—as in World War I—he preferred to work when he was unable to rest, writing in 1942: "…I could bear to sleep a little better, although the sleepless morning hours are very productive…" Whenever Rommel did manage to sleep soundly, he wrote about it to Lucie as if describing a noteworthy achievement. He happily mentioned occasions whenever he "overslept" and in another letter declared he was "sleeping like a log." On another occasion in 1941 he wrote: "I have been able to have my sleep…so now I'm ready for anything again." Obviously Rommel had problems getting rest and both he and his wife knew it.

His sleeping "achievements" tended to be the exception rather than the norm. Rommel's writings indicate he was awake and hyperactive most of the time even during World War II. Stress compounded his difficulties. He seemed to be chronically unable to rest. By 1943, his doctor was prescribing him sleeping medication in North Africa to force the issue.

While recovering from skull fracture injuries he sustained in Normandy in 1944, Rommel refused doctors' advice to sleep, instead waking and moving at an early hour. "He was too restless," his son Manfred wrote. Some of Rommel's contemporaries attributed his "restlessness" to his energetic personality. I believe Rommel's chronic insomnia was the symptom of a bigger underlying issue—PTSD—stemming from his traumas during the First World War.

The Ultimate Betrayal—Not Once, But Twice

Rommel smiles and shakes hands with one of his men during the interwar period. He was forcibly separated from his troops twice during his lifetime.

At the conclusion of "*Infantry Attacks!*," Rommel described a literal sense of "pain" at being ultimately separated from his men. There are lots of different words in German that Rommel could have used—for example, a word like *Leid*, meaning "sorrow." Rommel however used the word *Schmerz*, literally meaning "pain," to describe his feelings. This word is most often used to describe physical suffering from wounds or injuries. Rommel's use of this word emphasizes just how wounded he felt at being parted from his troops. Rommel stressed that he was "not allowed" by "higher powers" to return to his men. He seemed to have felt a sense of injustice about it.

History would repeat itself for Rommel during World War II when, recalled to Germany by Hitler on leave, he found himself removed from command of the Afrika Korps and separated from his troops forever. Rommel was extremely bitter over this final unexpected parting in World War II and expressed the wish that he could have stayed with his men to the end, even to be taken prisoner with them.

The final separation from his men in North Africa had a shattering impact on Rommel. Friends and family reported a drastic change in his mood and behavior that lasted until the end of his life in 1944. He was constantly angry, listless and depressed. He wrote that being parted from his troops had left him with "a bad scar."

This parting may well have reopened an old wound. We can only wonder if Rommel as an older man in World War II was reminded of the similar bitter experience he had when he was forcibly parted from his comrades in his youth.

My Views On Rommel

Rommel is a complex character. His critics call him an opportunist, an arrogant blunderer and even a Nazi. To others he is a hero and a great military genius. Sometimes people ask me what I think about him, personally. I've spent almost 20 years of my life studying Rommel and his campaigns. Working on this project gave me a new perspective about him and gave me a chance to reflect on Rommel in a new light. With that in mind, here are some opinions I offer to readers about Rommel.

Rommel and Nazism

I have done a lot of research about Nazi ideology, propaganda and war crimes; that includes on-site research in Germany and German-language research. In my research and articles, I attempt to expose Nazi war crimes as well as methods used by the Nazi regime to deceive people. Based on my research and experience, which has involved looking at both photographs, diaries, propaganda materials and other documents created by SS and Nazi war criminals, I believe that Rommel was not a Nazi.

- Rommel did not express racist, hateful or extremist views in any of his writings. Rommel did not make use of his World War II successes to enrich himself or gain extra privileges for himself or his family. Rommel did not steal works of art; he did not plunder territories he fought in nor take over Jewish-owned businesses. He did not use POWs or concentration camp inmates as his personal slaves. He did not save "trophy" photos of dead enemies from battlefields; his wartime photo collection is noteworthy for its lack of corpses and lack of images reflecting the ideals of Nazi ideology. Also, at a time when other Germans were divorcing inconvenient spouses and remarrying with other "Aryans," Rommel stayed married to his Polish wife who was having a lot of trouble providing documentation about her heritage to Third Reich authorities. According to his son Manfred, Rommel attempted to guarantee the safety of Lucie's uncle, a Polish Catholic priest, and to locate this man when he went missing in Poland; Rommel was unsuccessful in his efforts to find him and never found out what happened, but Manfred later learned after the war that this relative, who had shared a close bond with Lucie, had been shot by the Gestapo. Rommel was effectively fired from an assignment he had to train members of the Hitler Youth after he was insubordinate and raised complaints about the Nazi training program, with Baldur von Schirach allegedly complaining to Martin Bormann in 1935 that Rommel was "not a National Socialist." Rommel received no Nazi Party orders or decorations.

- The Nazi Propaganda Ministry created a fictional biography of Rommel during World War II, casting him as a Nazi, which they disseminated to the public. The Third Reich authorities would not have needed to invent an imaginary Nazi background for Rommel if he was indeed already a Nazi. Obviously even the Nazis thought their political views were missing from Rommel's life and career.
- Rommel had every opportunity in the world to be a Nazi and had no reason not to become one apart from his own inner convictions. Embracing Nazism openly in words and deeds would have opened many doors to him. He could have gained power, wealth and privileges, and even greater military distinctions and responsibilities. However he did not embrace Nazism to advance himself.

This photo shows Rommel in his Reichswehr uniform—before the rise of the Nazis. The uniform looks similar to World War II, but with a notable lack of swastikas. I don't believe it would ever have remotely occurred to Rommel to stop being a German soldier or to leave Germany regardless of who was in charge of the country.

Why, then, did Rommel willingly serve as a soldier of the Third Reich? This is a question I will answer in more detail in another book. However, I believe that being a German soldier was central to Rommel's identity and ethos. It probably would never have remotely occurred to Rommel to leave Germany or cease being a German soldier no matter what the circumstances. I also have the impression that Rommel was an independent thinker—someone who, a bit like fellow World War I veteran and *Pour le Mérite* recipient Ernst Jünger, wanted to be his own man and not simply "go along" with a political party (although unlike Rommel, Jünger was a vocal right-wing nationalist).

If Rommel had survived the war and had faced the Nuremberg Trials, I believe he would have been fairly exonerated of all charges that other high-ranking German military officers were found guilty of.

Rommel's Strengths & Weaknesses

Rommel could be myopic and self-centered. I don't believe he can be accused of being a cold-blooded careerist. Careerists exploit opportunities. They backstab, walk over bodies and turn to flattery to advance; you can see this in the lives of top Nazi officials. I can find no evidence that Rommel engaged in this kind of behavior. Yet what leads people to make that accusation is that they perceive a certain selfishness in him—and that selfishness was indeed present. Rommel tended to focus on matters of importance to himself and his own future.

Self-centeredness was one of Rommel's character flaws. It was noticeable when he was a child. When he disliked his lessons at school, he stopped studying and failed in his classes, regardless of the fact that his father was a school principal and that this situation was embarrassing to his family. Rommel was also selfish in his early relationships with women. He focused on his own feelings and desires, with seemingly little regard for the consequences his actions had on others. He was later sorry for his behavior and tried to make up for it. Rommel could also be self-centered about realizing his goals on the battlefield. He was so focused on meeting his set goals that he denied his men opportunities for food or rest. He also kept so focused on performing tasks that he wrecked his own health. Rommel can be fairly accused of "missing the bigger picture" around him at several key points in his life.

However Rommel was not totally selfish. As these memoirs demonstrate, he formed strong attachments to people around him and had goodwill towards others, including captured enemy soldiers. Rommel was capable of going against the "herd," stopping his own men from killing Italian POWs during World War I. His troops loved him because they could sense that he genuinely cared about them. His World War II letters show that he often reflected on how the war would end and what it would take to bring an end to it.

During World War II, Rommel felt forced to make a decision between being selfish and following his conscience. He chose to follow his conscience and act against Hitler by formulating a peace plan and preparing to negotiate with the Allies to end the war, and this ultimately cost him his life. Rommel's story ended in definite opposition to Hitler. Rommel should be given credit for that.

Rommel & the Resistance

Rommel's life was ended by the Nazis because he chose to stand against them. The details are still debated. Some argue that perhaps Rommel still felt some loyalty to Hitler even at the end of his life, but this makes no sense. If Rommel had in fact been loyal to Hitler, the Nazi dictator would have had zero reason to get rid of him—especially in view of his military abilities, which Germany desperately needed at that point in the war.

I therefore regard it as an injustice that Rommel does not appear among German

Resistance activists listed by the German Resistance Memorial Center (*Gedenkstätte Deutscher Widerstand*). I realize that is a bold assertion. Many people are uncomfortable with the fact that Rommel was a prominent military commander serving under the Third Reich and took an active part in the war.

However the German Resistance Memorial Center lists among its honorees numerous military men who fought for Germany during World War II, including former Afrika Korps members and men who fought on the Eastern Front; some were Nazis before changing their minds about

Rommel (left) is shown talking with his Chief of Staff Hans Speidel (right) and Gen. Carl Heinrich von Stülpnagel (center) in 1944. All three worked together to plot to end the war and remove Hitler from power. Both Speidel (who survived the war) and Stülpnagel (who committed numerous Nazi war crimes, and was executed by Hitler in a purge) have been honored publicly by the German Resistance Memorial Center by having their names listed as Resistance members. Rommel, however, is not named as a member of the Resistance despite the fact that he was killed for being part of it.

Hitler. Strikingly, the center has chosen to honor two figures implicated in war crimes: *Generaloberst* Erich Hoepner, a vocal antisemite and anti-Slavic racist who worked closely with the *Einsatzgruppen* death squads during Operation Barbarossa; and *General* Carl Heinrich von Stülpnagel, who took part in the invasion of the Soviet Union, assisted the SS in the arrests and executions of Jews, and also gave orders to shoot civilians in acts of reprisal. Both of these men have been publicly acknowledged—indeed, honored by having their biographies put up by the memorial website—as members of the German Resistance for participating in plotting against Hitler, an honor which I personally find distasteful given their criminal histories.

Rommel, on the other hand, committed no war crimes and was neither an antisemite nor a racist extremist, yet is not listed as a member of the German Resistance—despite the fact that even his Chief of Staff, Hans Speidel, who helped him plan to end the war against Hitler's wishes, is listed. Why these double standards? On the grounds of the current honorees alone, the decision to exclude Rommel is illogical.

The fact that Rommel died for his choice to oppose Hitler and Nazism should be enough to earn him a public acknowledgement as a member of the German Resistance—whether or not we know all the details. There is sufficient evidence already that he was working to bring an end to the Third Reich. I must ask: what greater proof can someone give of their convictions than forfeiting, and ultimately losing, one's own life? If people who committed heinous acts during the war can merit a place among Germany's heroes for connections to Stauffenberg and being executed by the Gestapo, then certainly Rommel, who behaved much more honorably and did not have an extremist or criminal history, deserves to be acknowledged.

Rommel was a flawed person but essentially a good man, which he proved with the choices that led to his death.

A Reflection about Rommel

I have visited the place where Rommel died. It is a rather long road, up a gradually rising hill, lined by trees. It is an isolated and silent place. Rommel had spent a lifetime focused on fighting in war but was unable to defend himself as he faced death. His death came suddenly. But it hadn't come from an enemy bullet—it had walked calmly into his house and announced itself with an icy sneer. He was not going to die on a battlefield, but in a strange and lonely spot less than 15 minutes from his home. And the people who were bringing an end to his life were not enemy soldiers, but fellow Germans. After a lifetime of priding himself on his military service, the last order he was given was simply to fold up and die. And, unless he wanted his family to be shot by Gestapo and SS who had surrounded the area, there was nothing he could do but obey. It was probably a fate he could never have imagined. One of the last things he told his son before going on that final, fatal drive up the road was that it was "hard" to be killed by his own people. The last few moments he spent on this earth, sitting in that car in smothering silence, surrounded by his blasé executioners while being acutely aware of the unfair hopelessness of his situation, were probably horrifying.

For all his flaws, I believe that Erwin Rommel was a good man, a good soldier and a good general. Yes, he was selfish and myopic. But take a closer look at Rommel and you will find a dedicated infantryman, struggling with inner demons from his First War, who never stopped feeling like he owed his country and his troops something. At a time when hatred was fashionable in his country—during both world wars—Rommel treated others with kindness and dignity, and encouraged soldiers around him to be humane. And he ultimately lost his life after taking action to overthrow an evil regime. That is not a myth.

Selected Sources and Reference Materials

A great deal of research went into this book over a period of more than five years, including visits to museums and historic sites in Germany. The following is a selection of some, but not all, of the sources and reference material I used for this project.

- *1944—Das Jahr der Entscheidung,* by Manfred Rommel, 2010.

- Collection of alleged Rommel letters to Lucie from Oct. 9, 1911 to Feb. 22, 1913 discussing their early relationship featured for auction on icollector.com in December 2013. Although I was not able to see the letters in person I believe them to be authentic based on the information in the transcripts, writing structure and views of the handwriting.

- "Desert Fox? Rommel Behaved More Like a Desert Rat: Nazi Covered Up Love Affair Which Led to Illegitimate Child For the Sake of His Career," by Alan Hall, in the *Daily Mail,* March 22, 2012.

- *Die Alpen im Krieg - Krieg in den Alpen: Die Anfänge der deutschen Gebirgstruppe 1915,* by Thomas Müller, 2015.

- "Die Enkel des Wüstenfuchses Erwin Rommel leben in Kempten," *Allgäuer Zeitung* newspaper, March 23, 2012.

- *In Stahlgewittern,* by Ernst Jünger, 1920.

- *Infantry Attacks!* by Erwin Rommel, 1937.

- *Invasion 1944,* by Hans Speidel, German language edition, 1950.

- "*Masking Gender: A German Carnival Custom in its Social Context,*" by Peter Tokofsky in Western Folklore sociology journal, Vol. 58 No. 3/4, 1999.

- *Mythos Rommel,* by Maurice Philip Remy, 2002.

- "Tender Love Letters Reveal Rommel's Romantic Side," Reuters, March 22, 2012.

- "The Nazi Wrote a Novel—Michael: by Joseph Goebbels" (translated by Joachim Neugroschel), by Tom Clark, Los Angeles Times, Oct. 25, 1987.

- *The Rommel Papers,* by Erwin Rommel, edited by B.H. Liddell Hart, 1952.

About the Author

Photo by Noël-Marie Fletcher

Zita Steele is the pen name of journalist, editor and military historian Zita Ballinger Fletcher. Zita is the author of more than 10 published books, including the first published collection of Field Marshal Erwin Rommel's wartime photography, *Bernard Montgomery's Art of War*, which won a Silver Medal Award from the Military Writers Society of America, and the historical fiction novel, *The Hidden Sphinx: A Novel of World War II Egypt*. She writes fiction and nonfiction and designs and illustrates her published work.

Zita is fluent in German and is pursuing a Master's of Arts degree in Military History. She has a Magna Cum Laude Bachelor's of Arts degree in Social Sciences from the Honors College at the University of South Florida. She was designated as a Commended Student by the National Merit Scholarship Corp. in 2008. She has won numerous awards for her writing.

Zita has written over 100 military history articles and also reviews books. She is the Editor of *Military History Quarterly* (MHQ) and *Vietnam* magazines published by HistoryNet, the world's largest publisher of history magazines. She lives in Washington, D.C.

www.ingramcontent.com/pod-product-compliance
Lightning Source LLC
Chambersburg PA
CBHW041241240426
43668CB00025B/2455